AF294884

Encyclopaedia of Mathematical Sciences

Volume 54

Editor-in-Chief: R. V. Gamkrelidze

G. M. Khenkin (Ed.)

Several Complex Variables V

Complex Analysis in Partial Differential Equations and Mathematical Physics

Springer-Verlag Berlin Heidelberg GmbH

Consulting Editors of the Series:
A. A. Agrachev, A. A. Gonchar, E. F. Mishchenko,
N. M. Ostianu, V. P. Sakharova, A. B. Zhishchenko

Title of the Russian edition:
Itogi nauki i tekhniki, Sovremennye problemy matematiki,
Fundamental'nye napravleniya, Vol. 54,
Kompleksnyj analiz - mnogie peremennye 5
Publisher VINITI, Moscow 1989

Mathematics Subject Classification (1991):
30Fxx, 32A45, 32C81, 32F99, 35Q40, 35Q75, 42B15
46F15, 53C07, 58E15, 81E30, 81T13, 83E50

ISBN 978-3-642-63433-8

Library of Congress Cataloging-in-Publication Data
Kompleksnyĭ analiz – mnogie peremennye 5. English.
Several complex variables V: complex analysis in partial differential
equations and mathematical physics / G. M. Khenkin (ed.).
p. cm. – (Encyclopaedia of mathematical sciences: v. 54)
Includes bibliographical references and indexes.
ISBN 978-3-642-63433-8   ISBN 978-3-642-58011-6 (eBook)
DOI 10.1007/978-3-642-58011-6
1. Functions of several complex variables. 2. Differential equations, Partial. 3. Mathematical physics.
I. Khenkin. G. M. II. Title. III. Title: Several complex variables 5. IV. Series.
QA331.7.K6713 1993 515'.94–dc20 93-4830

List of Editors and Authors

Editor-in-Chief

R.V. Gamkrelidze, Russian Academy of Sciences, Steklov Mathematical Institute,
ul. Vavilova 42, 117966 Moscow, Institute for Scientific Information (VINITI),
ul. Usievicha 20a, 125219 Moscow, Russia

Consulting Editor

G. M. Khenkin, Université de Paris VI, Pierre et Marie Curie, Mathématiques,
Tour 45-46, 4, place Jussieu, 75252 Paris Cedex 05, France

Authors

C. A. Berenstein, Department of Mathematics, University of Maryland, College
Park, MD 20742, USA

G. M. Khenkin, Université de Paris VI, Pierre et Marie Curie, Mathématiques,
Tour 45-46, 4, place Jussieu, 75252 Paris Cedex 05, France

A. Yu. Morozov, Institute for Theoretical and Experimental Physics,
117259 Moscow, Russia

R. G. Novikov, CNRS U.R.A. 758, Département de Mathématiques,
Université de Nantes, F-44072 Nantes Cedex 03, France

A. M. Perelomov, Institute for Theoretical and Experimental Physics,
117259 Moscow, Russia

D. C. Struppa, Department of Mathematics, George Mason University, Fairfax,
VA 22030-4444, USA

Contents

I. Complex Analysis and Convolution Equations

C.A. Berenstein, D.C. Struppa

Contents

Introduction

In this part, we present a survey of mean-periodicity phenomena which arise in connection with classical questions in complex analysis, partial differential equations, and more generally, convolution equations. A common feature of the problem we shall consider is the fact that their solutions depend on techniques and ideas from complex analysis. One finds in this way a remarkable and fruitful interplay between mean-periodicity and complex analysis. This is exactly what this part will try to explore.

It is probably appropriate to stress the classical flavor of all of our treatment. Even though we shall frequently refer to recent results and the latest theories (such as *algebraic analysis*, or the theory of *Bernstein–Sato polynomials*), it is important to observe that the roots of probably all the problems we discuss here are classical in spirit, since that is the approach we use. For instance, most of Chap. 2 is devoted to far-reaching generalizations of a result dating back to Euler, and it is soon discovered that the key tool for such generalizations was first introduced by Jacobi! As the reader will soon discover, similar arguments can be made for each of the subsequent chapters. Before we give a complete description of our work on a chapter-by-chapter basis, let us make a remark about the list of references. It is quite hard (maybe even impossible) to provide a complete list of references on such a vast topic. Nevertheless, we have aimed to give credit to all of the relevant work in this area, and we have also included some works in the references which are not directly quoted in the main text but which, nevertheless, appear to be of some interest. We sincerely hope that our effort will prove useful to the reader.

Now for the organization of this part. In Chap. 1, we describe the background of complex analysis, which lies at the basis of the modern approach, to mean-periodicity. The main tools needed are provided by the L^2-*theory for the existence of solutions with growth for the inhomogeneous $\bar{\partial}$-system* (the Cauchy–Riemann system). This theory is mostly due to L. Hörmander, but mention should also be made of the work of C.B. Morrey, J.J. Kohn, A. Andreotti, E. Vesentini, and S. Nakano. A key concept for this theory is that of *plurisubharmonicity*, which was only recently understood to be a fundamental one in several complex variables. Among the important results mentioned are theorems about the *extensions of holomorphic functions off algebraic (sometimes analytic) varieties*, as well as theorems about *division with bounds*. This chapter ends with a brief discussion of a different approach to these same questions, namely through the construction of *explicit kernels to solve the $\bar{\partial}$-system*. This idea is due to Khenkin, Grauert, Lieb, and Ramirez; and its modification to accomodate growth conditions was carried out by Berndtsson and Andersson.

Chapter 2 is the central one of this part of the book, and certainly the longest. In this chapter, we study *mean-periodic functions*, the main problem being the *existence of integral Fourier representations for such functions*. Here complex analysis alone is no longer sufficient, and functional analysis techniques have to be thrown in. The key result in this chapter is the renowned Ehrenpreis–Palamodov Fundamental Principle, which provides such a Fourier integral representation for solutions of linear partial differential operators with constant coefficients. The first part of the chapter is devoted to the preparation and explanation of such a result (no proof of it, of course), while the rest of the chapter takes care of its many different generalizations (to other spaces of functions, to convolution equations, and so on).

In Chap. 3, a change of perspective takes place, and the attention focuses on a more geometric type of mean-periodicity: mean-value theorems. The key topic of the chapter is the classical (1929) *Pompeiu problem*. This problem, besides its intrinsic interest, has recently shown its importance in concrete applications, with the creation of *computerized axial tomography*, based on similar principles for the *Radon transform*. This chapter does not touch on any of these applications (for which excellent surveys already exist) but, on the other hand, gives an overview of all the mathematical questions which arise from different generalizations of the Pompeiu problem, and in particular, we point out the strict link between these questions and mean-periodicity as studied in Chap. 2.

When dealing with the extension of the Fundamental Principle to the case of convolution equations, one is usually lead to consider *series of exponentials* (actually, one is faced with objects which are much less manageable, such as *grouped series of integrals, whose integrands are derivatives of exponentials*; but, at least in some simple instances, series of exponentials do indeed arise). These objects have been classically studied by many scholars in this field, either as *Taylor series* or as *Dirichlet series*, and many different questions have been asked at different times. We might just mention the question of *overconvergence* (Fabry, Ostrowski, Dienes, and so on), the problem of *quasianalyticity* (Beurling, Turán, Mandelbrojt, and others), or finally the questions of the *growth of series of exponentials*, with its many links to the theory of Dirichlet series (we shall just mention the name of V.I. Bernstein). The results of Chap. 2, together with a new perspective, now enable a different approach to these old and classical questions, and the description of this approach and of its results is the purpose of Chap. 4.

Chapter 5, on the other hand, is a brief description of a very recent method to prove the Fundamental Principle, based on the *theory of multidimensional residues*. This new approach is mainly due to the work of Berndtsson, Yger and Passare and, once again, relates to some very old questions. New proofs for the Ehrenpreis–Palamodov principle can thus be given (even for some very special convolution equations); and some classical problems related to *Fischer spaces* can also be given a novel treatment. As we shall see, the key tool in this construction is the so-called Bernstein–Sato polynomial.

Finally, in Chap. 6, we deal with a topic which alone could fill several volumes. The ambitious title of the chapter (to which the Bernstein–Sato polynomials is the suitable preface) is *Algebraic Analysis*. By this, we do not mean to say that a treatise on algebraic analysis is to be expected, but that we shall try to outline those aspects of this field that are mostly in line with the topics discussed before. Despite its revolutionary look, algebraic analysis is a very classical topic, which arose out of old problems, such as the surjectivity of differential operators, the propagation of singularities, and so on. These are the questions which underlie the rest of our contribution, and therefore it is of great interest to examine this new approach which, at least for the case of variable coefficients differential equations, seems to be working very well.

As the reader will see, at many points in this part, we refer to this last chapter, a fitting conclusion to the present survey.

Acknowledgements. The authors want to express their gratitude to Professor A.G. Vitushkin for his encouragement to write this survey. The work was started while the first author was visiting the L.A.M. of the University of Calabria. He thanks this institution and the Italian C.N.R. for their support. The second author is grateful to the Department of Mathematics of the University of Maryland for the hospitality during the period in which this text was completed, and for the excellent typing provided by Virginia Sauber. Finally, both authors point out that much of their research, on which this survey is based, was supported by grants from the National Science Foundation and the Ministero della Pubblica Istruzione.

Note from the editor. In this survey, the authors restrict themselves to the studies on convolution in spaces $\mathcal{D}$, $\mathcal{E}$, and their dual spaces. On the other hand, the Cauchy problem for convolution equations requires the consideration of functions (distributions) with polynomial decay (growth), which is also essentially connected to complex analysis. Precisely these questions are studied in the survey by L.R. Volerich and S.G. Gindikin, *The Cauchy Problem* (Vol. 32 of the Russian series, 1988, pp. 5–99).

Chapter 1
Complex Analysis

This chapter is devoted to some preliminary material in complex analysis which is basic for the sequel. One of the major problems in the theory of holomorphic functions in several variables is the following: Given an analytic variety V in $\mathbb{C}^n (n \geq 1)$, and a function f, holomorphic on V, is it possible to extend f to an entire function by keeping track, at the same time, of a given growth condition?

The case where no growth conditions are imposed is still significant, though the answer is easily seen to be affirmative. In fact, the space $H(V)$ of holomorphic functions on V is defined to be the space $H(\mathbb{C}^n)/\mathcal{J}$, with $\mathcal{J}$ being the ideal of functions vanishing on V, and so the surjectivity of the restriction map

$$\rho : H(\mathbb{C}^n) \to H(V)$$

is trivial. However, one can also give an intrinsic definition of $H(V)$, and in this case, the surjectivity of ρ is a consequence of Cartan's theorems A and B (Hörmander [1973]). On the other hand, if growth is imposed, the problem becomes significantly more difficult, and three major methods can be used in attempting to solve it. The first approach, originally due to Oka (Oka [1936; 1937]), consists of modifying Cartan's cohomological methods to the case in which "cohomology with bounds" is considered. Of course, the major difficulty lies exactly in defining the meaning of the term "cohomology with bounds," and the first examples of a consistent approach to this problem are to be found in Ehrenpreis's and Palamodov's monographs (Ehrenpreis [1970]; Palamodov [1970]), where they work out the case in which V is an algebraic variety in $\mathbb{C}^n$. In this situation, holomorphic functions with growth conditions do not constitute a sheaf, hence the usual definition of cohomology groups does not work. To sidestep this issue, Ehrenpreis (Ehrenpreis [1970]) introduces a suitable open covering of $\mathbb{C}^n$, whose open sets have decreasing sizes as they move to infinity. This compensates for the growth of the functions. Then Ehrenpreis proves directly the vanishing of what could be interpreted as the first cohomology group. The theory which Ehrenpreis and Palamodov developed is quite involved, but there are at least two cases which can be understood without much background: the case in which V is a hypersurface defined by the algebraic equation

$$P(z) = 0, \quad z \in \mathbb{C}^n,$$

where P is a polynomial which is a product of distinct irreducible factors, and, more generally, the case in which the variety V is defined by a (nonnecessarily) principal ideal. In both cases, their result can be understood as a quite refined version of Hilbert's Nullstellensatz.

6   C.A. Berenstein, D.C. Struppa

In the first case, the Ehrenpreis–Palamodov theorem states:

Theorem 1.1. *Let V be a hypersurface defined by the algebraic equation $P(z) = 0$, where P is a product of distinct irreducible polynomials. Then if f is a holomorphic function on V which, for some $A, B > 0$, $\rho > 0$, satisfies*

$$|f(z)| \leq A \exp(B|z|^\rho), \quad z \in V,$$

there exists $F \in H(\mathbb{C}^n), F|_V = f$, such that F satisfies, on all of $\mathbb{C}^n$, the same estimate. (For the same value of ρ, but with possibly different constants A, B).

The next simplest example one may want to describe deals with the problem of deciding the extendability of a function of finite order, on a variety V defined by the algebraic equation

$$[P(z)]^s = 0, \quad s \in \mathbb{N}, \quad z \in \mathbb{C},$$

with P being an irreducible polynomial. In this case, we may assume P to be monic in the variable z_n (for $z = (z_1, \ldots, z_n) \in \mathbb{C}^n$), and so a function on V is a *collection*

$$(g_1, \ldots, g_s)$$

of functions which are holomorphic on V and which satisfy the following compatibility condition:

There exists an entire function g such that

$$\frac{\partial^j g}{\partial z_n^j}\big|_V = g_j. \tag{1.1}$$

Under these conditions, one can extend Theorem 1.1 to prove that if all the g_j's are of order ρ and finite type, then g can be chosen to be of order ρ and finite type as well.

The operators $\frac{\partial^j}{\partial z_n^j}$ which appear in (1.1) are just a simple example of what is known in the general case under the name of "Noetherian operators." Let us give here the precise definition.

Definition 1.1. Let $\mathcal{F}$ be an ideal in the ring of polynomials $\mathbb{C}[z] = \mathbb{C}[z_1, \ldots, z_n]$. A *Noetherian operator* for $\mathcal{F}$ is a pair

$$\mathcal{A} = (W, A(z, D)),$$

where W is a Zariski closed subset of $\mathbb{C}^n$, and $A(z, D)$ is a differential operator with polynomial coefficients such that, for any $f \in \mathcal{F}$, one has

$$\mathcal{A}(f) := A(z, D)f|_W = 0.$$

The set W is said to be the *carrier* of $\mathcal{A}$.

As we mentioned before, the results of Ehrenpreis and Palamodov can be interpreted as a refined version of Hilbert's Nullstellensatz, and we can now exploit Definition 1.1 to provide a precise statement in this direction.

Theorem 1.2. *Let $\mathcal{I}$ be an ideal in the ring of polynomials, and let V be the locus of the common zeros of the generators of $\mathcal{I}$; then there exists a finite family $\mathcal{A}_1, \ldots, \mathcal{A}_s$ of Noetherian operators such that:*

(i) *the union of the carriers of $\mathcal{A}_1, \ldots, \mathcal{A}_s$ equals V;*

(ii) *a polynomial f belongs to the ideal $\mathcal{I}$ if and only if $\mathcal{A}_j(f) = 0$ for all $j = 1, \ldots, s$;*

(iii) *if g is an entire function such that each $\mathcal{A}_j(g) = g_j$ is of order ρ and finite type (i.e., $|g_j(z)| \leq A \exp(B|z|^\rho), z \in W_j$), then there exists $G \in H(\mathbb{C}^n)$ of order ρ and finite type, such that $G - g$ belongs to the ideal generated by $\mathcal{I}$ in $H(\mathbb{C}^n)$.*

The most general theorem of this kind deals with modules in $\mathbb{C}[z]^m$. In this case, the notion of Noetherian operator needs to be restated (Palamodov [1970]) as follows.

Definition 1.2. Let $P = (P_{jk}(z))$ be a polynomial matrix of size (J, K). In this case, consider standard the free $\mathbb{C}[z]$-module of rank K,

$$\mathcal{F} = \mathbb{C}[z]\varepsilon_1 \oplus \cdots \oplus \mathbb{C}[z]\varepsilon_K.$$

We associate to P the $\mathbb{C}[z]$-submodule $\mathcal{M}$ of $\mathcal{F}$ generated by

$$\psi_1, \ldots, \psi_J,$$

where

$$\psi_j(z) = \sum_{k=1}^{K} P_{jk}(z)\varepsilon_k, \quad 1 \leq j \leq J.$$

Definition 1.3. Let $\mathcal{M}$ be a submodule of $\mathcal{F}$. A *Noetherian operator* associated to $\mathcal{M}$ consists of a pair

$$\mathcal{A} = (W, A),$$

where W is a Zariski closed subset of $\mathbb{C}^n$, and A is a K-tuple $A = (A_1, \ldots, A_k)$ of differential operators $A_k = A_k(z, D)$ with polynomial coefficients, and

$$\sum_{k=1}^{K} A_k(z, D)Q_k(z) = 0$$

for all elements $m = \sum_{k=1}^{K} Q_k(z)\varepsilon_k$ in $\mathcal{M}$ and at all points z in W.

We can finally state the following extension of Theorem 1.2 to the case of modules.

Theorem 1.3. *Let $\mathcal{M}$ be a submodule of $\mathcal{F}$. Then there exists a finite set $\{A_j\}_{j=1}^s$ of Noetherian operators associated to $\mathcal{M}$ such that if $g = (g_1, \ldots, g_K)$ is some K-tuple of entire functions satisfying*

$$|A_j(z, D)g(z)| \leq A \exp(B|z|^\rho)$$

for all $z \in W_j$ and each $j = 1, \ldots, s$, then there exists another K-tuple of entire functions $(G_1, \ldots, G_K)$ such that each $G_k(k = 1, \ldots, K)$ is of order ρ and finite type, and such that

$$\sum_{k=1}^K G_k(z)\varepsilon_k - \sum_{k=1}^K g_k(z)\varepsilon_k \in H(\mathbb{C}^n) \bigotimes_{\mathbb{C}[z]} \mathcal{M}.$$

We may point out that this result now has several proofs. Its first statement appeared in 1961 (Ehrenpreis [1961]), and later on, complete proofs were given by Ehrenpreis and Palamodov (Ehrenpreis [1970]; Palamodov [1970]). Since then, many other proofs (essentially following the same principles) have appeared in the literature (see Berenstein [1970]; Berenstein and Dostal [1972]; Björk [1979]; Liess [1972]).

A key step in the evolution of this problem is the introduction of Hörmander's L^2-method in Hörmander [1965]. Now a totally new approach could be used to prove the vanishing of cohomology with bounds. The crucial tool of Hörmander's construction, of course, is the existence of solutions to the inhomogeneous Cauchy–Riemann equation, with control on the growth at infinity of the solutions.

Theorem 1.4. *Let φ be a plurisubharmonic function on $\mathbb{C}^n$, and let f be a $(0,1)$-form such that $\bar{\partial}f = 0$. If, moreover,*

$$\int_{\mathbb{C}^n} |f(z)|^2 e^{-2\varphi(z)} d\lambda(z) < +\infty,$$

with $d\lambda$ being the Lebesgue measure on $\mathbb{C}^n$, then there exists a function u on $\mathbb{C}^n$ such that

$$\int_{\mathbb{C}^n} \frac{|u(z)|^2 e^{-2\varphi(z)}}{1 + |z|^2)^2} d\lambda(z) \leq \int_{\mathbb{C}^n} |f(z)|^2 e^{-2\varphi(z)} d\lambda(z),$$

and which is a solution of the equation

$$\bar{\partial}u = f.$$

As Hörmander shows (Hörmander [1973]), Theorem 1.4 can easily be employed to prove the classical theorems A and B of Cartan by showing the vanishing of the cohomology with coefficients in a coherent sheaf on a Stein manifold. On the other hand, Hörmander's result can be used to improve on Hilbert's Nullstellensatz. As an example, in order to show the power of the

L^2-method, we wish to spend a few moments on the question of whether in a space of holomorphic functions with growth control, a function g belongs to the radical of a given ideal. Recall that if $\mathcal{I}$ is an ideal in a ring R, its radical $\sqrt{\mathcal{I}}$ is the ideal of those elements in R which raised to a suitable power belong to $\mathcal{I}$. With this terminology, the Nullstellensatz simply states that if $\mathcal{I}$ is an ideal in $\mathbb{C}[z]$ and if $f(z)$ is a polynomial which vanishes on the set of the common zeros of the generators of $\mathcal{I}$, then $f \in \sqrt{\mathcal{I}}$.

All of Hörmander's theory works in the general framework of plurisubharmonic functions, but in order to simplify our subsequent presentation, we will introduce a few restrictions. In the following, $p(z)$ will be a nonnegative plurisubharmonic function on $\mathbb{C}^n$ such that

(i) $\log(1 + |z|) = 0(p(z))$,
(ii) there exist constants $C, D > 0$ such that if $|z_1 - z_2| \leq 1$, then $p(z_1) \leq Cp(z_2) + D$.

Definition 1.4. For p, a plurisubharmonic function satisfying the above properties (i) and (ii), one defines

$$A_p = \{f \in H(\mathbb{C}^n) : |f(z)| \leq A \exp(Bp(z)), \text{ for some } A, B > 0\}.$$

Definition 1.5. The space of measurable functions f on $\mathbb{C}^n$, for which there exists $C > 0$ such that

$$\int_{\mathbb{C}^n} |f(z)|^2 e^{-Cp(z)} d\lambda(z) < +\infty, \tag{1.2}$$

is indicated by W_p.

By using properties (i) and (ii), one can easily show that $A_p = W_p \cap H(\mathbb{C}^n)$ and Theorem 1.4 can be reformulated in this framework. Namely, if we denote by L_r the space of $(0, r)$-differential forms whose coefficients are in W_p, one has:

Theorem 1.5. *For any $f \in L_{r+1}$ with $\bar{\partial}f = 0$, there exists $u \in L_r$ such that $\bar{\partial}u = f$.*

It may be worthwhile to point out the importance of the equality $A_p = W_p \cap H(\mathbb{C}^n)$. This is what makes the passage from L^2-bounds to supbounds possible, as it will appear clearly in the proof of Theorem 1.6.

One can use these tools to prove the following version of Hilbert's Nullstellensatz (due to Kelleher and Taylor [1967], but see also Kelleher and Taylor [1971a]; Hörmander [1967]; Gurevich [1972]).

Theorem 1.6. *Let $(f_1, \ldots, f_N)$ be an N-tuple of functions in A_p, let $g \in A_p$, and suppose that there exist $\varepsilon, C > 0$ such that, for $z \in \mathbb{C}^n$,*

$$|g(z)| \geq \varepsilon \exp(-Cp(z))[|f_1(z)| + \cdots + |f_N(z)|].$$

Then, for any $k \geq \min\{2n + 1, 2N - 1\}$, g^k belongs to the ideal generated in A_p by $f_1, \ldots, f_N$.

Proof. To give an idea of the techniques employed, we shall outline the proof for the case $n = 1$ (in which case g^3 belongs to the ideal generated by $f_1, \ldots, f_N$), since the case of $n > 1$ can be handled in a totally similar way, provided some natural algebraic techniques are employed (the so-called Koszul complex). Set therefore

$$\|f\|^2 = |f_1|^2 + \cdots + |f_N|^2,$$

and

$$\psi_j = g \frac{\bar{f}_j}{\|f\|^2} = g\varphi_j, \quad j = 1, \ldots, N.$$

It is then clear that $\sum \psi_j f_j = g$ and that the ψ_j are real-analytic (though not at all holomorphic). Now suppose we could solve, with $b_{j,k} \in W_p$ ($1 \leq j, k \leq N$), the equations

$$\frac{\partial b_{j,k}}{\partial \bar{z}} = g\psi_j \frac{\partial \psi_k}{\partial \bar{z}}; \tag{1.3}$$

then the theorem would follow immediately. Indeed, let

$$g_j = g^2\psi_j + \sum_{k=1}^{N}(b_{j,k} - b_{k,j})f_k;$$

then one has

$$\sum_{j=1}^{N} g_j f_j = g^2 \sum_{j=1}^{N} \psi_j f_j = g^3$$

and

$$\frac{\partial g_j}{\partial \bar{z}} = g^2 \frac{\partial \psi_j}{\partial \bar{z}} + \sum_{k=1}^{N} g f_k \left(\psi_j \frac{\partial \psi_k}{\partial \bar{z}} - \psi_k \frac{\partial \psi_j}{\partial \bar{z}} \right) = \cdots = 0.$$

Moreover, since $A_p = W_p \cap H(\mathbb{C}^n)$, it is clear that $g_j \in A_p$ as required. But now we note that a solution $b_{j,k} \in W_p$ for (1.3) exists in view of Theorem 1.5 (as $\bar{\partial}(g\psi_j \frac{\partial \psi_k}{\partial \bar{z}} d\bar{z}) = 0$ is automatic for $n = 1$), and therefore the result is proved in this case. $\qquad \square$

At this point, one should not forget other important work done in this area with respect to the variations on the classical *corona problem* by Kelleher, Taylor and Hörmander (Kelleher, Taylor and Hörmander [1967]; Hörmander [1967]; Kelleher [1966]), and the more recent work of Gentili and Struppa (Gentili and Struppa [1987]).

An even more refined result in this direction was obtained by H. Skoda, who essentially proved in Skoda [1972b] an integral version of Theorem 1.5, namely:

Theorem 1.7. *Let Ω be a pseudoconvex subset of $\mathbb{C}^n$, let p be a plurisubharmonic function on Ω, and let $f_1, \ldots, f_N$ be holomorphic functions on Ω. Set $q = \inf(n, N-1)$, and let V be the set of common zeros of $f_1, \ldots, f_N$. Then, for any function g, holomorphic on Ω, such that*

$$\int_{\Omega \setminus V} |g|^2 \|f\|^{-2q-2}(1 + \Delta \log \|f\|)e^{-p}d\lambda < +\infty,$$

there are N functions $h_1, \ldots, h_N$, holomorphic on Ω, such that $g = \sum_{i=1}^{N} f_i h_i$ and

$$\int_{\Omega} \|h\|^2 \|f\|^{-2q}(1 + |z|^2)^{-2}e^{-p}d\lambda \le 2 \int_{\Omega \setminus V} |g|^2 \|f\|^{-2q-2}(1 + \Delta \log \|f\|)e^{-p}d\lambda.$$

As we proceed to consider holomorphic functions instead of polynomials, it is clear that we gain in generality, though at the cost of some precision. In particular, the so-called division problem is better stated in this more general framework. Before we get to the holomorphic division problem, let us state a simple polynomial case.

Theorem 1.8. *Let f be an entire function of order ρ and finite type. Let $P_1, ..., P_N$ be polynomials in $\mathbb{C}^n$, and assume that f belongs to the ideal generated, in $H(\mathbb{C}^n)$, by $P_1, \ldots, P_N$. Then one can find entire functions $\lambda_1, \ldots, \lambda_N$, of order ρ and finite type, such that on $\mathbb{C}^n$,*

$$f(z) = \lambda_1(z)P_1(z) + \cdots + \lambda_N(z)P_N(z). \tag{1.4}$$

Notice, in particular, that the choice of the weight $|z|^\rho$ is not compulsory, since any radial weight (i.e., $p(z) = p(|z|)$) of not too rapid growth would work as well. Two quite natural questions arise at this moment: The first concerns the case in which f is also a polynomial. One can then find $\lambda_1, \ldots, \lambda_N$ to be polynomials, and now it is natural to inquire what can be said about the degrees of $\lambda_1, \ldots, \lambda_N$. This is already an interesting problem when $f = 1$ and $P_1, \ldots, P_N$ have no common zeros; up until quite recently, the only way to construct the polynomials $\lambda_1, \ldots, \lambda_N$ was to employ some kind of elimination theory. So, for example, if one assumes that $P_1, ..., P_N$ have no common zeros at infinity either (think of the P_j's as homogeneous polynomials in $\mathbb{CP}^n$), then the λ_j's can be chosen with

$$\deg(\lambda_j) \le n(D-1) + 1,$$

with $D = \max \deg(P_j)$. On the other hand, until 1987, the best result for the more general case was due to D.W. Masser and G. Wüstholz, who proved (Masser and Wüstholz [1983]) the existence of λ_j such that

$$\deg(\lambda_j) \le 2(2D)2^{n-1}, \quad j = 1, \ldots, N.$$

12 C.A. Berenstein, D.C. Struppa

Finally, D. Brownawell (Brownawell [1987]) employed some *analytical* results due to the authors (Berenstein and Struppa [1984]; Berenstein, Taylor and Yger [1983b]; Berenstein and Yger [1983]) to dramatically improve these bounds:

Theorem 1.9. *Let $P_1, \ldots, P_N \in \mathbb{C}[z_1, \ldots, z_n]$ have no common zeros, and let $D = \max_j \deg(P_j)$. Then the λ_j's in (1.4) can be taken so that*

$$\deg(\lambda_j P_j) \leq 3\mu n D^\mu,$$

with $\mu = \min(N, n)$.

The methods which were originally employed to prove Theorem 1.9 relied on an important development in complex analysis which (at least partially) supersedes Hörmander's L^2-technique; this is the work on integral representation formulas by Khenkin (Khenkin [1970]), Ramirez (Ramirez de Arellano [1970]), Grauert and Lieb (Grauert and Lieb [1970]), and its extensions by Andersson and Berndtsson (Andersson and Berndtsson [1982]), where weights are taken into consideration. We shall come back to this topic later. Very recently, J. Kollár (Kollár [1988]) found a purely algebraic proof of Theorem 1.9.

The second obvious question which arises from Theorem 1.8 is of a totally different nature; namely, what can we say when $P_1, \ldots, P_N$ are no longer polynomials but entire functions with some kind of growth which is not too fast (say $P_j \in A_p$)? In this case, unfortunately, no general result can be obtained, and some extra conditions on the N-tuple $(P_1, \ldots, P_N)$ have to be imposed.

The first comprehensive approach to this problem is probably that found in Berenstein and Taylor [1980b], though some of its basic ideas can be traced to Ehrenpreis [1970]. To simplify our description, let us assume for the time being that the variety $V = \{z \in \mathbb{C}^n : P_1(z) = \cdots = P_N(z) = 0\}$ is discrete (and also, to place ourselves in a generic situation, let us suppose $n = N$). Assume further that $P_j \in A_p$, for $j = 1, \ldots, n$. The first, crucial step consists of proving a *semilocal division lemma*: For this purpose, we shall consider a *bounded* component Ω of the semilocal neighborhood of V defined by

$$S(P, \varepsilon, C) = \{z \in \mathbb{C}^n : \sum_{j=1}^{n} |P_j(z)| < \varepsilon \exp(-Cp(z))\},$$

for some $\varepsilon, C > 0$. Now let λ be a function holomorphic in V, or, more simply, let $\lambda \in H(\mathbb{C}^n)$ and denote by $[\lambda]$ its equivalence class in $H(V) = H(\mathbb{C}^n)/\mathcal{I}$, with $\mathcal{I}$ being the ideal generated by $P_1, \ldots, P_N$ in $H(\mathbb{C}^n)$. Berenstein and Taylor (Berenstein and Taylor [1980b]) employ a classical formula due to Jacobi (Jacobi [1835]) to associate to each such function λ a new representative of $[\lambda]$, which enjoys particular properties. We refer the reader to Berenstein and Taylor [1980b] for the detailed construction of this Jacobi interpolation formula, which associates to $\lambda \in H(\Omega)$ a new function $I\lambda \in H(\Omega)$. We point out, on the other hand, the main properties of such a formula.

Theorem 1.10. *Let $\lambda \in H(\Omega)$, and consider its Jacobi interpolation $I\lambda$. Then:*

(i) *$I\lambda \in H(\Omega)$;*
(ii) *$[I\lambda] = [\lambda]$, i.e., $I\lambda - \lambda$ belongs to $\mathcal{I}$;*
(iii) *the modulus of $I\lambda(z)$ can be bound in terms of the size of $\lambda(z)$, the diameter of Ω, the length of $\partial\Omega$, and the size of $P_1, \ldots, P_n$ in Ω;*
(iv) *if λ belongs to $\mathcal{I}$, then $I\lambda \equiv 0$.*

With the use of such a formula, one is able to show that if $\lambda \in A_p(\mathbb{C}^n)$, and if $\lambda \in \mathcal{I}$, then the coefficients $g_1, \ldots, g_n$ in

$$\lambda = g_1 P_1 + \cdots + g_n P_n$$

can be chosen to satisfy the same kind of estimates depending on A_p on each compact connected component Ω of $S(P, \varepsilon, C)$. Of course, in order for the g_i to satisfy the same growth on all of $S(P, \varepsilon, C)$, one needs to know that each of its connected components is bounded and that their diameters are somehow uniform (in view of Theorem 1.10). These requirements, unfortunately, are not always satisfied, and one therefore needs to impose the following restrictive condition on the n-tuple $(P_1, \ldots, P_n)$:

Definition 1.6. An n-tuple $P = (P_1, \ldots, P_n)$ of elements in $A_p(\mathbb{C}^n)$ is said to be *slowly decreasing* if there are positive constants ε, C, C_1, C_2 such that

(i) each connected component of $S(P; \varepsilon, C)$ is relatively compact;
(ii) if z_1, z_2 belong to the same connected component of $S(P; \varepsilon, C)$, then $p(z_1) \leq C_1 p(z_2) + C_2$.

This definition, together with Theorem 1.10 allows the process of division with control of the bounds on a set such as $S(P; \varepsilon, C)$ (this is why the result is called semilocal division theorem); the passage from this semilocal result to the global one is finally another application (standard by now) of Hörmander's $\bar{\partial}$-techniques. Indeed (see also Theorem 1.6), one first extends the semilocal division in a C^∞ way and then uses the solvability of the inhomogeneous Cauchy–Riemann equation with bounds (Theorems 1.4 and 1.5) to correct this C^∞ solution to an entire one which satisfies the right growth conditions.

The final statement of the result goes as follows:

Theorem 1.11. *Let $P = (P_1, \ldots, P_n) \in [A_p(\mathbb{C}^n)]^n$ be slowly decreasing, and let $\lambda \in A_p(\mathbb{C}^n)$ belong to the ideal generated in $H(\mathbb{C}^n)$ by $P_1, \ldots, P_n$. Then there exist $\lambda_1, \ldots, \lambda_n \in A_p(\mathbb{C}^n)$ such that, on $\mathbb{C}^n$,*

$$\lambda = \lambda_1 P_1 + \cdots + \lambda_n P_n.$$

Up to now, we have dealt with the analytic division problem only for the case of discrete varieties; this, however, need not be so, and the nondiscrete case may be treated as well, with the only proviso that the variety V must be a *complete intersection*. We shall not get into the details of the notion of *slow decrease* for the nondiscrete case (the reader may see Berenstein and Taylor [1980b]), but we content ourselves with remarking that the Jacobi interpolation formula is typically a formula for discrete varieties; it is therefore necessary to *cut* our variety V with complex planes of a suitable dimension in order to have (on each plane) a discrete variety to which we can apply the Jacobi interpolation formula. It is then necessary to glue together these results to cover a full neighborhood $S(P; \varepsilon, C)$ of the variety V; one then proceeds as before to extend the result from $S(P; \varepsilon, C)$ to all of $\mathbb{C}^n$. In principle, therefore, two distinct vanishings of cohomology arguments are needed to complete the result for the nondiscrete case.

As we already mentioned, division and extension from a variety are almost two faces of the same problem; indeed, in Berenstein and Taylor [1980b], it is shown that under the slow-decrease conditions, extension with bounds can also be achieved, and the method follows the general ideas just described. One first constructs a semilocal extension to $S(P; \varepsilon, C)$, using the Jacobi interpolation formula, and then goes to all of $\mathbb{C}^n$ by means of the vanishing of the $\bar{\partial}$-cohomology.

The slowly decreasing condition, as expressed in Definition 1.6 and even more in the nondiscrete case, is somehow involved. Nevertheless, it is satisfied by a very large class of functions, as has been shown in Berenstein and Taylor [1980b], Berenstein and Yger [1986], and Yger [1988]. As is clear, the condition states that $\sum_{i=1}^{k} |P_i(z)|$ (for $1 \leq k \leq n$, in the general nondiscrete case) should not be too small too often. In most instances, this fact is a consequence, or a modification, of the minimum modulus theorem. Thus, for example, in the case of a single polynomial, the slow decrease is an immediate consequence of the celebrated Ehrenpreis–Malgrange lemma (Ehrenpreis [1970]) on the lower bound of polynomials. In the case of a single function of exponential type (the space $\mathrm{Exp}(\mathbb{C}^n)$ of functions of exponential type is exactly the space A_p, for $p(z) = |z|$), the division theorem is nothing other than the Lindelöf theorem, and the same argument holds whenever $p(z)$ is radial and does not grow too quickly at infinity (by this we mean $p(2z) = 0(p(z))$). Of course, when more than one function is considered, some extra conditions are necessary, even in the case of exponential-type functions which are, for example, satisfied by most n-tuples of exponential sums (i.e., when P is of the form

$$P(z) = \sum_{j=1}^{r} c_j e^{\alpha_j \cdot z}$$

for $c_j \in \mathbb{C}$, $\alpha_j \in \mathbb{C}^n$). On the other hand, if nonradial growth is considered, the situation immediately becomes more involved, even in one variable. An interesting case occurs if one takes $p(z) = |\mathrm{Im}\, z| + \log(1 + |z|)$. In this situation, A_p is isomorphic (actually, topologically isomorphic) to the space $\mathcal{E}'$ of compactly supported distributions (with the isomorphism being provided by the Fourier transform), and it can be easily shown that even in the case of a single function, division with bounds cannot always be achieved. Historically, this case was the first one which led to the notion of slow decrease (see the early papers of Ehrenpreis [1954–62; 1955b] and Hörmander [1968]). On the other hand, in Hörmander [1968], it was shown that any function in this space is close to actually being slowly decreasing. Let us now point out that all of these division results have a simple translation in terms of the ideal theory in the algebra A_p: Namely, one has that every slowly decreasing ideal is a closed ideal, and the search for conditions which ensure that ideals in weighted algebras of holomorphic functions are closed, has a rather long and instructive history. In this regard, we limit ourselves to refer the interested reader to some of the most relevant classical papers: Kelleher and Taylor [1972], Krasichkov-Ternovskij [1968b], Kahane [1959], Malgrange [1956], Ehrenpreis [1954–62], Schwartz [1943a; 1947], and the survey Nikol'skij [1968]. More details will be given in Chap. 2.

Another important outgrowth of Hörmander's work is the possibility of applying his methods to deal with spaces which had not been formerly studied by Ehrenpreis and Palamodov because of their topological structure, and which appear in a natural way when studying basic questions in quantum field theory. Let us shortly describe these spaces, following De Roever's approach as given in De Roever [1978]. These spaces arise in a standard way when generalizing the Ehrenpreis–Martineau theorem (Ehrenpreis [1959]; Martineau [1963]), which shows that the Fourier (or Fourier–Borel, as it is frequently called) transform is a topological isomorphism between the space of analytic functionals with compact convex carrier K and the space of exponential-type functions, whose growth is bounded by $A \exp(H_K(z))$, where H_K is the supporting function of K $(H_K(z) := \sup_{\zeta \in K} \operatorname{Re}(z \cdot \zeta))$. The interesting object which arises from quantum field theory is the space of those analytic functions which are carried by a closed convex set $\Omega \subseteq \mathbb{C}^n$ (the case in which Ω is bounded is the one treated by the Ehrenpreis–Martineau theorem). In this situation, it is possible to show the existence of an open convex cone Γ in $\mathbb{C}^n$, and of a convex function g on Γ, which is homogeneous of degree one, such that a topological isomorphism exists between the space of analytic functionals and the space $H(\Gamma, g)$ of holomorphic functions of exponential type g in Γ (see De Roever [1978] for precise definitions). This isomorphism is of particular interest if Γ is the open tube domain

$$\Gamma^c = \mathbb{R}^n + iC,$$

where C is the open convex cone in $\mathbb{R}^n$. Then the functions holomorphic in Γ^c are the Fourier transforms of analytic functionals carried by some convex sets which may be unbounded in the imaginary direction. What is more interesting to us is the fact that De Roever uses Hörmander's L^2-methods to prove, for polynomial ideals in $H(\Gamma, g)$, a division theorem exactly as Theorem 1.8, where entire functions of order ρ and finite type are replaced by holomorphic functions in $H(\Gamma, g)$.

A final interesting aspect of De Roever's theory is the fact that for a tube domain Γ, these holomorphic functions have boundary values which are Fourier hyperfunctions, in the sense defined by T. Kawai (Kawai [1970]). Therefore, this naturally leads us to consider the different approach to the cohomology with bounds, which was introduced by the Japanese school of M. Sato, beginning with Kawai's master's thesis (Kawai [1970]), where holomorphic functions with growth conditions are made into a sheaf by suitably compactifying the space $\mathbb{R}^n$ in a radial way. To describe these ideas, we need some preliminary definitions.

Definition 1.7. Denote by $\mathbb{D}^n$ the compactification $\mathbb{R}^n \amalg S_\infty^{n-1}$ of $\mathbb{R}^n$, where S_∞^{n-1} is an $(n-1)$-dimensional sphere at infinity. $\mathbb{D}^n$ is given its natural topology.

Now consider the space $\mathbb{D}^n \times i\mathbb{R}^n$ and define on it the *sheaf of slowly increasing holomorphic functions* as follows ($\mathcal{O}$ denotes the sheaf of holomorphic functions on $\mathbb{C}^n$).

Definition 1.8. The sheaf $\tilde{\mathcal{O}}$ of slowly increasing holomorphic functions is the sheaf whose section module $\tilde{\mathcal{O}}(\Omega)$ over an open set $\Omega \subseteq \mathbb{D}^n \times i\mathbb{R}^n$ is the set of all holomorphic functions $f \in \mathcal{O}(\Omega \cap \mathbb{C}^n)$ which, for any positive ε and any compact set K in Ω, satisfy

$$\sup_{z \in K \cap \mathbb{C}^n} |f(z)e^{-\varepsilon|z|}| < +\infty$$

Definition 1.9. An open set $\Omega \subseteq \mathbb{D}^n \times i\mathbb{R}^n$ is said to be $\tilde{\mathcal{O}}$-pseudoconvex if the following conditions are satisfied:

(i) $\sup_{z \in V} |\operatorname{Im} z| < +\infty$, for $V = \Omega \cap \mathbb{C}^n$;
(ii) there exists a plurisubharmonic function $\varphi(z)$ on V which satisfies

$$\{z : \varphi(z) < c\} \subset\subset V \quad \text{for any} \quad c,$$

and
$$\sup L \cap \mathbb{C}^n \varphi(z) \leq M_L \quad \text{for any} \quad L \subset\subset \Omega.$$

Using Hörmander's L^2-technique and after some natural modifications, Kawai proved the vanishing of cohomology in this new framework.

Theorem 1.12. *For any $\tilde{\mathcal{O}}$-pseudoconvex domain Ω in $\mathbb{D}^n \times i\mathbb{R}^n$, we have $H^s(\Omega, \tilde{\mathcal{O}}) = 0$, for $s \geq 1$.*

This important result was then employed by A. Kaneko (Kaneko [1970; 1972]) to treat overdetermined systems in the space of hyperfunctions. We shall return to this point later. Note, however, that condition (i) in Definition 1.9 can be weakened and one can obtain, for example, that

$$H^s(\mathbb{D}^n \times i\mathbb{R}^n, \tilde{\mathcal{O}}) = 0, \quad s \geq 1,$$

while clearly $\mathbb{D}^n \times i\mathbb{R}^n$ is not $\tilde{\mathcal{O}}$-pseudoconvex.

This approach of Kawai's shows that the L^2-methods can be used towards proving the vanishing of other kinds of cohomology with bounds. This idea has been carefully exploited by Saburi (Saburi [1978]) and by Meril in a series of papers (Meril [1983a,b,c]; see also Berenstein and Struppa [1983], strongly influenced by Meril's work). In particular, in order to deal with the spaces studied by De Roever, Meril defines a different sheaf of holomorphic functions satisfying a growth condition. In this first work on the topic (Meril [1983a]), Meril considers $S^{2n-1}_\infty = \{U_\infty : \text{set directions in } U \subseteq S^{2n-1}\}$ and the *spherical compactification of* $\mathbb{C}^n$ defined by

$$\mathbb{D}^{2n} = \mathbb{C}^n \amalg S^{2n-1}_\infty,$$

with its natural topology. Then let g be the supporting function of a fixed convex subset of $\mathbb{C}^n$ which contains the origin. The compact will be chosen so that $g \in \mathcal{C}^1(\mathbb{C}^n \backslash \{0\})$. One then defines the sheaf ${}^g\mathcal{O}$ as follows: For every open set $\Omega \subseteq \mathbb{C}^2 n$, set

$${}^g\mathcal{O}(\Omega) = \big\{ f \in H(\Omega \cap \mathbb{C}^n) : \forall K \subset\subset \Omega, \forall \varepsilon > 0,$$
$$\sup_{K \cap \mathbb{C}^n} |f(z) \exp(-g(z) - \varepsilon|z|)| < +\infty \big\}.$$

It is obvious that such a sheaf coincides with the sheaf $\mathcal{O}$ when restricted to $\mathbb{C}^n$, and if g is the supporting function of a compact K_0, then

$$\exp(K_0) = \Gamma(\mathbb{D}^{2n}, {}^g\mathcal{O}).$$

In Meril's paper, the sheaf of p-forms ${}^g\mathcal{O}^p$ is then given a soft resolution by means of the sheaves ${}^g\mathcal{L}_2^{p,q}$, where p, q are integers denoting the bidegrees of differential forms and where ${}^g\mathcal{L}_2$ is defined in the same way as ${}^g\mathcal{O}$, replacing sup norms with L^2-norms. Vanishing of cohomology is then proved, essentially following Kawai's paper (Kawai [1970]), and division and extension are proved for ideals generated by holomorphic functions in De Roever's spaces. As is probably clear, all of the previous works rely more or less directly on the solvability of the $\bar{\partial}$-equation proved by Hörmander. Quite recently though, a totally new method was developed by Khenkin, Grauert, Lieb, and Ramirez to provide explicit solutions to the $\bar{\partial}$-equation (Khenkin [1970; 1969]; Ramirez de Arellano [1970]; Grauert and Lieb [1970]). As could be expected, this method has led to great improvements with respect to most of the problems discussed earlier. Let us give a partial description of the developments in this area: For further detail, we refer the reader to the monographs Henkin (= Khenkin) and Leiterer [1984], and Range [1986] and to the survey Khenkin [1990].

One of the main problems in complex analysis is finding explicit integral representations. The first such formula is probably the Cauchy–Green formula (also known as the Pompeiu formula), in which, for *a C^1-function f on the closure of a bounded set $D \subseteq \mathbb{C}$ with C^1-boundary, one has*

$$f(z) = \frac{1}{2\pi i} \int_{\partial D} \frac{f(\zeta)d\zeta}{\zeta - z} - \frac{1}{2\pi i} \int_D \frac{\bar{\partial}f(\zeta) \wedge d\zeta}{\zeta - z}, \quad z \in D. \tag{1.5}$$

Extensions of (1.5) to domains of $\mathbb{C}^n$ have been the main concern of complex analysts for quite a while, giving rise to a whole lot of different generalizations. For the sake of compactness of exposition, we follow Khenkin and Leiterer (Henkin and Leiterer [1984]) and shall give some preliminary definitions and notations. For fixed z, we define,

$$\omega(\zeta) := d\zeta_1 \wedge \cdots \wedge d\zeta_n,$$

$$\omega'_\zeta(\bar\zeta - \bar z) := \sum_{j=1}^n (-1)^{j+1}(\bar\zeta_j - \bar z_j)\left(\bigwedge_{k \neq j} d\bar\zeta_k\right).$$

If $D \subset\subset \mathbb{C}^n$ is an open set and f is a bounded 1-form on D, then we define

$$(B_D f)(z) := \frac{(n-1)!}{(2\pi i)^n} \int_D f(\zeta) \wedge \frac{\omega'_\zeta(\bar\zeta - \bar z) \wedge \omega(\zeta)}{|\zeta - z|^{2n}}, \quad z \in D.$$

If $D \subset\subset \mathbb{C}^n$ is an open set with piecewise C^1-boundary and f is a bounded measurable function on ∂D, then we define

$$(B_{\partial D} f)(z) := \frac{(n-1)!}{2\pi i)^n} \int_{\partial D} f(\zeta) \wedge \frac{\omega'_\zeta(\bar\zeta - \bar z) \wedge \omega(\zeta)}{|\zeta - z|^{2n}}, \quad z \in D.$$

Note that for $n = 1$, $B_D f$ and $B_{\partial D} f$ are the operators of the Pompeiu formula (1.5). In fact, the definitions of $B_D, B_{\partial D}$ can be extended to differential forms of arbitrary degree in an obvious way, which we will therefore not repeat.

Let D be a bounded open set in $\mathbb{C}^n$. A $\mathbb{C}^n$-valued C^1-function $w(z, \zeta) = (w_1(z, \zeta), \ldots, w_n(z, \zeta))$ defined for ζ in some neighborhood of ∂D and $z \in D$ is called a *Leray map for D* if ($\langle\,,\rangle$ denoting the usual bilinear product in $\mathbb{C}^n$)

$$\langle w(z, \zeta), \zeta - z\rangle \neq 0 \quad \text{for all} \quad (z, \zeta) \in D \times \partial D. \tag{1.6}$$

Let $w(z, \zeta)$ be a Leray map for D and suppose that ∂D is piecewise C^1 and define

$$\eta^w(z, \zeta, \lambda) := (1 - \lambda)\frac{w(z, \zeta)}{\langle w(z, \zeta), \zeta - z\rangle} + \lambda\frac{\bar\zeta - \bar z}{\langle\bar\zeta - \bar z, \zeta - z\rangle}$$

for all $z \in D, 0 \leq \lambda \leq 1$, and ζ in some neighborhood of ∂D. Then, by (1.6) for every fixed $z \in D$, the differential form

$$\frac{\omega'_\zeta(w(z, \zeta)) \wedge \omega(\zeta)}{\langle w(z, \zeta), \zeta - z\rangle} := \sum_{j=1}^n (-1)^{j+1}\frac{w_j(z, \zeta)}{\langle w(z, \zeta), \zeta - z\rangle^n}\bigwedge_{k \neq j} d_\zeta w_k(z, \zeta) \wedge \omega(\zeta)$$

is continuous for ζ in some neighborhood of ∂D, and the differential form

$$\omega'_{\zeta,\lambda}(\eta^w(z, \zeta, \lambda)) \wedge \omega(\zeta) := \sum_{j=1}^n (-1)^{j+1}\eta^w_j(z, \zeta, \lambda)\bigwedge_{k \neq j} d_{\zeta,\lambda}\eta^w_k(z, \zeta, \lambda) \wedge \omega(\zeta)$$

is continuous for ζ in some neighborhood of ∂D and for $0 \leq \lambda \leq 1$. Therefore, one can define for f a bounded measurable complex-valued function on ∂D,

$$(L^w_{\partial D} f)(z) := \frac{(n-1)!}{2\pi i)^n} \int_{\partial D} f(\zeta) \frac{\omega'_{zeta}(w(z,\zeta)) \wedge \omega(\zeta)}{\langle w(z,\zeta), \zeta - z \rangle^n}, \quad z \in D,$$

and, for f, a bounded 1-form on ∂D,

$$(R^w_{\partial D} f)(z) := \frac{(n-1)!}{2\pi i)^n} \int_{\substack{\partial D \\ 0 \leq \lambda \leq 1}} f(\zeta) \wedge \omega'_{\zeta,\lambda}(\eta^w(z,\zeta,\lambda)) \wedge \omega(\zeta), \quad z \in D.$$

As before, these definitions can be immediately extended to differential forms of arbitrary degree. We can now state the first important generalization of the Cauchy–Green formula, namely the Bochner–Martinelli formula (Martinelli [1938]; May [1941]; Bochner [1943]).

Theorem 1.13. *Let $D \subset\subset \mathbb{C}^n$ be an open set with piecewise C^1 boundary, and let f be a continuous function on $\bar{D}$ such that $\bar{\partial} f$ is also continuous on $\bar{D}$. Then on D,*

$$f = B_{\partial D} f - B_D \bar{\partial} f.$$

For $n \geq 2$, the kernel in the Bochner–Martinelli formula does not depend holomorphically on z, so that in 1956 Leray developed a different formula in which $\bar{\zeta} - \bar{z}$ is replaced by a more general Leray map $w(z,\zeta)$ (the original idea actually dates back to 1943 and is due to Fantappiè (Fantappiè [1943]). This formula really lies at the base of all the subsequent results. The following theorem was obtained by Khenkin (Khenkin [1970]) in 1969.

Theorem 1.14. *Let $D \subset\subset \mathbb{C}^n$ be an open set with piecewise C^1 boundary, and let $w(z,\zeta)$ be a Leray map for D. Then, for every continuous function f on $\bar{D}$ such that $\bar{\partial} f$ is also continous on $\bar{D}$, we have on D,*

$$f = L^w_{\partial D} f - R^w_{\partial D} \bar{\partial} f - B_D \bar{\partial} f.$$

Note that the Leray formula reduces to the Bochner–Martinelli one when $w(z,\zeta) = \bar{\zeta} - \bar{z}$. It is often of interest to have extensions of Theorems 1.13 and 1.14 to the case of differential forms, and not just functions. These extensions are known, respectively, under the names of Koppelman formula (Theorem 1.15 was first proved by Koppelman in 1967; see Koppelman [1967]) and of Koppelman–Leray formula (actually first proved by Lieb in 1970 (Lieb [1970–71]), and by Polyakov (Polyakov [1971]) and Ovrelid in 1971 (Ovrelid [1971a])).

Theorem 1.15. *Let $D \subset\subset \mathbb{C}^n$ be an open set with piecewise C^1 boundary, and let f be a continuous $(0,q)$-form on $\bar{D}$ such that $\bar{\partial} f$ is also continuous on $\bar{D}, 0 \leq q \leq n$. Then the forms $B_{\partial D} f, B_D \bar{\partial} f, B_D f,$ and $\bar{\partial} B_D f$ are continuous in D, and we have on D,*

$$(-1)^q f = B_{\partial D} f - B_D \bar{\partial} f + \bar{\partial} B_D f.$$

Theorem 1.16. *Let $D \subset\subset \mathbb{C}^n$ be an open set with piecewise C^1 boundary, and let $w(z, \zeta)$ be a Leray map for D. Then, for every continuous $(0, q)$-form on $\bar{D}$, $0 \le q \le n$, the forms $L^w_{\partial D} f$, $R^w_{\partial D} f$, $B_D \bar{\partial} f$, $R^w_{\partial D} \bar{\partial} f$, $B_D f$, $\bar{\partial} R^w_{\partial D} f$, and $\bar{\partial} B_D f$ are continuous in D, and we have on D,*

$$(-1)^q f = L^w_{\partial D} f - (R^w_{\partial D} + B_D) \bar{\partial} f + \bar{\partial}(R^w_{\partial D} + B_D) f.$$

All of these formulas are representation formulas for continuous functions and forms. However, the important results from our point of view relate to the explicit construction of solutions to the equations $\bar{\partial} u = f$. Here we provide some of these constructions in different situations before we get to the weighted results of Berndtsson and Andersson. Let $D \subset\subset \mathbb{C}^n$ be a strictly convex open set with C^2-boundary, defined by

$$D = \{z \in \mathbb{C}^n, \rho(z) < 0\},$$

where ρ is a real-valued $\mathbb{C}^2$ function in $\mathbb{C}^n$ such that, for some $\alpha > 0$,

$$\sum_{j,k=1}^{2n} \frac{\partial^2 \rho(z)}{\partial x_j \partial x_k} t_j t_k \ge \alpha |t|^2, \quad z \in \partial D, \quad t \in \mathbb{R}^{2n}.$$

Set

$$w_\rho(\zeta) := 2 \left(\frac{\partial \rho}{\partial \zeta_1}(\zeta), \dots, \frac{\partial \rho}{\partial \zeta_n}(\zeta) \right),$$

which can be shown to be a Leray map for D. The following result follows from the Koppelman–Leray formula (Theorem 1.16).

Theorem 1.17. *Let D be as above, and let f be a continuous $(0, q)$-form on $\bar{D}$ such that $\bar{\partial} f = 0$ in D, $1 \le q \le n$. Then*

$$u := (-1)^q (R^{w_\rho}_{\partial D} f + B_D f)$$

is a solution of $\bar{\partial} u = f$ in D. This solution belongs to $C^\alpha_{(0,q-1)}(D)$ for all $0 < \alpha \le 1/2$.

Modifications of this result which allow for Hölder estimates can also be given (see e.g., Henkin and Leiterer [1984]; Range [1986]). Also, analogous results on the $\bar{\partial}$-equation can be carried over to a much more general situation when one considers $(0, q)$-forms with values in holomorphic vector bundles over complex manifolds.

Let us now give a more detailed description of the work of Andersson and Berndtsson (Andersson and Berndtsson [1982]). Their starting point is the Leray formula (Theorem 1.14), which represents every function as a sum of three terms, each of which contains a kernel. One wants to construct those kernels in such a way that they provide the required growth. The following version of such a formula has been given in Andersson and Berndtsson [1982].

Let $D \subset\subset \mathbb{C}^n$ be open and with C^2 boundary. Assume the existence of a C^1 function

$$s : \bar{D} \times \bar{D} \to \mathbb{C}^n,$$

such that for every compact subset $K \subset D$, there are constants $C_1, C_2 > 0$ for which the following conditions hold for every $z \in \bar{D}$, $\zeta \in K$:

$$\|s(z,\zeta)\| \leq C_1 \|z - \zeta\|,$$

$$|\langle s, z - \zeta \rangle| \geq C_2 \|z - \zeta\|^2,$$

and define

$$S := \sum_{j=1}^{n} s_j \, d\zeta_j.$$

We now consider pairs of functions (q, Γ) with the following properties: The function

$$q : \bar{D} \times \bar{D} \to \mathbb{C}^n$$

is C^1, and the map $z \mapsto q(z, \zeta)$ is holomorphic in D for every $\zeta \in \bar{D}$. Let Φ be the function defined by

$$\Phi(z, \zeta) := \langle q, z - \zeta \rangle + 1,$$

and let Q be the differential form

$$Q := \sum_{j=1}^{n} s_j(z, \zeta) \, d\zeta_j$$

(note that $\bar{\partial} Q = \bar{\partial}_{\zeta, z} Q = \bar{\partial}_\zeta Q$). Moreover, let Γ be a holomorphic function of a single complex variable, defined in a neighborhood of $\Phi(\bar{D} \times \bar{D})$ and such that

$$\Gamma(1) = 1,$$

and for α being a nonnegative integer, denote by $\Gamma^{(\alpha)}$ the function defined in $D \times D$ by

$$\Gamma^{(\alpha)} := \frac{d^\alpha \Gamma}{dt^\alpha}\Big|_{t = \Phi(z, \zeta)}.$$

Given $M + 1$ pairs $(q_1, \Gamma_1), \ldots, (q_{M+1}, \Gamma_{M+1})$ and a multiindex $\alpha = (\alpha_1, \ldots, \alpha_{M+1})$, we set

$$\Gamma^{(\alpha)} = \Gamma_1^{(\alpha_1)} \cdot \cdots \cdot \Gamma_{M+1}^{(\alpha_{M+1})}$$

and

$$(\bar{\partial} Q)^\alpha = (\bar{\partial} Q_1)^{\alpha_1} \wedge \cdots \wedge \bar{\partial} Q_1)^{\alpha_{M+1}}.$$

With this notation, we can define the explicit kernels P and K which are used in the Koppelman formula. We let

$$P := \sum_{|\alpha|=n} \frac{1}{\alpha!} \Gamma^{(\alpha)} (\bar{\partial} Q)^\alpha,$$

and for α being a multiindex of length $M + 1$ and α_0 a nonnegative integer, set

$$K := \sum_{\alpha_0 + |\alpha| = M-1} \frac{1}{\alpha!} \Gamma^{(\alpha)} \frac{S \wedge (\bar{\partial}S)^{\alpha_0} \wedge (\bar{\partial}Q)^{\alpha}}{\langle s, z - \zeta \rangle^{\alpha_0 + 1}}.$$

Then, in Andersson and Berndtsson [1982], the following representation is proved:

Theorem 1.18. *Every $f \in C^1(\bar{D})$ can be represented as follows: for $z \in D$,*

$$f(z0 = \frac{1}{(2\pi i)^n} \left\{ \int_{\partial D} f(\zeta) K(z, \zeta) - \int_D \bar{\partial} f(\zeta) \wedge K(z, \zeta) + \int_D f(\zeta) P(z, \zeta) \right\}.$$

One can actually apply this result to give explicit versions for the division problem, as done in Berndtsson [1983] and Berenstein and Yger [1983]. In view of the technical complexity of such formulas, we refer the reader to the original papers. We now want to conclude this chapter by mentioning several related developments which, while not exactly in the same spirit as our main topics, do indeed provide some information on closely related matters.

The first one is related to the work of Demailly (Demailly [1982b]) and to the recent work of Ohsawa, Forster, Takegoshi, and Nakano (Ohsawa and Takegoshi [1987]; Forster and Ohsawa [1987]; Nakano [1986]) on the extension of holomorphic functions with various growth conditions. The most classical extension problems deal with the extension from a variety in $\mathbb{C}^n$ to all of $\mathbb{C}^n$.

Without going into details, we want to point out that solutions of the $\bar{\partial}$-system on vector bundles have been employed in Demailly [1982b] to provide more refined results on the extension of holomorphic functions with growth conditions. Let us give the flavor of his work by mentioning one such result: Let X be an n-dimensional, weakly pseudoconvex Kähler manifold, and let E, M be two hermitian vector bundles over X, with $rk(M) = m$. Let $Y = F^{-1}(0)$ be an analytic subset of X, defined by a holomorphic section F of M, and set

$$U = \{z \in X : |F(z)| < 1\}.$$

One has the following extension theorem (see also an earlier paper, Jennane [1978]).

Theorem 1.19. *Let η be a positive real number and $q = \inf(n, m)$. Let Y be X-negligible and suppose that the curvature form $c(E)$ of E satisfies the following inequality (to be interpreted in the sense of Nakano semipositiveness):*

$$i\, c(E) \geq \left(\frac{\eta}{1 + |F|^2} + \frac{q}{|F|^2} \right) \langle i\, c(M) \cdot F, F \rangle - i\, \mathrm{Ricci}\,(X).$$

Then, for every section g of E over U such that

$$\int_U |g|^2 dV < +\infty,$$

there exists a section G of E over X which coincides with g over Y and such that

$$\int_X \frac{|G|^2 dV}{(1+|F|^2)^{q+\eta}} \le C(q,\eta) \int_U |g|^2 dV,$$

with $C(q,\eta) = 1 + \frac{(q+1)^2}{\eta}$.

A recent work of Ohsawa and Takegoshi deals with the extension of L^2-holomorphic functions on a variety to bounded domains in $\mathbb{C}^n$. We quote here the following important result from Ohsawa and Takegoshi [1987].

Theorem 1.20. *Let Ω be a bounded pseudoconvex domain in $\mathbb{C}^n$, let φ be a plurisubharmonic function in Ω, and $H \subset \mathbb{C}^n$ a complex hyperplane. Then there exists a constant C depending only on the diameter of Ω such that, for any holomorphic function f on $\Omega \cap H$ verifying*

$$\int_{\Omega \cap H} e^{-\varphi} |f|^2 d\lambda < +\infty,$$

(where $d\lambda$ is the $(2n-2)$-dimensional Lebesgue measure), there is a holomorphic function F on Ω, satisfying $F|_{\Omega \cap H} = f$ and

$$\int_\Omega e^{-\varphi} |F|^2 d\lambda \le C \int_{\Omega \cap H} e^{-\varphi} |f|^2 d\lambda.$$

This is a generalization of Bombieri's theorem (Hörmander [1973], Theorem 4.4.4). The result is based on the Andreotti–Vesentini theory (Andreotti and Vesentini [1965]), which, in this case, proves to be more flexible than Hörmander's method.

A second topic which we have not mentioned is the study of division, extension and, more generally, of the ideal theory in the algebra $A^\infty(\Omega)$ of functions which are holomorphic on Ω and smooth up to the boundary. Quite recently, a nice survey on this topic has been published (Saerens [1987]), and we refer to it for details. Let us just mention the work of Bartolomeis and Tomassini (Bartolomeis and Tomassini [1981; 1982]), on finitely generated ideals in $A^\infty(\Omega)$, of Bruna and Ortega (Bruna and Ortega [1984a,b]), Amar (Amar [1980; 1983; 1984; 1985; 1987]), and of Gay and Sebbar (Gay and Sebbar [1985]; Sebbar [1985]).

All of the previous results concern the existence of extensions of given holomorphic functions with suitable growth conditions, but a closely related question, first discussed in Khenkin and Mityagin [1971] and Djakov and Mityagin [1980], is the *linearity* of such an extension. This problem was eventually reduced (see Khenkin and Mityagin [1971]) to the question of the linearity of the solvability of the $\bar{\partial}$-system in various spaces and was finally approached via functional analytic tools by Taylor (Taylor [1982]) and eventually by Meise, Taylor, and Vogt (see Meise and Taylor [1988; 1989]; Meise, Taylor and Vogt [1987a,b; 1988]).

Let us therefore state two of their most interesting results. Let $\{z_k\}$ be a discrete sequence of points in $\mathbb{C}^n$; let $\hat{\mathcal{E}}'(\mathbb{R}^n)$ denote the space of entire functions on $\mathbb{C}^n$ which are Fourier transforms of compactly supported distributions on $\mathbb{R}^n$, and let Γ be the space of all the sequences $\{\gamma_k\}$ of complex numbers such that

$$|\gamma_k| \leq B(1 + |z_k|)^C \exp(D|\mathrm{Im}\, z_k|),$$

for some $B, C, D > 0$. One therefore has a natural restriction map

$$\rho : \hat{\mathcal{E}}'(\mathbb{R}^n) \to \Gamma,$$

and $\{z_k\}$ is said to be *interpolating* if ρ is surjective. In Taylor [1982], the following result was proved.

Theorem 1.21. *The restriction ρ admits a continuous linear right inverse if and only if*

$$\lim_{k \to +\infty} \sup \frac{|\mathrm{Im}\, z_k|}{\log(1 + |z_k|)} < +\infty.$$

A more powerful result is proved in Meise and Taylor [1988], from which we quote the following extension theorem which relies on the topological property (DN), introduced by Vogt in Vogt and Wagner [1980], to which we refer for details.

Theorem 1.22. *Let p be a plurisubharmonic function on $\mathbb{C}^n$, such that the strong dual $[A_p(\mathbb{C}^n)]'$ of $A_p(\mathbb{C}^n)$ has the (DN) property, and let $F = (F_1, \ldots, F_k)$ be a slowly decreasing k-tuple in $A_p(\mathbb{C}^n)$. Then, if V is the multiplicity variety of the ideal generated by $F_1, \ldots, F_k$, there exists a continuous linear extension operator*

$$E : A_p(V) \to A_p(\mathbb{C}^n),$$

where $A_p(V) = \{f \in H(V) : \sup |f(z)| \exp(-Ap(z))| < +\infty \text{ for some } A > 0\}$.

Chapter 2
Mean-Periodicity and Representation Theorems

The expression *mean-periodic* was first used by J. Delsarte in Delsarte [1935] to indicate functions which were solutions of very special convolution equations (we will soon describe the concrete examples). In recent years, and due to the work of a large group of mathematicians (Delsarte, Kahane, Ehrenpreis, Malgrange, Martineau, Schwartz, Hörmander, and others), we have now reached a fairly complete understanding of the structure of such equations and of the links existing between these problems and many (apparently unrelated) other questions in complex analysis, functional analysis, and partial differential equations. In this chapter (the central one of this contribution), we shall try to put all of these questions into the correct perspective and to link the chapter with the developments touched upon in the other chapters of this work.

However, before we begin, we need to put forward a word of caution: In all of this theory, a leading role has been played by L. Hörmander. His work is so pervasive—and so clearly discussed in the second volume of his monograph (Hörmander [1962])—that we should probably be quoting him every other line; for this reason, we decided to point this out in advance and to refer the reader to Hörmander [1983; 1985] for a full appreciation of his contribution (it is really hard to decide which of his many ideas seems to be the most decisive of all).

Let us now start by fixing some notations: We already met the topological vector space $\mathcal{E} = \mathcal{E}(\mathbb{R}^n)$ of C^∞ functions, endowed with the topology of uniform convergence of all derivatives on compact subsets of $\mathbb{R}^n$, which makes it a Fréchet–Montel space. We denote the space of Schwartz test functions by $\mathcal{D} = \mathcal{D}(\mathbb{R}^n)$: As a set, $\mathcal{D}$ consists of those C^∞ functions on $\mathbb{R}^n$ whose support is compact, while its usual topology (Schwartz [1950–51]) is such that the natural inclusion $\mathcal{D} \hookrightarrow \mathcal{E}$ is *dense and continuous*, and the topology of $\mathcal{D}$ is strictly finer than the one induced by $\mathcal{E}$. The space $\mathcal{D}'$ of Schwartz distributions then appears as the dual of $\mathcal{D}$ and will be considered with the *strong dual* topology, i.e., with the topology of uniform convergence on the *bounded* subsets of $\mathcal{D}$ (cf. Schwartz [1950–51]). Since the inclusion $\mathcal{D} \hookrightarrow \mathcal{E}$ has a dense image, the dual $\mathcal{E}'$ of $\mathcal{E}$ can be identified with a vector subspace of $\mathcal{D}'$ and if $\mathcal{E}'$ is also given the strong topology, then $\mathcal{E}' \hookrightarrow \mathcal{D}'$ becomes a continuous injection; as we know, $\mathcal{E}'$ turns out to be the subspace of $\mathcal{D}'$ made of compactly supported distributions. The spaces $\mathcal{E}$ and $\mathcal{D}$ are reflexive, that is, $\mathcal{D} = (\mathcal{D}')'$ and $\mathcal{E} = (\mathcal{E}')'$ (always taking strong topologies). For Ω, an open subset of $\mathbb{C}^n$, on the other hand, one can define the space $H(\Omega)$ of holomorphic functions of Ω as the closed subspace of $\mathcal{E}(\Omega)$ defined by the Cauchy–Riemann equations $\frac{\partial f}{\partial \bar{z}_j} = 0$, for $j = 1, \dots, n$, so that $H(\Omega)$ inherits from $\mathcal{E}(\Omega)$ a structure of Fréchet space (by independent arguments, one also knows that $H(\Omega)$ is Montel and therefore reflexive). There are two important ways of relating the spaces H and $\mathcal{E}$ (this is somehow underlying—though not explicitly—most of the arguments in Chap. 4) which we want to recall. First, one can take $f \in H(\mathbb{C}^n)$ and restrict it to $\mathbb{R}^n = \mathrm{Re}(\mathbb{C}^n)$. This restriction is a real-analytic function and, in particular, belongs to $\mathcal{E}(\mathbb{R}^n)$; the restriction

$$\begin{cases} H(\mathbb{C}^n) \longrightarrow \mathcal{E}(\mathbb{R}^n) \\ f \longmapsto f_{|\mathrm{Re}(\mathbb{C}^n)} \end{cases} \tag{2.1}$$

is a continuous linear map (as the topology of H is equivalent to the one induced by $\mathcal{E}(\mathbb{R}^{2n})$), which is easily seen to be one-to-one and with dense image. On the other hand, the identification of $\mathbb{C}^n$ with $\mathbb{R}^{2n}$ provides a natural injection

$$H(\mathbb{C}^n) \hookrightarrow \mathcal{E}(\mathbb{R}^{2n}); \qquad (2.2)$$

such an injection clearly has a closed image. Let us now define $H'(\mathbb{C}^n)$, the strong dual of $H(\mathbb{C}^n)$, as the space of analytic functionals on $\mathbb{C}^n$ (all of these functionals are *compactly carried*, unlike those studied by De Roever, and which we mentioned in the previous chapter). We can take the transposes of (2.1) and (2.2) to obtain two continuous linear restriction mappings

$$\mathcal{E}'(\mathbb{R}^{2n}) \to H'(\mathbb{C}^n) \quad \text{and} \quad \mathcal{E}'(\mathbb{R}^n) \to H'(\mathbb{C}^n),$$

the first of which (by Hahn–Banach) is a surjection which shows that every analytic functional can be extended to a distribution in $\mathcal{E}'(\mathbb{R}^{2n})$. As is well known, the collection $\{H'(\Omega)\}_{\Omega \text{ open}}$ does *not* form a sheaf, so that we cannot define a notion of support for analytic functionals. This is one of the reasons for the differences that will appear when we study mean-periodicity with respect to distributions and with respect to analytic functionals. If $\Omega \subseteq \mathbb{C}^n$ is open, let r_Ω be the restriction

$$r_\Omega : H(\mathbb{C}^n) \longrightarrow H(\Omega)$$
$$f \longmapsto f_{|\Omega}$$

and let $^t r_\Omega$ be its transpose (generally speaking, $^t r_\Omega$ is *not* one-to-one). We say that $\mu \in H'(\mathbb{C}^n)$ is *carried* by Ω if there exists $\tilde{\mu} \in H'(\Omega)$ such that $^t r_\Omega(\tilde{\mu}) = \mu$; this is equivalent to saying that there exists a distribution with compact support in Ω, which represents μ. We finally say that K, a compact subset of $\mathbb{C}^n$, is a *carrier* for μ (or that μ is *carried* by K) if μ is carried by every neighborhood of K.

The link between problems on mean-periodicity and the complex analysis we have developed in Chap. 1 is given by the Fourier transform; for $\varphi \in \mathcal{D}$, its Fourier transform $\hat{\varphi}$ is the real-analytic function defined on $\mathbb{R}^n$ by

$$\hat{\varphi}(\xi) = \int_{\mathbb{R}^n} \varphi(\xi) \exp(-ix \cdot \xi), \quad \xi \in \mathbb{R}^n. \qquad (2.3)$$

Of course, $\hat{\varphi}(\xi)$ extends naturally to all of $\mathbb{C}^n$ as an entire function of exponential type. The Fourier transform cannot be defined on $\mathcal{D}'$, whereas, if T is in $\mathcal{E}'$, its Fourier transform is defined by

$$\hat{T}(\xi) = \langle T_x, \exp(-ix \cdot \xi) \rangle, \quad \xi \in \mathbb{R}^n \qquad (2.4)$$

(T_x means that we look upon T as a distribution on the x-variable, so that ξ is just a parameter). Also, in this case, the Fourier transform $\hat{T}$ can be extended as an entire function on $\mathbb{C}^n$, and (2.3), (2.4) can be simultaneously replaced (recall that $\mathcal{D} \hookrightarrow \mathcal{E}'$) by

$$\mathcal{F}T(\zeta) = \hat{T}(\zeta) = \langle T_x, \exp(-ix \cdot \zeta) \rangle, \quad \zeta \in \mathbb{C}^n. \qquad (2.5)$$

The image of $\mathcal{F}$ is described in a precise way by the Paley–Wiener–Schwartz theorem: Let Ω be an open convex subset of $\mathbb{R}^n$, and let $\{K_j\}_{j\geq 1}$ be a fixed increasing sequence of compact convex subsets of Ω, with $K_j \subset \text{int}(K_{j+1})$, which exhausts Ω, and denote by $H_j = H_{K_j}$ the supporting function of K_j, i.e.,

$$H_j(x) = \sup_{y \in K_j} (x \cdot y), \quad x \in \mathbb{R}^n.$$

Now let $\sigma_j(z) = j\log(1 + |z|) + H_j(\text{Im } z)$; for $\Sigma = \{\sigma_j\}_{j\geq 1}$ let $A[\Sigma]$ be the space of all entire functions g which satisfy

$$\sup_{\mathbb{C}^n} |g(z)|e^{-\sigma_j(z)} < +\infty$$

for some j, equipped with the natural locally convex inductive limit topology.

Theorem 2.1. *The Fourier transform establishes a topological isomorphism between $\mathcal{E}'(\Omega)$ and $A[\Sigma]$.*

A similar result can be obtained for analytic functionals, the pioneer in this field being A. Martineau, who introduced, in Martineau [1963], the Fourier–Borel transform of an analytic functional $\mu \in H'(\mathbb{C}^n)$ to be the entire function

$$\hat{\mu}(\zeta) = \langle \mu_z, \exp(z \cdot \bar{\zeta})\rangle.$$

To state the analogue of Theorem 2.1, let $\Omega \subseteq \mathbb{C}^n$ be open and convex, and let $\{K_j\}_{j\geq 1}$ be a sequence as above: Set $\sigma_j(z) = \sup_{K_j} \text{Re}(z \cdot \bar{\zeta}) = H_j(z)$ and define $A[\Sigma]$ in the same way as before.

Theorem 2.2. *Let $\mu \in H'(\mathbb{C}^n)$. Then μ is carried by the compact convex set $K \subseteq \mathbb{C}^n$ if and only if, for every $\varepsilon > 0$, there exists $A_\varepsilon > 0$ such that*

$$|\hat{\mu}(z)| \leq A_\varepsilon \exp(H_K(z) + \varepsilon|z|), \quad z \in \mathbb{C}^n.$$

One can show that the Fourier–Borel transform is a topological isomorphism between $H'(\Omega)$ and $A[\Sigma]$.

The same considerations work for the space $\hat{\mathcal{D}}$ as shown in Ehrenpreis [1970], except that the description of the topology is more complicated (see also Hörmander [1983; 1985]).

Let us now go back to the problem of mean-periodicity, which is our main concern at the moment. As is well known, every translation invariant on $\mathcal{E}$, $\mathcal{D}'$, or H is a convolution operator, in the sense that if T is a *continuous translation invariant operator* on one of the spaces above, then there exists μ (in $\mathcal{E}'$, $\mathcal{E}'$, $\mathcal{H}'$, respectively) such that $T(-) = \mu * -$. On the other hand, we know that among partial differential operators, those with constant coefficients are completely characterized by the fact that they are translation invariant; it is therefore natural to consider convolution operators as the first extension of linear partial differential operators with constant coefficients.

Here one should point out the strong relationship linking convolution with the Fourier transform: Indeed, $\mathcal{E}'$ is an algebra under convolution, while $A[\Sigma]$

(see Theorem 2.1) is an algebra under the usual pointwise multiplication of functions. The isomorphism given by the Fourier transform respects this extra structure in the sense that

$$\mathcal{F}(\mu * \nu) = \mathcal{F}(\mu) \cdot \mathcal{F}(\nu).$$

If one looks back at the history of the development of the theory of partial differential equations (which, in this chapter, will always be linear and with constant coefficients), it clearly stands out that a crucial point was the creation, in 1945, of the theory of distributions by L. Schwartz (Schwartz [1950–51]). Distributions, and the machinery of Fourier transform, gave great freedom to analysis, which, as Gel'fand and Shilov point out in Gel'fand and Shilov [1968], is essentially based on the fact that for most of the spaces of test functions and their duals, the duality bracket is just a natural generalization of the usual integral $\int f(x)g(x)dx$.

Let $P = P(\xi_1, \ldots, \xi_n)$ be a polynomial in $\mathbb{C}[\xi_1, \ldots, \xi_n]$, set

$$D = \left(-i\frac{\partial}{\partial x_1}, \ldots, -i\frac{\partial}{\partial x_n} \right)$$

and, for $\alpha = (\alpha_1, \ldots, \alpha_n)$ any multiindex of nonnegative integers, with $|\alpha| = \alpha_1 + \cdots + \alpha_n$, set

$$D^\alpha = (-i)^{|\alpha|} \frac{\partial^{|\alpha|}}{\partial x_1^{\alpha_1} \cdot \cdots \cdot \partial x_1^{\alpha_1}}, \qquad \xi^\alpha = \xi_1^{\alpha_1} \cdot \cdots \cdot \xi_n^{\alpha_n}.$$

In 1950, Schwartz asked the following three basic questions: Does every operator $P(D)$ admit a *fundamental solution* (i.e., a distribution E such that $P(D)E = \delta$)? When is $P(D)$ *hypoelliptic* (i.e., such that any distribution f solution of $P(D)f = 0$ must be C^∞)? Which partial differential operators $P(D)$ have a continuous right inverse?

These problems have many connections with our main topic, so we should spend a few words on them: The fact that *every nontrivial $P(D)$ has a fundamental solution* was proved, independently and, roughly speaking, at the same time, by B. Malgrange (in his thesis, Malgrange [1956], which is a basic paper both on convolution equations and on partial differential equations) and by L. Ehrenpreis (Ehrenpreis [1954–62]). Their proofs are different, but in both cases the key argument is a clever use of the Hahn–Banach theorem: In the hard part of their proofs, they employ Fourier transform techniques to obtain an explicit inequality which, as a consequence, shows that the map

$$P(D)u \mapsto u(0)$$

from $P(D)\mathcal{D}$ to $\mathbb{C}$ is continuous (from a given Banach space in which $\mathcal{D}$ is dense). Hence, by Hahn–Banach, there exists $E \in \mathcal{D}'(\mathbb{R}^n)$ such that $\langle E, P(D)u \rangle = u(0)$ for any $u \in \mathcal{D}$, i.e.,

$$P(D)E(-x) = \delta(x).$$

One might notice the *nonconstructive* character of this proof; the problem of *constructing* fundamental solutions was first tackled by Hadamard (Hadamard [1923]), then by Petrowski (Petrowski [1939]) for very special cases of elliptic operators, and later by Hörmander (Hörmander [1983; 1985]) for hypoelliptic operators, and by Treves for all operators (see Treves [1975] for detailed references, and also Chaps. 5 and 6 for a different approach to this same problem). The fact that there is always a fundamental solution E in the space S' of tempered distributions turned out to be a much deeper problem that was solved by Hörmander (Hörmander [1958]) and Lojasiewicz (Lojasiewicz [1959]). A different solution is provided by the methods indicated in Chap. 5 (Bernstein [1971]; Atiyah [1970]).

In Hörmander's thesis (Hörmander [1955b]), a characterization of hypoellipticity is given, using the same kind of arguments as those used by Malgrange and Ehrenpreis for the problem of the existence of the fundamental solution. The result, by the way, relies on Hörmander's explicit construction of a fundamental solution for a hypoelliptic operator.

The success of the method of using abstract functional analysis to obtain precise inequalities was now clear and was exploited in a consistent way; thus, Malgrange (Malgrange [1956]) used the Hahn–Banach theorem to prove the *approximation theorem* for solutions of homogeneous partial differential equations. Let $P(D)$ be a partial differential operator: If $z_0 \in \mathbb{C}^n$ satisfies $P(z_0) = 0$, then a solution $u \in \mathcal{E}$ of the homogeneous equation

$$P(D)u = 0 \qquad (2.6)$$

is $u(x) = \exp(2\pi i x \cdot z_0)$. Depending on the multiplicity of z_0 as a zero of P, one can also find polynomials $Q(x)$ such that $Q(x) \exp(2\pi i x \cdot z_0)$ is still a solution of (2.6) (the choice of $Q(x)$ is strictly related to the Noetherian operators which we have already described in Chap. 1). This kind of solution is very important and will be called an *exponential-polynomial* solution of (2.6). By the well-known principle due to Euler (Euler [1743]), if $n = 1$ (i.e., the case of linear ordinary differential equations with constant coefficients), then the (*finite*) linear combinations of exponential polynomial solutions yield all possible solutions of (2.6). This is clearly false if $n > 1$; still, one might ask whether the linear span W of all such solutions is *dense* in the space $\mathcal{E}^P$ of *all* solutions of (2.6). In Malgrange [1956], Malgrange proves this to be true even when (2.6) is replaced by

$$\mu * f = 0,$$

with $\mu \in \mathcal{E}'(\mathbb{R}^n)$, $f \in \mathcal{E}(\mathbb{R}^n)$: Recall that the definition of convolution is

$$\mu * f(x) := \langle \mu_t, t \mapsto f(x - t) \rangle.$$

But the most remarkable advance along these lines was obtained by Ehrenpreis (Ehrenpreis and Mautner [1955]; Ehrenpreis [1970]) and, independently, by Palamodov (Palamodov [1970]). Their result, to which we will always refer by the name of *Fundamental Principle* (according to Ehrenpreis's own terminology) is a much stronger generalization of Euler's principle than the approximation theorem is. Actually, this result is the exact analogue of Euler's principle for $n > 1$, as it gives the explicit form of a general solution to (2.6), with only the use of integrals and sums of exponential polynomial solutions:

To be precise, we shall state the Fundamental Principle for the case of a single partial differential operator acting on $\mathcal{E} = \mathcal{E}(\mathbb{R}^n)$; but its proof (as we shall soon see) holds in a much wider situation.

Theorem 2.3. *Let $f \in \mathcal{E}$ and let $P(D)$ be a linear differential operator with constant coefficients. Then f satisfies $P(D)f = 0$ if and only if it can be represented as*

$$f(x) = \sum_{j=1}^{J} \int_{V_j} \partial_j (\exp(ix \cdot z)) d\mu_j(z), \qquad (2.7)$$

where J is some positive integer, ∂_j are partial differential operators on $\mathbb{C}^n$ with constant coefficients, and the V_j's are algebraic varieties contained in the characteristic variety of P,

$$V = \{z \in \mathbb{C}^n : P(z) = 0\}.$$

Moreover, the Radon measures $d\mu_j$ have supports contained in V_j and satisfy, for a suitable positive continuous function k on $\mathbb{C}^n$, the growth condition

$$\int_{V_j} k(z)|d\mu_j|(z) < +\infty,$$

which ensures the convergence of the integrals which appear in (2.7), for all $x \in \mathbb{R}^n$.

To be very precise, we should have $\check{P}$ instead of P in the theorem. This is due to the fact that ${}^t P(D) = P(-D)$. Later on, a similar remark applies to convolution operators.

Some remarks are necessary to clarify the meaning and the relevance of this result. First, one can see that if $n = 1$, then V is a finite set of points, and (2.7) becomes the Euler representation. Moreover, for $n > 1$, the family $(V_j, \delta_j)_{j=1}^{J}$ is naturally associated to P, and it describes what Ehrenpreis calls the *multiplicity variety* associated to P (in terms of Palamodov's approach, and following Chap. 1, we are confronted with nothing but a set of Noetherian operators associated to P). Indeed, in Ehrenpreis [1970], it is shown that for any polynomial P, one can construct a (nonnecessarily unique) multiplicity variety $\mathcal{B}_P = \{(V_j, \delta_j)_{j=1}^{J}\}$ such that an entire function $F \in H(\mathbb{C}^n)$ belongs to the ideal $I(P)$ generated by P in $H(\mathbb{C}^n)$ if and only if $\partial_j F_{|V_j} \equiv 0$, for all $j = 1, \ldots, J$.

As to Theorem 2.3, we need to mention that it holds (suitably modified) when $P(D)$ is replaced by a *matrix* of differential operators. In this case, we naturally deal with a *system* of differential equations, and the multiplicity varity $\mathcal{B}_P$ is replaced by a *vector* of multiplicity varieties (V becomes the set of points where the matrix $[P_{ij}(z)]$ has nonmaximal rank). The difficulties which surface in this case are highly nontrivial, and the attempt to overcome them constitutes a good deal of both Ehrenpreis [1970] and Palamodov [1970]; in particular, the differential operators ∂_j (as Palamodov showed in a famous example) turn out to have polynomial coefficients.

Another important point which can be made is the fact that Ehrenpreis and Palamodov proved Theorem 2.3 for a very large class of spaces. Ehrenpreis introduced a very interesting family of spaces, for which Theorem 2.3 could be proved: He called these spaces LAU-*spaces* (localizable analytically uniform); they appear as a natural generalization of the spaces A_p introduced by Hörmander, which were already discussed in Chap. 1. Roughly speaking, a topological vector space is an LAU-space if its strong dual is isomorphic (via a map which is usually a variation of the classical Fourier transform) to a space of entire functions whose growth at infinity is controlled by a family $\mathcal{K}$ of weights, satisfying suitable technical properties: Such properties are intended to replace the conditions on the plurisubharmonic weight p, which Hörmander imposed for his A_p-spaces, and imply, for example, the closure under derivation, and the possibility of applying $\bar{\partial}$-techniques (as we saw in Chap. 1, such techniques can be successfully applied if factors such as $(1+|z|)^n$ can be absorbed in our weights; this is the case for LAU-spaces). The link between A_p-spaces and LAU-spaces is stronger than it may seem at first sight, and it is clearly pointed out in Hansen [1981], where a different characterization for LAU-spaces is given. This new characterization was strongly exploited in Struppa [1983a], about which we shall talk later on. Among the main examples of such spaces, we recall $\mathcal{E}(\Omega)$ (Ω being an open convex set in $\mathbb{R}^n$), $\mathcal{D}'(\mathbb{R}^n)$, $\mathcal{D}'(\Omega)$ (Ω as before), $H(\mathbb{C}^n)$, $H(\Omega)$ (now Ω is an open convex set in $\mathbb{C}^n$). Note that when distributions are involved, formula (2.7) has to be interpreted symbolically, i.e., it must be obviously replaced by

$$\langle f, \varphi \rangle = \sum_{j=1}^{J} \int_{V_j} \partial_j(\hat{\varphi}(z)) d\mu_j(z),$$

where φ is any test function. One should also notice that Kaneko, in three interesting papers (Kaneko [1970; 1972; 1973]), has succeeded in employing a modification of Hörmander's techniques to prove a version of Theorem 2.3 for hyperfunction solutions of a system of differential equations; we refer to Chap. 6 for more details on the notion of hyperfunctions. Kaneko's work also provides generalization of Hartogs' theorem on the removability of compact singularities for holomorphic functions.

Let us now briefly attempt to sketch a proof for Theorem 2.3.

The first step consists of showing that the Fourier representation (2.7) can be reduced to the theorems in complex analysis which we discussed in Chap. 1

(actually, in view of the simplified version of the Fundamental Principle which we are proving, a very simple form of the results of Chap. 1 is needed). Let $f \in \mathcal{E}(\mathbb{C}^n)$ such that $P(D)f = 0$. Then, for all $\mu \in \mathcal{E}'(\mathbb{R}^n)$, one has

$$0 = \langle P(D)f, \mu \rangle = \langle f, {}^t P(D)\mu \rangle, \tag{2.8}$$

where ${}^t P(D)$ is the transpose of $P(D)$ and acts from $\mathcal{E}'(\mathbb{R}^n)$ to $\mathcal{E}'(\mathbb{R}^n)$. Taking Fourier transforms in (2.8), one deduces that f defines a continuous linear functional on $\frac{\hat{\mathcal{E}}'}{P \cdot \hat{\mathcal{E}}'}$ (vice versa, it is easily seen that every continuous linear functional on $\frac{\hat{\mathcal{E}}'}{P \cdot \hat{\mathcal{E}}'}$ defines a solution of $P(D)f = 0$). Therefore, in order to prove Theorem 2.3, one needs to give a handy *topological* description of the quotient space $\frac{\hat{\mathcal{E}}'}{P \cdot \hat{\mathcal{E}}'}$. To reach such a description, one notices that if two entire functions F_1, F_2 in $\hat{\mathcal{E}}'$ belong to the same coset module $P \cdot \hat{\mathcal{E}}'$, then their restriction to the variety

$$V = \{ z \in \mathbb{C}^n : P(z) = 0 \}$$

must coincide. In fact, the equality $F_{1|V} = F_{2|V}$ must be interpreted in a more subtle way, since we can actually say that $F_1 - F_2$ vanishes (at each point of V) with order larger than or equal to the order of P; it is therefore natural to try to split V into a *finite* number of pieces V_j $(j = 1, \ldots, J)$, each of which is endowed with a differential operator ∂_j, such that

$$\partial_j (F_1 - F_2)_{|V_j} = 0.$$

Theorem 2.3 is nothing but a corollary of the fact that this construction can be reversed; to state precisely this result, let $\mathcal{B}_P = \{(V_j, \partial_j)\}$ be the multiplicity variety associated to P. A system of functions (see also Chap. 1) $\{F_j = \partial_j F_{|V_j}\}$, where F is an analytic function on $V = \bigcup_{j=1}^{J} V_j$ (the *support* of $\mathcal{B}_P$), is called an analytic function on $\mathcal{B}_P$. We then define the vector space $\hat{\mathcal{E}}'(\mathcal{B}_P)$ as the space of all analytic functions of $\mathcal{B}_P$ which satisfy, on V, the bounds that define $\hat{\mathcal{E}}'$ (the actual definition of $\hat{\mathcal{E}}'(\mathcal{B}_P)$ is more complicated, but we shall content ourselves with this rough description). These bounds define a natural *locally convex topology* on $\hat{\mathcal{E}}'(\mathcal{B}_P)$, and the Fundamental Principle is now a trivial application of the Riesz representation theorem in view of the following result.

Theorem 2.4. *The spaces $\frac{\hat{\mathcal{E}}'}{P \cdot \hat{\mathcal{E}}'}$ and $\hat{\mathcal{E}}'(\mathcal{B}_P)$ are topologically isomorphic,* with the isomorphism being provided by the restriction of an entire function to V.

The proof of Theorem 2.4, on the other hand, relies on the deep results in complex analysis which we mentioned in Chap. 1. To be precise, one needs to divide and to extend functions in the algebra $\hat{\mathcal{E}}'$ with respect to the algebraic variety associated to P. As we know from Theorem 2.1, the growth in $\hat{\mathcal{E}}'$ can be described by means of suitable plurisubharmonic weights on $\mathbb{C}^n$, as it is the case for all LAU-spaces, so that we can give the following form for our complex analysis result.

Theorem 2.5. *Using the same notation as above, the following statements hold true for all plurisubharmonic functions φ on $\mathbb{C}^n$; there is a constant $M > 0$ such that*

(i) *for every entire function $v : \mathbb{C}^n \to \mathbb{C}$,*

$$\partial_j(P \cdot v)_{|V_j} = 0, \quad j = 1, \ldots, J;$$

Conversely, if an entire function $g : \mathbb{C}^n \to \mathbb{C}$ satisfies

$$\partial_j(g)_{|V_j} = 0, \quad j = 1, \ldots, J,$$

then there exists an entire function v such that $g = P \cdot v$ on $\mathbb{C}^n$, and

$$\sup_{\mathbb{C}^n} |v(z)| e^{-\varphi_M(z)} (1 + |z|)^{-M} \leq \sup |g(z)| e^{-\varphi(z)},$$

where

$$\varphi_M(z) = \sup\{\varphi(z + z') : z' \in \mathbb{C}^n, |z'| \leq M\} \quad \text{on } \mathbb{C}^n;$$

(ii) *for any entire function $g : \mathbb{C}^n \to \mathbb{C}$, there is another entire function $f : \mathbb{C}^n \to \mathbb{C}$ such that*

$$\partial_j(f - g)_{V_j} = 0, \quad j = 1, \ldots, J,$$

and

$$\sup_{\mathbb{C}^n} |f(z)| e^{-\varphi_M(z)} (1 + |z|)^{-M} \leq \max_j \sup_{V_j} |\partial_j g(z)| e^{-\varphi(z)}.$$

These considerations lead us naturally to the end of the story for partial differential equations (note that we are talking about *constant coefficients*; as to the case of *variable coefficients*, we refer the reader to Chap. 6) and to the beginning of the study of convolution equations, which, as we mentioned, are the most natural generalizations of linear partial differential equations with constant coefficients. Then consider $\mu \in \mathcal{E}'$ and the convolution operator

$$\mu* : \mathcal{E} \to \mathcal{E} \tag{2.9}$$

which it defines; when taking its transpose, and later its Fourier transform, (2.9) becomes multiplication in $\hat{\mathcal{E}}'$; indeed

$$\mu * {}^t : \mathcal{E}' \to \mathcal{E}'$$

is defined by

$$\langle \mu * {}^t\nu, f \rangle = \langle \nu, \mu * f \rangle$$

for any $\nu \in \hat{\mathcal{E}}'$, $f \in \mathcal{E}$; if we now take the Fourier transformation, we obtain a multiplication operator

$$\hat{\mu}\cdot : \hat{\mathcal{E}}' \to \hat{\mathcal{E}}'$$

which is easily shown to be a continuous endomorphism of $\hat{\mathcal{E}}'$. This remark is then used to define convolution operators on every LAU-space; let X be an LAU-space: a *multiplier* for $\hat{X}'$ is any continuous endomorphism F of $\hat{X}'$; a *convolutor* on X, on the other hand, is defined via a multiplier F, by the adjoint of the mapping

$$f \mapsto \mathcal{F}^{-1}(F(\mathcal{F}f)), \quad f \in X'.$$

This definition shows, for example, that $H'(\mathbb{C}^n)$ is the space of convolutors for $H(\mathbb{C}^n)$, while $\hat{\mathcal{E}}'(\mathbb{R}^n)$ is the space of multipliers of $\hat{\mathcal{D}}(\mathbb{R}^n)$, and therefore $\mathcal{E}'(\mathbb{R}^n)$ is the space of convolutors for $\mathcal{D}(\mathbb{R}^n)$.

Going back to Malgrange's thesis (Malgrange [1956]) and Ehrenpreis's division papers (Ehrenpreis [1954–62]), we find their approach (for X either $\mathcal{E}(\mathbb{R}^n)$ or $\mathcal{D}'(\mathbb{R}^n)$) to both the approximation problem and the question of the surjectivity of

$$\mu* : \mathcal{D}'(\mathbb{R}^n) \to \mathcal{D}'(\mathbb{R}^n),$$

with $\mu \in \mathcal{E}'(\mathbb{R}^n)$. As to the first problem, they proved the following result.

Theorem 2.6. *Let $0 \neq \mu \in \mathcal{E}'(\mathbb{R}^n)$ and let $\Omega \subseteq \mathbb{R}^n$ be open and convex. Let $\mathcal{E}_\mu(\Omega)$ be the closed linear subspace of $\mathcal{E}(\Omega)$ given by those $f \in \mathcal{E}(\Omega)$ for which $\mu * f = 0$, and let W_μ be the set of all linear combinations of exponential polynomial solutions of $\mu * f = 0$ in $\mathbb{R}^n$. Then the restriction to Ω of the elements in W_μ are dense in $\mathcal{E}_\mu(\Omega)$, with the topology induced by $\mathcal{E}(\Omega)$.*

Let us point out that the modern proof of Theorem 2.6 actually relies on estimates for plurisubharmonic functions which, in spirit, are quite close to Theorem 2.5 (see Hörmander [1983, Chap. 16; 1985] for complete details; in Hörmander's book, actually, the original argument is simplified by the systematic use of the theory of analytic functionals). Let us take the opportunity to mention an open problem in this area: Up to now, it is not known which conditions should be placed on $\mu \in H'(\mathbb{C}^n)$ in order to have a positive answer for the approximation problem for $H(\Omega)$, with Ω being a convex open subset of $\mathbb{C}^n$; partial answers are contained in some works of Napalkov, Epifanov, and Morzhakov (Epifanov [1969]; Morzhakov [1974]; Napalkov [1979a]).

More interesting, from our point of view, is the study of the surjectivity of convolution operators; the main and earliest results in this field are certainly due to the pioneering work of Ehrenpreis (Ehrenpreis [1954–62; 1970]). He found a remarkably simple necessary and sufficient condition for the surjectivity of convolution operators in different spaces; let us begin by quoting an important result from Hörmander [1983; 1985], but which is essentially contained in Ehrenpreis's work:

Theorem 2.7. *Let $\mu \in \mathcal{E}'(\mathbb{R}^n)$; then the following conditions are equivalent:*

(i) *there is a constant $A > 0$ such that for every $x \in \mathbb{R}^n$,*

$$\sup\{|\hat{\mu}(z)| : z \in \mathbb{C}^n, |z - x| < A\log(1 + |x|)\} > (A + |x|)^{-A};$$

(ii) *for every $a > 0$, there is a constant $A > 0$ such that for every $X \in \mathbb{R}$,*

$$\sup\{|\hat{\mu}(x + y)| : y \in \mathbb{C}^n, |y| < a\log(1 + |x|)\} > (A + |x|)^{-A};$$

(iii) *for every $a > 0$, there is a constant $A > 0$ such that for every $x \in \mathbb{R}^n, |x| > 2$,*

$$\int_{|z|<a} \log|\mu(x + z\log|x|)|d\lambda(z) > -A\log|x|,$$

with $d\lambda(z)$ being the Legesgue measure;

(iv) *if $\nu \in \mathcal{E}'$ and $\hat{\nu}/\hat{\mu}$ is an entire function, then it is the Fourier transform of a distribution in $\mathcal{E}'$.*

In accordance with this result, Ehrenpreis defined a compactly supported distribution $\mu \in \mathcal{E}'(\mathbb{R}^n)$ to be *invertible* if it satisfies any one of the equivalent conditions in Theorem 2.7; the obvious reason for this definition is the following result (independently obtained by Ehrenpreis and Malgrange).

Theorem 2.8. *The convolution operator $\mu* : \mathcal{E} \to \mathcal{E}$ is surjective if and only if μ is invertible.*

We sketch here the proof of this result, since it will lead us to another important topic we have not yet touched upon. By standard functional analysis, the operator $\mu* : \mathcal{E} \to \mathcal{E}$ is surjective if and only if its transpose $\mu*\prime : \mathcal{E}' \to \mathcal{E}'$ is injective and has closed range. Since the Fourier transform is a *topological* isomorphism, this is equivalent to requiring that multiplication by $\hat{\mu}$ (acting from $\hat{\mathcal{E}}'$ to itself) be injective and have closed range. Injectivity is always trivially satisfied, since (if $\mu \neq 0$) $\hat{\mu}$ is a nontrivial holomorphic function in $\hat{\mathcal{E}}'$, and the product in $\hat{\mathcal{E}}'$ is clearly nondegenerate. Thus, $\mu*$ is onto if and only if the image of $\hat{\mu}\cdot$, i.e., the *ideal $(\hat{\mu})$ generated by $\hat{\mu}$ in $\hat{\mathcal{E}}'$ is closed in $\hat{\mathcal{E}}'$*. Now, if one can prove the identity

$$\hat{\mu} \cdot \hat{\mathcal{E}}' = (\hat{\mu} \cdot H) \cap \hat{\mathcal{E}}' \tag{2.10}$$

then the closedness of the ideal $\hat{\mu} \cdot \hat{\mathcal{E}}'$ follows immediately, since by Cartan's Theorem B, $\hat{\mu} \cdot H$ is closed. Now we just have to notice the equivalence of (2.10) with (iv) in Theorem 2.7.

We can therefore see that the question of surjectivity reduces to the question of closure of ideals; this is clearly intimately related to the so-called problem of spectral synthesis: Given an ideal I in an algebra of holomorphic functions, one wants to know whether *the ideal localizes*, i.e., whether the ideal I coincides with the local ideal I_{loc} which it generates. (We remind the reader that I_{loc} is the collection of global sections of the sheaf of ideals associated to I.) It has to be mentioned that, by using Fourier transform and a duality argument, spectral synthesis is equivalent to the positive solution of the approximation problem. Hence, we know that we have spectral synthesis for principal ideals in $\hat{\mathcal{E}}'$ (by Theorem 2.6) for polynomial ideals in $\hat{\mathcal{E}}'$ (as a consequence of the Fundamental Principle). For $n = 1$, the spectral synthesis in $\hat{\mathcal{E}}'$, $\hat{H}'$ was proved by Schwartz in his well-known work Schwartz [1947], to which we shall come back later. For $n > 1$, it is false in general (Gurevich [1975]). The bibliography on closed ideals, as well as on localizing ideals, is enormous; we refer the reader to the reference list in Nikol'skij [1974].

Returning to the question of surjectivity, one finds the more refined results of Hörmander [1983; 1985], who completely solved the surjectivity problem for distributions: Let $\mu \in \mathcal{E}'(\mathbb{R}^n)$, $\mathrm{supp}(\mu) = K$, and let Ω_1, Ω_2 be two open subsets of $\mathbb{R}^n$ such that

$$\Omega_2 - K \subseteq \Omega_1, \tag{2.11}$$

so that, for any $f \in \mathcal{D}'(\Omega_2)$, $\mu * f$ is a well-defined distribution in $\mathcal{E}'(\Omega_1)$. The following definitions are due to Hörmander and modeled on the notion of P-convexity introduced for similar purposes in Malgrange [1956].

Definition 2.1. A pair (Ω_1, Ω_2) of open subsets of $\mathbb{R}^n$ satisfying (2.11) is said to be μ-*convex for supports* if it satisfies either of the following equivalent conditions:

(i) $\mathrm{dist}(\mathrm{supp}(\nu), \mathbb{R}^n \backslash \Omega_2) = \mathrm{dist}(\mathrm{supp}(\mu * \nu), \mathbb{R}^n \backslash \Omega_1)$ for any $\nu \in \mathcal{E}'(\Omega_2)$;
(ii) for every compact $K_1 \subset \Omega_1$, there exists a compact $K_2 \subset \Omega_2$ such that $\mathrm{supp}(\nu) \subset K_2$ if $\nu \in \mathcal{E}'(\Omega_2)$ and $\mathrm{supp}(\mu * \nu) \subset K_1$.

Definition 2.2. A pair of open sets (Ω_1, Ω_2) in $\mathbb{R}^n$ satisfying (2.11) is said to be μ-*convex for singular supports* if it satisfies either of the following (equivalent) conditions:

(i) $\mathrm{dist}(\mathrm{sing}\ \mathrm{supp}(\nu), \mathbb{R}^n \backslash \Omega_2) = \mathrm{dist}(\mathrm{sing}\ \mathrm{supp}(\check{\mu} * \nu), \mathbb{R}^n \backslash \Omega_1)$ for any $\nu \in \mathcal{E}'(\Omega_2)$;
(ii) for every compact $K_1 \subset \Omega_1$, there exists a compact $K_2 \in \Omega_2$ such that $\mathrm{sing}\ \mathrm{supp}(\nu) \subset K_2$ if $\nu \in \mathcal{E}'(\Omega_2)$ and $\mathrm{sing}\ \mathrm{supp}(\check{\mu} * \nu) \subset K_1$.

The relevance of these definitions lies in the following theorem, whose proof combines the division theorems of Chap. 1 with the duality theory for Fréchet spaces (and limits of the same).

Theorem 2.9. *Let $\mu \in \mathcal{E}'(\mathbb{R}^n)$ and let Ω_1, Ω_2 be open sets in $\mathbb{R}^n$ satisfying* (2.11). *Then $\mu * \mathcal{D}'(\Omega_2) = \mathcal{D}'(\Omega_1)$ if and only if μ is invertible and the pair (Ω_1, Ω_2) is μ-convex for supports and for singular supports.*

Many other related results (e.g., when $\mathcal{D}'$ is replaced by $\mathcal{D}'_F$, the space of finite order distribution) are known, but we shall not give any details here. We might mention here the work of Berenstein and Dostal concerning the invertibility of some distributions arising from simple geometrical objects. Thus, for example, in their paper Berenstein and Dostal [1976], they show that the characteristic function of a closed convex C^∞ surface with positive

Gaussian curvature *is invertible* (think, e.g., of the sphere), although it *does not propagate singularities* (this is a more subtle condition than invertibility, for which we refer the reader to Berenstein and Dostal [1973b]). In particular, surjectivity might fail even for half-spaces Ω_1, Ω_2.

The situation, however, becomes much more delicate if one tries to look for the possibility of achieving some kind of Fundamental Principle for convolution equations, or systems of them. The first crucial paper in this area is due to Schwartz, who proved (in 1947) the following Fundamental Principle for convolution equations for functions of one variable (Schwartz [1947]).

Theorem 2.10. *Let* $\mu \in \mathcal{E}'(\mathbb{R})$, $V = \{z \in \mathbb{C} : \hat{\mu}(z) = 0\}$, *and let* $f \in \mathcal{E}(\mathbb{R})$ *be such that* $\mu * f = 0$. *Then* f *admits a unique formal (in the sense that the series might not converge) representation*

$$\sum_{\alpha \in V} c_\alpha(x) e^{i\alpha x},$$

where the c_α*'s are polynomials of degree strictly less than the multiplicity of* α *as a root of* $\hat{\mu}(z) = 0$. *Moreover, it is possible to construct infinitely many disjoint subsets* $V_k \subset V$ *such that* $\bigcup_{k=1}^{+\infty} V_k = V$, *and the series*

$$\sum_{k=1}^{+\infty} \left(\sum_{\alpha \in V_k} c_\alpha(x) e^{i\alpha x} \right)$$

converges in $\mathcal{E}$ *to* f, *after performing a certain "abelian summation" process, which can be interpreted as*

$$f(x) = \lim_{\varepsilon \to 0} \sum_{k=1}^{+\infty} \left(\sum_{\alpha \in V_k} c_{\alpha,\varepsilon}(x) e^{i\alpha x} \right),$$

where both the limit and the series have to be interpreted in the topology of $\mathcal{E}$.

We point out that both the grouping of terms, with the V_k, and the use of Abel convergence factors, are not always necessary. In particular, Ehrenpreis has shown in Ehrenpreis [1954–62] that if μ is invertible, then the Abel factors are unnecessary, while in Ehrenpreis and Malliavin [1974] and Berenstein and Taylor [1979], the respective authors gave a precise characterization of those invertible convolutors μ for which no groupings are necessary. We also point out that the phenomenon of grouping was known since the early papers of Gel'fond and Leont'ev, where a detailed study of difference-differential equations had been carried out (the original papers Gel'fond [1951] and Leont'ev [1949; 1951] contain the germ of their ideas that led to the most important results which we quote).

Of crucial importance in Theorem 2.10 is the *uniqueness* of the series representation, which immediately implies the following corollary, usually known under the name of *spectral analysis theorem*.

Theorem 2.11. *Let* $\mu_1, \ldots, \mu_r \in \mathcal{E}'(\mathbb{R})$, *and suppose that* $f \in \mathcal{E}(\mathbb{R})$ *satisfies the system*

$$\mu_1 * f = \cdots = \mu_r * f = 0.$$

Then, if $V = \{z \in \mathbb{C} : \hat{\mu}_1(z) = \cdots = \hat{\mu}_r(z) = 0\} = \emptyset, \quad f \equiv 0.$

Let us mention here the important work of Kahane (Kahane [1957; 1959]), who gave a different proof of Schwartz's work (just as far as spectral synthesis is concerned) by generalizing the so-called Carleman transform. This was originally described in Carleman [1944] as an effective way to provide a kind of Fourier transform in one variable for functions of polynomial growth and was later understood to anticipate the idea of hyperfunctions (see Martineau [1964a]), where the several-variables case is also considered. It might be interesting to evaluate Kahane's work in this new light, examining the possibility of using Martineau's generalization of Carleman's work to provide a similar approach to spectral synthesis in more than one variable (which, for a single equation, was shown to be true by Malgrange).

In Schwartz [1947], the author proposed that his result could be extended to the case when $n > 1$ (convolution equations in $\mathbb{R}^n$) by first trying to obtain an analogue of Theorem 2.11. As we mentioned, Ehrenpreis's Fundamental Principle (first announced in 1960) showed that this was the case whenever $\hat{\mu}_1, \ldots, \hat{\mu}_r$ are *polynomials*. However, it seemed very difficult to extend his result to the case in which the $\hat{\mu}_j$'s are more general *analytic functions*. The first significant extension to this case is due to Berenstein [1970] (see also Berenstein and Dostal [1972]) for the case of a single convolution equation $\mu * f = 0$, in which $\hat{\mu}$ is a distinguished polynomial (instead of being a polynomial), i.e., for $z = (z_1, z'), z' = (z_2, \ldots, z_n)$ and some integer m,

$$\hat{\mu}(z_1, z') = z_1^m + P_1(z')z_1^{m-1} + \cdots + P_m(z'),$$

with $P_j(z')$ being holomorphic in $\mathbb{C}^{n-1}$. The idea of the proof for this Fundamental Principle follows rather closely the one given by Ehrenpreis, but some extra care is needed in view of the occurrence of some unexpected problems concerning the fact that the roots of $\hat{\mu}$ could coalesce very rapidly. This led to the concept of *slowly decreasing* given in Berenstein and Taylor [1980b].

The hidden reason for the difficulty in proving a more powerful and general extension became clear only in 1975 when Gurevich proved in Gurevich [1975] that Theorem 2.11 is, in general, false for $n \geq 2$. More precisely, Gurevich proved the existence of six distributions $\mu_1, \ldots, \mu_6 \in \mathcal{E}'(\mathbb{R}^n)$ for which

$$\{z \in \mathbb{C}^n : \hat{\mu}_1(z) = \cdots = \hat{\mu}_6(z) = 0\} = \emptyset$$

(so that no exponential polynomial solutions exist), but

$$\{f \in \mathcal{E}(\mathbb{R}^n) : \mu_1 * f = \cdots = \mu_6 * f = 0\} \neq \{0\}.$$

At this point, it becomes necessary to understand how to eliminate this unpleasant phenomenon; this understanding has been fully reached in Berenstein and Taylor [1980b], but it is quite interesting to follow how the clarification was reached. A seminal paper in this direction was written in 1960 by J. Delsarte (Delsarte [1960]), though the proofs in it are not completely correct.

Theorem 2.12. *Let $\delta(a, b)$ denote the Dirac delta mass at the points (a, b) of $\mathbb{R}^2$. Let Q be the square $[0, 1] \times [0, 1]$ and let $\varphi_1, \varphi_2 \in L^1(\mathbb{R}^2)$ be supported in Q; take eight complex numbers $a_1, a_2, b_1, b_2, c_1, c_2, d_1, d_2$, such that $c_1 d_2 - c_2 d_1, b_1 d_2 - b_2 d_1, a_1 b_2 - a_2 b_1, a_1 c_2 - a_2 c_1$ are all different from zero; for $j = 1, 2$, set*

$$\mu_j = a_j \delta(0, 0) + b_j \delta(1, 0) + c_j \delta(0, 1) + d_j \delta(1, 1) + \varphi_j dx dy.$$

If $\{z \in \mathbb{C}^2 : \hat{\mu}_1(z) = \hat{\mu}_2(z) = 0\}$ has only simple points, then the system $\mu_1 * f = \mu_2 * f = 0$ admits spectral synthesis.

The interest of this paper is, at least, twofold. On one side it clearly points out some examples in which things work as hoped, on the other its proof is extremely instructive, since it rediscovers the old interpolation formula due to Jacobi, which we briefly described in Chap. 1. It was soon clear how to make use of these indications: In Yger [1979; 1977], the author generalized Delsarte's example to the case in which Q is replaced by a convex compact nontrivial polygon P, and where φ_1, φ_2 are measures supported in the interior of P (some technical conditions on the sides of P are also needed); actually, Yger followed Ehrenpreis's proof for the Fundamental Principle, together with a compactness argument (based on Douady's preferred neighborhoods theorem) to prove not just spectral synthesis, but even a representation theorem for the solutions of the system (thus generalizing a previous result of Y. Meyer in Meyer [1976]). On the other hand, simultaneously and independently, Berenstein and Taylor, in a series of papers (Berenstein and Taylor [1979; 1980a,b]), made the most important advances on this question. Using the Jacobi interpolation formula and following some ideas on closed ideals which had appeared in Kelleher and Taylor [1971a] as well as in Gurevich's paper Gurevich [1972], they finally found the right conditions on $\mu_1, \ldots, \mu_r$ in $\mathcal{E}'(\mathbb{R}^n)$ for spectral synthesis to hold, and even more important, they were now able to prove an integral representation theorem for solutions of *slowly decreasing* systems of convolution equations, which was a genuine extension of the Fundamental Principle ($\mu_1, \ldots, \mu_r$ are said to be *jointly slowly decreasing* if $\hat{\mu}_1, \ldots, \hat{\mu}_r$ are slowly decreasing in the sense of Chap. 1).

Theorem 2.13. *Let $u_1, \ldots, \mu_r \in \mathcal{E}'(\mathbb{R}^n)$, $1 \le r \le n$, be jointly slowly decreasing. Then there exists a locally finite family of closed sets $\{V_j : j \in J\}$, a partition of J into finite subsets J_k, and partial differential operators ∂_j on $\mathbb{C}^n$ (with analytic coefficients) such that*

(i) $\bigcup_{j \in J} V_j \subseteq V = \{z \in \mathbb{C}^n : \hat{\mu}_1(z) = \cdots = \hat{\mu}_r(z) = 0\}$;
(ii) *the functions $x \mapsto \partial_j(\exp(ix \cdot z))$, $z \in V_j$ are solutions of the system*

$$\mu_1 * f = \cdots = \mu_r * f = 0; \tag{2.12}$$

(iii) *any solution $f \in \mathcal{E}(\mathbb{R}^n)$ of (2.12) can be represented as*

$$f(x) = \sum_{k=1}^{+\infty} \left(\sum_{j \in J_k} \int_{V_j} \partial_j(\exp(ix \cdot z)) d\nu_j(z) \right),$$

where the $d\nu_j$'s are Radon measures supported in V_j, and where the series and the integrals converge in $\mathcal{E}(\mathbb{R}^n)$.

This theorem was originally proved by Berenstein and Taylor for $\mathcal{E}(\mathbb{R}^n)$ and $H(\mathbb{C}^n)$, but it was soon generalized in Struppa's thesis (Struppa [1983a]) to a list of LAU-spaces (essentially, all LAU-spaces defined on all of $\mathbb{R}^n$ or $\mathbb{C}^n$). Indeed, the original proof of Theorem 2.13 uses the fact that $\hat{\mathcal{E}}'$ and $\hat{H}'$ are A_p-spaces in the sense of Hörmander (and of Chap. 1). On the other hand, in Struppa [1983a], the author used Hansen's new characterization of LAU-spaces (Hansen [1981]), together with the approach of Berenstein and Dostal [1972], to extend the result to $\mathcal{D}'(\mathbb{R}^n)$ and the other LAU-spaces. Another extension of Theorem 2.13 which is carried out in Struppa [1983a] is to a rectangular matrix of convolution operators (i.e., an $n \times m$ system of convolution equations). This extension, which relies on some algebraic machinery first developed in Buchsbaum [1964], now seems of relevant interest in some questions in algebraic analysis.

Before turning to other developments in mean-periodicity, let us point out that for a single convolution equation, the Fundamental Principle always holds (Berenstein and Taylor [1980b]) if we allow the abelian summation, as done for $n = 1$.

In view of Gurevich's example, this is the most general extension of Malgrange's thesis we could hope for.

Given that for more than one equation one needs the slowly decreasing condition, this condition, though generic, is hard to grasp. It therefore becomes important to find concrete examples of systems of convolution equations which satisfy the slowly decreasing condition. In fact, many open questions still lie in this direction, despite the very active work of Berenstein and Yger in this area. Before we quote their results, let us mention some of these open questions:

(a) *Consider m exponential polynomials of n variables with frequencies in $\mathbb{Q}^n$ and algebraic coefficients, which define a discrete (or empty) variety; are they jointly slowly decreasing?*

(b) *Consider m exponential polynomials as above, which define an empty variety. Can one show that the ideal they generate in $\hat{\mathcal{E}}'$ is all of $\hat{\mathcal{E}}'$?*

(c) *Let F be an exponential polynomial in one variable with algebraic coefficients and real algebraic frequencies: Are there positive constants c, N such that $F(z_1) = F(z_2) = 0$ implies (when $z_1 \neq z_2$)*

$$|z_1 - z_2| > c \exp(-N|\mathrm{Im}\, z_1|) \cdot (1 + |z_1|)^{-N} ?$$

Among the positive results, on the other hand, one has the following ones.

Theorem 2.14. *Let $P(z)$ be an exponential polynomial in $\mathbb{C}^n$, $P(z) = \sum_{j=1}^{N} c_j(z) \exp(\alpha_j \cdot z)$ (the α_j's are said to be the frequencies of $P(z)$) with $\alpha_j \in \mathbb{R}^n$. Then $P(z)$ is slowly decreasing.*

If we then restrict our attention to exponential sums (i.e., exponential polynomials with constants c_j's), we can prove a result concerning any complete intersection variety.

Theorem 2.15. *Let $P_1, \ldots, P_m$ $(m \leq n)$ be exponential sums with integral frequencies such that $V = \bigcap_{j=1}^{m} P_j^{-1}(0)$ is of codimension m in $\mathbb{C}^n$. Then $(P_1, \ldots, P_m)$ are jointly slowly decreasing.*

In view of these first results, one can consider the following general question. (However, it can easily be seen that such a question, arising in harmonic analysis when taking Fourier transforms of distributions with finite support is $\mathbb{R}^n$, i.e., arising from the general study of *difference-differential operators*, actually is tied to interesting problems in transcendental number theory.) *Given m exponential polynomials $F_1, \ldots, F_m$ with real frequencies and such that the set $\{z \in \mathbb{C}^n : F_1(z) = \cdots = F_m(z) = 0\}$ is either empty or discrete, is it possible to estimate the size of*

$$\{z \in \mathbb{C}^n : \sum_{i=1}^{m} |F_i(z)| < \varepsilon \exp(-Cp(z))\}$$

for $p(z) = |\mathrm{Im}\ z| + \log(1 + |z|)$?

As Berenstein and Yger pointed out, this result would correspond to a refined type of transcendency of the exponential functions with respect to the algebraic functions.

In Berenstein, Taylor and Yger [1983a] and Berenstein and Yger [1986a], a positive answer was given to a particular case of this question:

Theorem 2.16. *Let $F_1(z)$ and $F_2(z)$ be two exponential polynomials in $\mathbb{C}^2$ whose frequencies lie in $\mathbb{Q}^2$, and such that $\{z \in \mathbb{C}^n : F_1(z) = F_2(z) = 0\}$ is discrete. Then for any $\varepsilon, C > 0$, there are $\varepsilon_1, C_1 > 0$ such that*

(i) *the connected components of*

$$\{z \in \mathbb{C}^2 : |F_1(z)| + |F_2(z)| < \varepsilon_1 \exp(-C_1 p(z))\}$$

are bounded;

(ii) *if two points z_1, z_2 belong to the same connected component, then*

$$d(z_1, z_2) < \varepsilon \exp(-Cp(z_1)).$$

In Berenstein and Yger [1986a], this result was extended to $m \geq 2$ exponential polynomials in two variables, but also the case $m = 3$ (in any number of variables) is exhaustively studied. Many other related results are proved in Berenstein and Yger [1986b], to which we refer the interested reader. We content ourselves by mentioning two results, as well as *Ehrenpreis's conjecture (Berenstein and Yger [1986a]): Let $F_1, \ldots, F_m$ be exponential polynomials whose frequencies are algebraic numbers; then the family $(F_1, \ldots, F_m)$ is slowly decreasing if $V = \{z \in \mathbb{C}^n : F_1(z) = \cdots = F_m(z) = 0\}$ has codimension m.*

Theorem 2.17. *Given an exponential polynomial F with real frequencies, and a family $P_1, \ldots, P_m$ of polynomials defining an algebraic variety of dimension smaller than or equal to one, if the variety*

$$\{z \in \mathbb{C}^n : F(z) = P_1(z) = \cdots = P_m(z) = 0\}$$

is discrete, then the $(m+1)$-tuple $(F, P_1, \ldots, P_m)$ is slowly decreasing.

Theorem 2.18. *If F, G are exponential polynomials (with frequencies in $\mathbb{C}^2$) such that F/G is an entire function, then there is an exponential polynomial H and a polynomial P, factorizable in affine factors, such that for any $z \in \mathbb{C}^n$,*

$$\frac{F(z)}{G(z)} = \frac{H(z)}{P(z)}.$$

Without doubt, the reader will notice the strong link which exists between this result and the famous Ritt theorem; more details on this can be found in the earlier paper Berenstein and Dostal [1974a], as well as in Ritt [1929].

Also in connection with this work on (integral) representations of mean periodic functions, we wish to mention R. Gay's papers (Gay [1976a,b; 1980]), which deal with the case of entire functions of exponential type. Gay's work has had an important outgrowth in a recent paper by L. Gruman. In Gruman [1984], he uses the methods of Berenstein and Taylor, as well as ideas from Gay's work, to study *difference-differential equations with variable coefficients.* He is able to prove some surjectivity results in terms of some (quite complicated) compatibility conditions; we refer the reader to the original paper for the statements, which are too involved to be reproduced here.

Further developments came out of Meril's work (which we already quoted in Chap. 1 when we talked about division problems); in a series of papers (Meril [1983a,b,c]; Berenstein and Struppa [1983]) by Meril and the authors, the problems of spectral synthesis and of Fourier representation for mean-periodic functions in spaces of functions which are just holomorphic in cones (with exponential-type growth conditions) were completely solved. The methods are a combination of Meril's original approach, together with the basic tools developed in Berenstein and Taylor [1980b], except for the $\bar{\partial}$-methods, where one could not use Hörmander's results directly, but has to resort to De Roever's modifications (De Roever [1978]) of the method (these modifications are quite akin to those developed by Kaneko for the closely related problem of finding a fundamental principle for hyperfunctions; see Kaneko [1973]).

More difficulties appeared when we tried to extend the study of mean-periodicity to spaces such as $\mathcal{E}(\Omega)$ and $H(\Omega)$ for Ω, an open convex set in

$\mathbb{R}^n$ and an open convex set in $\mathbb{C}^n$, respectively. Even though the general philosophy does not need to be changed, the study of mean-periodicity in $\mathcal{E}(\Omega)$ is more delicate in view of the LAU-structure of the space $\mathcal{E}(\Omega)$, which is more subtle than the case $\Omega = \mathbb{R}^n$. In Berenstein and Struppa [1987a], the authors considered the following case: Let $\Omega_1, \Omega_2 \subseteq \mathbb{R}^n$ be two open convex sets and let $\mu_1, \ldots, \mu_r \in \mathcal{E}'(\mathbb{R}^n)$, $1 \le r \le n$, be such that $\mathrm{ch}(\mathrm{supp}\,\breve{\mu}_j) = K$ for a compact convex set K, such that $\Omega_1 = \Omega_2 + K$. In order to establish an analogue of Theorem 2.13 for the system

$$\mu_1 * f = \cdots = \mu_r * f = 0, \quad f \in \mathcal{E}(\Omega_1),$$

it is unfortunately necessary to impose *stronger* conditions on $(\mu_1, \ldots, \mu_r)$ than for the case of $\Omega = \mathbb{R}^n$ (this fact is later employed and discussed in the authors' paper, Berenstein and Struppa [1987b]). To give an idea of the situation, let us provide the precise definition for the discrete case (in the more general situation, one has to resort to a cutting up argument as hinted at in Chap. 1).

Definition 2.3. An n-tuple $(\mu_1, \ldots, \mu_n)$ in $\mathcal{E}'(\mathbb{R}^n)$ of distributions as above is said to be *slowly decreasing* (for convex pairs) if there exist positive constants C, D, m such that for $V = \{z \in \mathbb{C}^n : \hat{\mu}_1(z) = \cdots = \hat{\mu}_n(z) = 0\}$, $V_i = \{z \in \mathbb{C}^n : \hat{\mu}_i(z) = 0\}$, $d(z, V) = \min(1, (\mathrm{dist}(z, V))$, and similar definitions for $d(z, V_i)$, one has

$$|\hat{\mu}_i(z) \ge C(1 + |z|)^{-D} d(z, V_i)^m \exp(H_K(\mathrm{Im}\ z)), \quad (i = 1, \ldots, n,\ \forall\ z \in \mathbb{C}^n)$$

and

$$|\hat{\mu}(z)| \ge C(1 + |z|)^{-D} d(z, V)^m \exp(H_K(\mathrm{Im}\ z)) \quad (\forall\ z \in \mathbb{C}^n),$$

and for any $\varepsilon > 0$, there are constants $A, B > 0$ such that the set

$$S(\varepsilon, A, B) = \{z \in \mathbb{C}^n : d(z, V_i) \le A(1 + |z|)^{-B} \exp(-\varepsilon|\,\mathrm{Im}\ z|), \forall\ i\}$$

has connected components with uniformly bounded diameters.

It is then proved in Berenstein and Struppa [1987a] that spectral synthesis and integral Fourier representations hold for these systems. We omit the cumbersome statement of the theorem. On the other hand, we want to mention a different result in this direction, due to Berenstein and Gay; in their paper Berenstein and Gay [1986a] (which we shall subsequently quote in Chap. 3 when dealing with the Pompeiu problem), the authors are interested in obtaining a series development of mean-periodic functions in $\mathcal{E}(\Omega)$ when the μ_j's are characteristic functions of spheres; in Berenstein and Dostal [1976], Berenstein and Dostal had shown that, unfortunately, such distributions will not fulfill the requirements of Def. 2.3, so that the results of Berenstein and Struppa [1987a] cannot be used. On the other hand, in Berenstein and Struppa [1987a], we had also obtained some knowledge of the behavior of the terms of the series expansion, and this was not necessary for the result Berenstein and Gay were looking for. In Berenstein and Gay [1986a], they managed to prove the following series development (without any information on the terms of the series).

Theorem 2.19. *Let Ω be an open convex subset in $\mathbb{R}^n (n \geq 2)$, $\mu \in \mathcal{E}'(\mathbb{R}^n)$ an invertible distribution, and $K = \mathrm{ch}\,(\mathrm{supp}\,\check{\mu})$. Any function $f \in \mathcal{E}(\Omega + K)$, satisfying $\mu * f = 0$, can be written as*

$$f(x) = \sum_{j \geq 1} P_j(x), \quad (x \in \Omega + K),$$

with P_j being exponential polynomials mean-periodic with respect to μ, and the series being convergent in the C^∞ topology of $\Omega + K$. Furthermore, given a sequence $\{s_j\}_{j \geq 1}$ of positive numbers, letting $P_0 = 0$, we can choose the P_j so that the absolute value of all frequencies in P_{j+1} exceeds the largest absolute value of the frequencies in P_j by at least s_{j+1}.

Before turning to the case of functions in $H(\Omega)$, let us point out a problem which still needs to be solved: In Berenstein and Struppa [1987a], besides the slowly decreasing condition, it is necessary that all of the μ_j's have the same support K; this is not the case in many concrete situations. *Can one prove a representation theorem for the systems $\mu_1 * f = \cdots = \mu_r * f = 0$, $f \in \mathcal{E}(\Omega)$ in which the μ_j's have different supports?*

The extension and modifications of the approach to deal with the case of $H(\Omega)$, with Ω being an open convex subset of $\mathbb{C}^n$, was carried on in Meril and Struppa [1987a] by Meril and Struppa (we already mentioned at the beginning of this chapter the reason for the different behavior of analytic functionals and distributions). In Meril and Struppa [1987a], one studies the following situation: For $\Omega \subseteq \mathbb{C}^n$ being a convex open set, and $\mu \in H'(\mathbb{C}^n)$ carried by a convex compact set K, one has

$$\mu* : H(\Omega + K) \to H(\Omega),$$

and three problems can be studied:

(i) When is $\mu*$ a surjective operator?
(ii) Prove a Fundamental Principle for systems such as

$$\mu_1 * f = \cdots = \mu_r * f = 0, \quad f \in H(\Omega + K).$$

(iii) When can one extend solutions of such systems beyond $\Omega + K$?

For $n = 1$, Ephrom (Ephrom [1983]) and, earlier, Napalkov (Napalkov [1974; 1979a]) have shown that $\mu*$ is onto if and only if $\hat{\mu}$ is of completely regular growth (Levin [1964]). For $n \geq 2$, Morzhakov has shown that surjectivity follows from

(a) $\hat{\mu}$ is of completely regular growth, and
(b) the radial regularized indicatrix $\wedge*_{\hat{\mu}}$ of $\hat{\mu}$ coincides with H_K. More recently, Gruman and Lelong have shown in Gruman and Lelong [1986]

(a reference book for most of the complex analysis which appears in this contribution) that if Ω is strictly convex, with a C^2 boundary, then conditions (a) and (b) are necessary and sufficient for the surjectivity of $\mu*$. On the other hand, in Meril and Struppa [1987a], it is shown that *if Ω is bounded and if K is strictly convex with a C^1 boundary, and if $\mu*$ is onto, then* (b) *holds; moreover, independently of any regularity for K, if $\mu*$ is onto, then K is a support for μ (and not just a carrier).*

With respect to problem (ii), the proof from Berenstein and Struppa [1987a] can be modified to suit the holomorphic case, and a Fundamental Principle is proved. It is with the use of this principle that one can prove some interesting characterizations of hyperbolicity (see both Meril and Struppa [1987a] and the related Berenstein and Struppa [1987b]); this question, however, will be taken up later in Chap. 4.

Since hyperbolicity is one kind of answer to question (iii), a different kind of extension of solutions was considered in Meril and Struppa [1987a]:

Find good conditions so that every solution of a convolution equation $\mu *$ $f = 0$ ($\mu \in H'(K_0)$, $f \in H(\Omega + K_0)$) can be extended to a larger open set $\Omega_1 \supseteq \Omega + K_0$ which depends on the asymptotic cone of the zero set of $\hat{\mu}$.

The paper Meril and Struppa [1987a] is inspired by previous results of Kiselman (Kiselman [1969]) and Sebbar (Sebbar [1980]). However, the techniques of algebraic analysis have enabled Aoki (Aoki [1988]) to give a simpler treatment of this question (see also Kashiwara and Schapira [1979] and Chap. 6 below). The results of Kiselman and Sebbar deal with the case in which $\mu*$ is a constant coefficient partial differential operator and the case in which $K_0 = \{0\}$, respectively. In Meril and Struppa [1987a], an extension to any compact K_0 is given.

The extension problem leads us to another classical question, that of the removability of singularities. The well-known *Hartog's extension theorem* shows the removability of compact singularities for holomorphic functions in $\mathbb{C}^n$, $n \geq 2$. In 1960, Ehrenpreis explicitly pointed out the relation between Hartog theorem and the theory of overdetermined systems of partial equations (Ehrenpreis [1961b]). This idea was further developed for convolution equations in the spaces $\mathcal{E}(\Omega), \mathcal{D}'(\Omega), \mathcal{E}_\omega(\Omega)$ by Meril and Struppa (Meril and Struppa [1987b]; Struppa [1988b]), and for infinite-order differential equations in $\mathcal{B}(\Omega)$ (hyperfunctions) by Kawai and Struppa (Kawai and Struppa [1990]). For the details, the interested reader is referred to the fairly complete survey (Struppa [1988a]).

Two more directions of research should be discussed before ending this chapter. The first one is mainly due to Meise, Taylor, and Vogt (Meise, Momm and Taylor [1987]; Meise, Schwerdtfeger and Taylor [1986]; Meise and Taylor [1987]; Meise, Taylor and Vogt [1987b]) and has a functional analysis flavor. Without going into details, for which we refer the reader to the original papers, we wish to mention that the core of their approach is the construction of a suitable Schauder basis for the kernel of a given convolution operator. Following an idea of Meise (Meise [1985]), they prove that for any surjective convolution operator on $\mathcal{E}_\omega(\mathbb{R})$ (this is the Fréchet space of all ω-ultradifferentiable functions on $\mathbb{R}$, in the sense of Beurling and Björck; see Beurling [1957] and Björck [1966]) with an infinite-dimensional kernel, its "local" zero solutions admit a local Fourier expansion. More precisely, *there exists a Schauder basis $\{g_j\}_{j\geq 1}$ of $ker(\mu*)$, $\mu \in \mathcal{E}'_\omega(\mathbb{R})$, consisting of exponential polynomials such that for each $\rho > 0$, there exist $r > \rho$, $R > r$, and $h_j \in \mathcal{E}'_\omega((-R,R))$ such that for each $\varphi \in \mathcal{E}_\omega(\mathbb{R})$ with $\mu * \varphi|[-r,r] = 0$, we have*

$$\varphi(x) = \sum_{j=1}^{+\infty} f_j(x)h_j(\varphi), \quad x \in (-\rho,\rho);$$

this expansion converges in $\mathcal{E}_\omega((-\rho,\rho))$, while if φ belongs to $\ker(\mu*)$, it converges in $\mathcal{E}_\omega(\mathbb{R})$, and the coefficients are uniquely determined.

The last topic we want to mention in this chapter is the extension of all the previous material to the case of symmetric spaces. The case of partial differential equations has been studied by Kuchment, using Helgason's general results (see Helgason [1962; 1965; 1970]; Kuchment [1981; 1985a,b]). In the case of convolutino equations, much work has recently been done, mainly by Berenstein and Gay, Berenstein and Zalcman (see their papers quoted in Chap. 3 in relation with the Pompeiu problem on symmetric spaces) as well as by Kuchment (Kuchment [1982]); he gives a representation theorem for mean-periodic functions in this context by essentially following the papers Berenstein and Taylor [1979; 1980b] and Wawrzynczyk [1985; 1987]). Let us briefly fix the notations and then proceed to state the most relevant results: Let G be a unimodular semisimple connected noncompact Lie group, with finite center, and let K be a maximal compact subgroup of G. Then we have that $X = G/K$, the homogeneous space of right cosets gK, with the natural projection $\pi : G \mapsto X$, is a symmetric space of noncompact type and every symmetric space of this type can be realized in this way. If dg (resp. dk) is the Haar measure of G (resp. the normalized Haar measure of K, with $\int_k dk = 1$), one can define a measure dx on X by setting

$$\int_X f(x)dx = \int_G (f \circ \pi)(g)dg, \quad \text{for} \quad f \in \mathcal{D}(X),$$

where, as for any second countable smooth manifold M, $\mathcal{D}(M)$, $\mathcal{D}'(M)$, $\mathcal{E}(M)$, $\mathcal{E}'(M)$ have their customary meanings. Compactly supported functions on G can be convolved by setting

$$(f * \varphi)(g) = \int_G f(gh^{-1})\varphi(h)dh = \int_G f(h)\varphi(h^{-1}g)dh,$$

and as usual, this extends to distributions (the last equality being a consequence of the unimodularity of G). A measure on G can be introduced by setting

$$\delta_K : f \mapsto \int_G f(k)dk,$$

so that every function $f \in \mathcal{E}(G)$ defines a function $f_\pi \in \mathcal{E}(X)$ by

$$f_\pi \circ \pi = f * \delta_K;$$

hence, every $T \in \mathcal{D}'(X)$ lifts to a distribution $\tilde{T}$ on G defined by

$$\tilde{T}(f) = T(f_\pi),$$

for any $f \in \mathcal{D}(G)$. Now, if a function φ on X is regarded as a distribution, then

$$\tilde{\varphi}(g) = \varphi(\pi(g)), \quad (\tilde{\varphi})_\pi = \varphi,$$

so that *the functions (and the distributions) on X can be identified with functions (and distributions) on G which are right-invariant under K, i.e., which satisfy $\varphi(gk) = \varphi(g)$ for all $k \in K$.* The lifting described before also induces a notion of convolution in $\mathcal{D}'(X)$ by

$$\mu * \nu := \tilde{\mu} * \tilde{\nu}$$

if at least one of the two distributions is compactly supported. This convolution is associative and satisfies $\mu * \delta = \delta * \mu = \mu$ if δ is the distribution on X given by $\delta(f) = f(\pi(e))$, where e is the identity in G. If $f \in \mathcal{D}_0(G)$, i.e., if $f \in \mathcal{D}(G)$ is K-biinvariant, one can define a *spherical Fourier transform* $\mathcal{F}$ given by

$$\mathcal{F}(f)(\lambda) = \int_G f(g)\varphi_\lambda(g^{-1})dg, \qquad (2.13)$$

for all $\lambda \in \mathcal{A}'$, where $\mathcal{A}'$ is the dual of a real vector space $\mathcal{A}$, whose dimension n is the rank of X (or equivalently, the real rank of G). The φ_λ which appear in (2.13) are the spherical functions of G (Helgason [1984]). Denoting by $d\lambda$ the Lebesgue measure divided by $(2\pi)^{n/2}$, Helgason proved (Helgason [1962; 1984]) the *inversion formula*

$$f(g) = \frac{1}{N} \int_{\mathcal{A}'} \mathcal{F}(f)(\lambda)\varphi_\lambda(g)|c(\lambda)|^{-2}d\lambda,$$

where $c(\lambda)^{-1}$ is an analytic function on $\mathcal{A}'$, and N is the order of the *Weyl group* W of G. If we denote by $\mathcal{A}'_{\mathbb{C}}$ the complexification of $\mathcal{A}'$, which can be identified with $\mathbb{C}^n$, then the functions φ_λ are defined for any $\lambda \in \mathcal{A}'_{\mathbb{C}}$, and $\mathcal{F}(f)(\lambda)$ extends to an entire function on $\mathbb{C}^n$; the following analogue of the Paley–Wiener theorem can be given (Helgason [1962; 1984]).

Theorem 2.20. *The spherical Fourier transform $\mathcal{F}$ provides a bijection between $\mathcal{D}_0(G)$ and the space of entire functions of exponential type, W-invariant on $\mathcal{A}'_{\mathbb{C}}$ and rapidly decreasing on $\mathcal{A}'$. This bijection extends to a linear isomorphism between the space $\mathcal{E}'_0(G)$ of K-biinvariants compactly supported distributions on G, and the space $\mathcal{F}(\mathcal{E}'_0(G))$ of entire functions of exponential type, W-invariant and of polynomial growth on $\mathcal{A}'$.*

As Berenstein and Zalcman proved in Berenstein and Zalcman [1980], $\mathcal{F}$ is actually a *topological algebra isomorphism*. For the sake of simplicity, let us now indicate $\mathcal{A}'_{\mathbb{C}}$ by $\mathbb{C}^n$ so that, in the sequel, $H(\mathbb{C}^n), \exp(\mathbb{C}^n)$, and $\hat{\mathcal{E}}'(\mathbb{R}^n)$ will represent $H(\mathcal{A}'_{\mathbb{C}}), \exp(\mathcal{A}'_{\mathbb{C}})$, respectively, and the space of functions of exponential type in $\mathrm{Exp}(\mathcal{A}'_{\mathbb{C}})$ which, on $\mathcal{A}'$, have polynomial growth. Similarly, we denote by $[\mathrm{Exp}(\mathbb{C}^n)]_W$ and by $[\hat{\mathcal{E}}'(\mathbb{R}^n)]_W$ the subalgebras consisting of W-invariant functions, where W is a finite group of linear transformations on $\mathbb{C}^n$ (real, in the second case). We can now conclude by quoting the following results (Berenstein and Gay [1986b]):

Let $\frac{2}{3} < k < 1$ and consider two functions f_1, f_2 in $A_{q_k}(\mathbb{C}^n)$, $q_k(z) = |z_1|^k + \cdots + |z_n|^k$, with no common zeros, and which generate a nonlocalizable ideal I in A_{q_k} (these functions exist by Gurevich's basic paper, Gurevich [1975], and the same construction holds in $\mathrm{Exp}(\mathbb{C}^n)$). (By this we mean, as we already explained, that the closure of I in A_{q_k} does not coincide with I_{loc}. In this case, since f_1, f_2 have no common zeros, $I_{\mathrm{loc}} = A_{q_k}$.) Now, let W be a finite group of unitary transformations, whose order we denote by N. Set $r = N + 1$ and consider r distinct complex numbers, $\lambda_1, \ldots, \lambda_r$, to which we associate the r functions

$$\varphi_j(z) = \prod_{s \in W} (f_1(s(z)) + \lambda_j f_2(s(z))), \quad z \in \mathbb{C}^n, \quad j = 1, \ldots, r.$$

Theorem 2.21. *The functions $\varphi_1, \ldots, \varphi_r$ have no common zeros and generate a nonlocalizable ideal in the subspace $A_{q_k}(\mathbb{C}^n)_W$ of W-invariant functions of $A_{q_k}(\mathbb{C}^n)$.*

Even more interesting is the fact that Berenstein and Gay succeed, with a totally different method based on ideas of Sheppard and Todd [1954], in proving that the same lack of spectral synthesis holds even for very general algebras. Let $q(z) = |z_1|^{a_1} + \cdots + |z_n|^{a_n}, a_i > 0, i = 1, \ldots, n$, and consider the algebra $A_q(\mathbb{C}^n)$.

Theorem 2.22. *There exists an integer N (depending on $a_1, \ldots, a_n$) such that there exists a nonlocalizable ideal $(f_1, \ldots, f_N)$, generated by functions with no common zeros, in $A_q(\mathbb{C}^n)$.*

To conclude, we give one positive result on mean periodicity, due to Kuchment [1982].

Theorem 2.23. *Let $f \in \mathcal{E}(X)$, $\mu \in \mathcal{E}'(X)$, such that $\mu * f = 0$, and suppose $\mathcal{F}(\mu)$ to be slowly decreasing (the definition being as in the case of $X = \mathbb{R}^n$). Then (with the same notations as in Theorem 2.13), one can represent f as follows*

$$f(x) = \sum_{i,j,\delta} \sum_{\ell} \left[\sum_{k \in J_\ell} \int_{V_k} \partial_k \varphi_{i,j,\delta}(x,\lambda) d\mu_{i,j,\delta,k}(\lambda) \right],$$

where the sum with respect to i, j, δ is finite, the μ's are Radon measures with supports in V_k, the series (with respect to ℓ) and the integrals converge in

$\mathcal{E}(X)$, and the functions $\varphi_{i,j,\delta}$ (defined in Kuchment [1982]) are the precise analogue of the exponential polynomials.

The proof of this result is based on Helgason's theory of the spherical Fourier transform (Helgason [1981; 1984]), and on the work of Berenstein and Taylor (Berenstein and Taylor [1980b]); as Kuchment himself mentions in Kuchment [1982], it is also possible to obtain analogous results for K-finite functions on the tangent space to X, which are mean-periodic, by replacing the functions $\varphi_{i,j,\delta}$ by generalized Bessel functions.

Chapter 3
The Pompeiu Problem

The key topic of this chapter is the so-called Pompeiu problem, which (in a very particular case) can be stated as follows:

Problem. Let $\mathcal{F} = \{\mu_1, \ldots, \mu_m\}$ be a finite family of compactly supported distributions in $\mathbb{R}^n$. One can associate to $\mathcal{F}$ the *Pompeiu transform*, which acts as follows:

$$P = P_{\mathcal{F}} : \mathcal{E}(\mathbb{R}^n) \to [\mathcal{E}(M(n))]^m,$$

where $M(n)$ is the semidirect sum of $\mathbb{R}^n$ with $SO(n)$, the group of rigid motions of $\mathbb{R}^n$. The Pompeiu transform is defined on each $f \in \mathcal{E}(\mathbb{R}^n)$ by

$$P_{\mathcal{F}}(f)(\sigma) = \{\langle \sigma \mu_j, f \rangle\} := \langle \mu_j, f \circ \sigma^{-1} \rangle : j = 1, \ldots, m\},$$

for $\sigma \in M(n)$. One can pose three basic questions about the Pompeiu transform:

(i) Is $P_{\mathcal{F}}$ injective?
(ii) If not, characterize its kernel.
(iii) If $P_{\mathcal{F}}$ is injective, find its inverse (where defined).

The fancy notation which we have just introduced actually includes a number of well-known examples, the first of which was posed and studied by the Romanian mathematician D. Pompeiu in 1929 (Pompeiu [1929a,b]). In this classical case, the family $\mathcal{F}$ is simply the characteristic function of a single open bounded set $\Omega \subset \mathbb{R}^n$, $n \geq 2$. Then

$$P_{\mathcal{F}}(f)(\sigma) = \int_{\sigma(\Omega)} f(x)dx, \tag{3.1}$$

for any $\sigma \in M(n)$, $f \in \mathcal{E}(\mathbb{R}^n)$. One says that Ω has the Pompeiu property if $P_{\mathcal{F}}$ is injective. It can be easily seen that, even in the plane, not every set has the Pompeiu property. Indeed, let $\Omega = \Delta(0, R)$ for fixed $R > 0$, and $f = f(x_1, x_2) = \exp[i(\alpha_1 x_1 + \alpha_2 x_2)]$. Then, since $\sigma(\Omega)$ is just a disk of radius

R and center $P_0 = (\xi_0, \eta_0)$, with these last coordinates depending on σ, the integral (3.1) amounts to

$$\int_{\Delta(P_0, R)} \exp[i(\alpha_1 x_1 + \alpha_2 x_2)] dx_1 dx_2 = 2\pi R \exp[i\langle P_0, \alpha\rangle] \cdot \frac{J_1(R|\alpha|)}{|\alpha|},$$

where J_1 is the Bessel function of the first kind, $\alpha = (\alpha_1, \alpha_2)$, and $|\alpha|^2 = |\alpha_1|^2 + |\alpha_2|^2$. Therefore, whenever $R|\alpha|$ is a zero of the Bessel function J_1, we have that f belongs to the kernel of $P_{\mathcal{F}}$ which, therefore, is not injective. Using the theorems of Chap. 2, one can actually prove in this case that the exponentials $\exp[i(\alpha_1 x_1 + \alpha_2 x_2)]$ generate $\ker(P_{\mathcal{F}})$.

An important remark can be made to this point: Indeed, by using the decomposition of $M(n)$ as the semidirect product of $\mathbb{R}^n$ and $\mathrm{SO}(n)$, the kernel of $P_{\mathcal{F}}$ can be understood as the space of solutions of a (possibly infinite) system of homogeneous convolution equations. In the case in which $\mathcal{F}$ contains just the characteristic function of a single fixed radius disk, this infinite system can actually be reduced to a single *nontrivial* equation. On the other hand, if we had two disks, say $\Delta(0, R_1)$ and $\Delta(0, R_2)$, then the system could be reduced to a different one made of two equations, and now the previous computations would show that if R_1/R_2 does not belong to the set of quotients of zeros of J_1, then there would be no exponential solutions. It can be proved that the Pompeiu transform would be injective in this case (see Brown, Schreiber and Taylor [1973]; Berenstein and Zalcman [1980]; Zalcman [1972]).

In many concrete cases, one can determine by means of the following theorem whether or not a set Ω has the Pompeiu property (Berenstein and Zalcman [1980]; Brown, Schreiber and Taylor [1973]).

Theorem 3.1. *Let Ω be an open bounded set in $\mathbb{R}^n$, with Lipschitz boundary, and whose complement is connected. Then Ω does not have the Pompeiu property if and only if there exists a constant $\alpha > 0$, and a function u on Ω, satisfying the following overdetermined Neumann boundary problem:*

$$\begin{cases} \Delta u + \alpha u = 0 & \text{on} \quad \Omega \\ \dfrac{\partial u}{\partial n} = 0, \ u = 1 & \text{on} \quad \partial\Omega. \end{cases} \tag{3.2}$$

A most important corollary of this result is the following.

Theorem 3.2. *Under the same geometrical assumptions of* Theorem 3.1, *if Ω does not have the Pompeiu property, then $\partial\Omega$ is a real-analytic submanifold of $\mathbb{R}^n$.*

This result, which is proved in Williams [1976], in particular shows that any domain whose boundary has some "corner" does indeed enjoy the Pompeiu property.

Because Theorem 3.1 gives an equivalence between the failure of the Pompeiu property and the existence of a solution to the system (3.2), one can immediately see why the disk fails to have the Pompeiu property, since if Ω were a disk, problem (3.2) would have infinitely many solutions. This last fact indeed characterizes the balls of $\mathbb{R}^n$ among the bounded open domains which satisfy the geometrical conditions of (3.1) (Berenstein [1980]; Berenstein and Yang [1982]). It is an old conjecture (known as the *Schiffer problem* and dating back to Lord Rayleigh) that the only domain for which (3.2) has a solution is the ball.

As we mentioned before, the work on this kind of problem (which we formulated in the C^∞ category, but which we can rephrase with suitable interpretations in the case of distributions) goes back to Pompeiu, but we refer the interested reader to the beautiful paper by Zalcman (Zalcman [1980]) for a fairly complete list of references.

Let us now show how the Pompeiu problem relates to the well-known Morera theorem. The basic result which one has in mind is the following.

Theorem 3.3. *In $\mathbb{C}^n$, consider the $(n, n-1)$-differential forms*

$$\omega_j = dz_1 \wedge \cdots \wedge dz_n \wedge d\bar{z}_1 \wedge \cdots \wedge d\bar{z}_{j-1} \wedge d\bar{z}_{j+1} \wedge \cdots \wedge d\bar{z}_n,$$

let Ω be an open set with piecewise C^1 boundary, and suppose that Ω has the Pompeiu property with respect to $\mathbb{R}^{2n}$. Let $f \in C(\mathbb{C}^n)$. Then f is an entire holomorphic function if and only if

$$\int_{\sigma(\partial\Omega)} f(z)\omega_j = 0$$

for every $\sigma \in M(n)$ and $j = 1, \ldots, n$.

The proof of this fact is just a standard application of Stokes' formula to reduce the problem to Pompeiu's. Indeed, the open sets for which the conclusions of the Morera theorem hold are precisely those for which the Pompeiu property holds.

We shall later see that this theorem can be further generalized when we talk about local Pompeiu properties. Before we turn to other generalizations of the Pompeiu problem, we wish to point out that we can replace the triple $(\mathbb{R}^n, M(n), dx)$ in the formulation of this problem by a triple $(X, \mathcal{G}, d\mu)$, where X is the manifold and $\mathcal{G}$ is a Lie group which acts transitively on X and which leaves the measure $d\mu$ on X invariant (this is shown in Berenstein and Zalcman [1980]). In this general setting, we know very little about the validity of the general Pompeiu properties, except in the case in which X is an irreducible symmetric space of the noncompact type, $X = G/K$, and $\operatorname{rk} k(X) = \operatorname{real\ rank}(G) = 1$ (the case of a compact Lie group can also be easily analyzed). This more general analysis is based on properties of ideals which are invariant under rotations (or, more generally, under K). In fact, the statement of the equivalence between the Pompeiu problem and the overdetermined boundary-value problem remains valid under the natural interpretation in which Δ in (3.2) is the Laplace–Beltrami operator (for details, see Berenstein and Shashahani [1983] and the related papers Bagchi and Sitaram [1979]; Shashahani and Sitaram [1987]).

Let us now turn to a slight variation of the definition of the Pompeiu transform by simply allowing the distributions $\mu_1, \ldots, \mu_m$ which constitute $\mathcal{F}$ to be in $\mathcal{S}'$ ($\mathcal{D}'$, respectively) on $\mathbb{R}^n$. Then the corresponding map acts as follows:

$$P_{\mathcal{F}} : \mathcal{S}(\mathbb{R}^n) \to [\mathcal{S}(M(n))]^m$$

(or, respectively, $P_{\mathcal{F}} : \mathcal{D}(\mathbb{R}^n) \to [\mathcal{D}(M(n))]^m$), where the definition of $\mathcal{S}(M(n))$ can be easily supplied by the reader. A particular example of this situation is the classical *Radon transform*; in this case, just take the family $\mathcal{F}$ of the single-density measure $\mu = dx_1 \cdot \cdots \cdot dx_{n-1}$ on the hyperplane $x_n = 0$ (such a measure μ induces a surface measure dA on every hyperplane). Inversion formulas in the theory of the Radon transform are very well known, see, e.g., Helgason [1965], but slightly less known is the following *hole theorem* due to Cormack and Helgason (see Helgason [1980] and Natterer [1986]).

Theorem 3.4. *Let K be a compact convex set in $\mathbb{R}^n$, and let f be a function in $\mathcal{S}(\mathbb{R}^n)$ such that*

$$\int_H f \, dA = 0,$$

for every hyperplane H such that $H \cap K = \emptyset$; then $\operatorname{supp}(f) \subset K$.

When talking about the Radon transform, we would like to mention the recent work of Ehrenpreis (Ehrenpreis [1992]), in which Fourier analysis on $\mathbb{R}^n$ is employed in a very imaginative way to study the Radon transform. The roots of Ehrenpreis's work, as well as many of the results mentioned in this chapter, can be traced to the thesis of F. John (John [1955]).

Closely related to the Radon transform (it can actually be reduced to it, via projective transformations, as Solmon showed in Solmon [1976]) is the so-called *Radon transform on spheres*, which is defined as follows: Denote by $S(x)$ the sphere of center x and radius $\frac{|x|}{2}$ in $\mathbb{R}^n$ with its normalized surface measure $d\sigma$. Then we can define, following Cormack and Quinto [1980], a transformation

$$S : \mathcal{E}(\mathbb{R}^n) \to \mathcal{E}(\mathbb{R}^n)$$

given by

$$S(f)(x) := \int_{S(x)} f \, d\sigma.$$

It turns out that this transformation is invertible, and in fact, there are explicit inversion formulas in Cormack and Quinto [1980] which imply the following variation of the hole theorem (i.e., of Theorem 3.4).

Theorem 3.5. *Let F be C^∞ on a ball centered in the origin of $\mathbb{R}^n$, and of radius R, and suppose that $S(f)(x)$ vanishes for $|x| < \frac{R}{2}$. Then $f(x)$ vanishes for $|x| < R$.*

Note that though the definition of the transform S only requires f to be continuous, this theorem requires that f is C^∞ in a full neighborhood of the origin. Examples show that the result is actually false for functions which are not C^∞.

More general problems arise when integrating on families of sets which are just invariant under rotations. The most complete theory in this direction corresponds to what is known as the *local Pompeiu problem*. Given a distribution μ of compact support in $\mathbb{R}^n$ and an open subset U, we consider the set

$$G(U,\mu) = \{g \in M(n) : \operatorname{supp}(g\mu) \subset U\}.$$

The *local Pompeiu transform* for a family $\mathcal{F} = \{\mu_1, \dots, \mu_m\}$ then maps

$$\mathcal{E}(U) \to \bigoplus_{i=1}^{m} G(U, \mu_j).$$

Its definition is the same one we gave for the (global) Pompeiu transform. We shall now say that the family $\mathcal{F}$ has the *local Pompeiu property* with respect to U if this map is injective.

Among the many different results in this area, we would like to mention the following one in which the exceptional set (first described in Zalcman [1972] and, independently, in Brown, Schreiber and Taylor [1973]) is given by

$$E_n = \{\text{quotients of two zeros of the Bessel function } J_{n/2}\}.$$

Theorem 3.6 (Berenstein and Gay [1986a]). *Let r_1, r_2 be positive numbers such that $r_1, r_2 \notin E_n$, and let Ω be an open subset of $\mathbb{R}^n$ such that every point lies in some open ball contained in Ω, of radius strictly larger than $r_1 + r_2$. If f is a continuous function on Ω such that*

$$\int_{B(x,r_j)} f(y)dy = 0$$

for every $\bar{B}(x, r_j) \subseteq \Omega$, $j = 1, 2$, then $f \equiv 0$.

Under appropriate arithmetical restrictions, the conditions on r_1, r_2 described in Theorem 3.6 are actually necessary and sufficient.

This particular theorem has also been given a different proof by Berenstein, Gay, and Yger in Berenstein, Gay and Yger [1990]. The proof consists of the explicit construction of the inverse, as long as r_1/r_2 does not belong to the exceptional set described before, and provided that $r_1 + r_2 < R$. This answers, in particular, a question posed by Zalcman in Zalcman [1980].

Theorem 3.7 (Berenstein and Gay [1986a]). *Define*

$$\sigma(t) = 2^{(n-2)/2}\Gamma(n/2)J_{(n-2)/2}(t)/t^{(n-2)/2},$$

and set

$$H_n = \{\xi/\eta : \xi, \eta \in (0, +\infty), \sigma(\xi) = \sigma(\eta) = 1\}.$$

If $R > r_1 + r_2$, $r_1/r_2 \notin H_n$, and f is a continuous function in $B(0, R)$ satisfying (where $d\sigma$ is the normalized Lebesgue measure on S^{n-1})

$$f(x) = \int_{S^{n-1}} f(x + r_1 y) d\sigma(y), \quad \forall x : |x| < R - r_1$$

and

$$f(x) = \int_{S^{n-1}} f(x + r_2 y) d\sigma(y), \quad \forall x : |x| < R - r_2,$$

then f is harmonic in $B(0, R)$.

Let us point out that in this theorem, the centers about which one checks the mean-value property are quite limited. It is interesting to compare this with an old theorem of Volterra, showing that for functions continuous on the closed disk, the mean-value property on a single circle about each point already implies harmonicity (for details and its relation to an open conjecture of Littlewood, the reader is referred to the two papers Zalcman [1973; 1980]).

Following Berenstein and Gay [1989], we can also give a local version of the Morera theorem, which we stated above as Theorem 3.3.

Theorem 3.8. *Let Ω be a bounded open set in $\mathbb{C}^n$ with sufficiently regular boundary and suppose that Ω satisfies the Pompeiu property. Let R^* be a positive number such that $\bar{\Omega}$ is contained in the open ball $B(0, R^*)$ and let $R \geq 2R^*$. Then, with the same notations employed in Theorem 3.3, a continuous function f in $B(0, R)$ is holomorphic in $B(0, R)$ if and only if*

$$\int_{\sigma(\partial\Omega)} f(x)\omega_j = 0,$$

for every $j = 1, \ldots, n$ and every $\sigma \in G(B(0, R), \chi_\Omega)$.

Finally, one can use this local version of the Pompeiu problem to prove an interesting form of the hole theorem (i.e., Theorem 3.4), which we can state as follows.

Theorem 3.9. *Let Ω be a bounded open set in $\mathbb{R}^n$ with sufficiently regular boundary and which satisfies the Pompeiu property. Let K be a convex set in $\mathbb{R}^n$. Suppose that f is a continuous function in the complement of K such that*

$$\int_{\sigma(\Omega)} f dx = 0,$$

for every $\sigma \in M(n)$, such that $\sigma(\Omega) \cap K = \emptyset$; then f vanishes identically outside K.

The reader may notice that the theorems above are given for families of functions which are invariant under rotations, and their proofs (see the original references for the details) are based on the fact that the set $G(U, \chi_\Omega$ is both invariant under the subgroup of the rotations and *sufficiently large*. If instead, one would just have families of distributions which are invariant under rotations, then not everything would work any longer, as the following example shows.

Example 3.1. Take, in $\mathbb{C}$, the family of all circles centered at the point $e^{i\theta}$ and of radius less than 1, i.e., $\mathcal{F} = \{S_{\theta,r} : 0 \leq \theta < 2\pi, \ 0 < r < 1\}$. Now consider the transformation from the space of C^∞ functions on the punctured disk $\Delta(0,2)\backslash\{(0,0)\}$ into $\mathcal{E}(S^1 \times [0,1])$, defined by

$$f \mapsto \int_{S_{\theta,r}} f(z)d\sigma;$$

then it is possible to show that the transformation thus defined is not injective, since, e.g., the function $f(z) = \log|z|$ is in its kernel.

Example 3.2. Take a smooth Jordan curve Γ with the origin in its interior and fix r such that $0 < r < \text{dist}(0, \Gamma)$. Consider the set

$$\Omega = \bigcup_{z \in \Gamma} \Delta(z,r)$$

and let f be a continuous function in $\bar{\Omega}$ with the property that for every $z \in \Gamma$, its restriction $f|_{\partial\Delta(z,r)}$ has an analytic extension to $\Delta(z,r)$. The natural question we may ask is whether we can deduce from this information that f is holomorphic all over Ω. A positive answer in this direction was given in Globevnik [1983], at least when Γ is a circle (see also Globevnik [1990]).

Recently Globevnik and Rudin (Globevnik and Rudin [1990]) obtained a characterization of harmonicity which is akin to Volterra's theorem as well as to Theorem 3.7.

Theorem 3.10. *Consider a function f, continuous on $\Delta(0, R)$. Then f is harmonic if and only if, for every convex set U contained in $\Delta(0, R)$ with a smooth boundary and such that the origin belongs to U,*

$$\int_{\partial U} f d\omega_U = f(0),$$

where $d\omega_U$ is the harmonic measure of ∂U with respect to the origin.

On the other hand, in Zalcman [1973], the author generalized the equivalence between harmonicity and the mean-value theorems in a different direction. He proved that for any measure, one can find a system of homogeneous linear partial differential operators with constant coefficients such that a continuous function f satisfies a generalized mean-value condition with respect to the given measure if and only if it is annihilated by the system. It is interesting to point out that the proof of this result in its most general form relies on the Fundamental Principle of Ehrenpreis and Palamodov, which we discussed in Chap. 2. Thus, even though we apparently moved away from our original point, this shows how close all of these results are to one another. Before we give some examples of Zalcman's results in this direction, let us at least state a special case of his theorems.

Theorem 3.11. *Let Ω be some open convex set in $\mathrm{I\!R}^n$. For each homogeneous polynomial P, there exists a finite measure μ of compact support such that any function f continuous on Ω is a weak solution of*

$$P(D)f = 0$$

if and only if

$$\int f(x + rt)d\mu(t) = 0 \tag{3.3}$$

for all $x \in \Omega$ and r such that $0 < r < \mathrm{dist}(x, \partial\Omega)$. In fact, any measure of the form $P(D)T$, where T is a distribution of compact support such that $\hat{T}(0) \neq 0$, has the required property.

Example 3.3. The weak solutions to

$$\frac{\partial^n f}{\partial z^n} \pm \frac{\partial^n f}{\partial z^{-n}} = 0$$

are characterized by the condition

$$\int_0^{2\pi} f(x + r\cos\theta, y + r\sin\theta) \left\{ \begin{array}{c} \cos n\theta \\ \sin n\theta \end{array} \right\} d\theta = 0.$$

When $n = 2$, equation (3.4) becomes the classical D'Alembert equation $f_{xx} - f_{yy} = 0$, and the corresponding mean-value condition was already known, e.g., to V.L. Shapiro (Shapiro [1969]).

As we already mentioned, this kind of result is by no means restricted to the case of a single homogeneous polynomial.

Theorem 3.12 (Zalcman [1973]). *Let Ω, μ be as in the previous theorem, and consider $f \in C(\Omega)$. Then f satisfies (3.3) for every $x \in \Omega$ and every r such that $0 < r < \mathrm{dist}(x, \partial\Omega)$ if and only if f is a weak solution in Ω of the system*

$$Q_n(D)f = 0, \quad n = 0, 1, 2, \ldots,$$

where $Q_n(D) = \sum_{|\alpha|=n} A_\alpha D^\alpha$, and $\hat{\mu}(z) = \sum_\alpha A_\alpha z^\alpha$.

Well-known, maybe we should say *classical*, examples for Theorem 3.12 are provided by Weber's relation, which holds for solutions of the Helmholtz equation, or the two-dimensional Euler–Poisson–Darboux equation (for these and other examples, see Zalcman [1973]).

To conclude this chapter, we would like to mention that the question of explicit inversion to the Pompeiu transform is related to the *deconvolution problem*, a problem that arises in many applications in signal processing, acoustics, and so forth. We refer the reader to some of our recent work for details: Berenstein, Krishnaprasad and Taylor [1984]; Berenstein and Yger [1983; 1989; 1991b].

Chapter 4
Series of Exponentials

In this chapter, we once again consider the problem of mean-periodic functions, but from a totally different point of view. Actually, the idea is to deal with objects which *a priori* do not seem to be mean-periodic functions and to show that they actually are. Once this is done, one tries to exploit the general properties of mean-periodic functions to deduce further information on the original object. To understand in a more concrete way the meaning of this philosophy, we should begin by recalling the importance that *exponentials with complex frequencies* play in the theory of mean-periodic functions. Indeed, if $\alpha \in \mathbb{C}^n$, $x \in \mathbb{R}^n$, and if we take $\mu \in \mathcal{E}'(\mathbb{R}^n)$, then the function

$$f(x) = e^{i\alpha \cdot x} \tag{4.1}$$

is a solution of the convolution equation

$$\mu * f(x) = 0 \tag{4.2}$$

if and only if $\hat{\mu}(\alpha) = 0$. More generally, as we have seen in Chap. 2, the particular solutions (4.1) of equation (4.2) are particularly important, since they generate (together with suitable derivatives) the space of all C^∞ solutions to equation (4.2) (for the precise statement, see Theorem 2.13). On the other hand, something more precise is true: As we have seen in Chap. 2, the Fundamental Principle of Ehrenpreis and Palamodov shows that all solutions to all systems of homogeneous linear partial differential equations with constant coefficients in $\mathcal{E}(\mathbb{R}^n)$ (as well as in many other spaces) can be written as "linear combinations" of such elementary solutions where, actually, infinitely many of them appear so that they have to be integrated along the characteristic variety of the system itself. In the special case in which such a variety is discrete, all the solutions appear as exponential polynomials (i.e., as we already mentioned, finite sums of exponentials, each one of which may be multiplied by a polynomial). We also mentioned the fact that such a Fundamental Principle can be extended to suitable systems of convolution equations (the so-called slowly decreasing systems). In the case in which we have a slowly decreasing system of convolution equations (let us still think in terms of distributions $\mu_1, \ldots, \mu_n \in \mathcal{E}'(\mathbb{R}^n)$ and of functions $f \in \mathcal{E}(\mathbb{R}^n)$, as we shall later turn to other spaces) of the kind

$$\mu_1 * f = \cdots = \mu_n * f = 0, \tag{4.3}$$

then (since the number of equations equals the number n of space variables) the characteristic variety

$$V = \{z \in \mathbb{C}^n : \hat{\mu}_1(z) = \cdots = \hat{\mu}_n(z) = 0\}$$

is generically discrete, and all the solutions to (4.3) can be written as infinite series of the form

$$\sum_{K=1}^{+\infty}\sum_{k=1}^{J_K} c_k(x)e^{i\alpha_k\cdot x}. \tag{4.4}$$

We therefore see that *series of exponentials* are just a particular case of (4.4) which arise when $c_k \equiv$ const and when $J_K = 1$, hence they are particular examples of mean-periodic functions as long as the frequencies belong to the characteristic variety of a system. This is the point of view which we want to adopt in dealing with series of exponentials, and we should probably mention the fact that it was originally Ehrenpreis who in Ehrenpreis [1970] suggested this approach: " ... we could extend the concept of exponential sums even further. Namely, we can regard (formally) a sum

$$\sum c^j \exp(ia^j x) = f(x)$$

as a solution of the convolution equation $S * f = 0$, where the Fourier transform of S is just $\Pi[1 = z/a^j]$ (perhaps with some convergence factors added) .... However, our methods are now powerful enough to treat this situation ... ". We remark that the difficulty which Ehrenpreis mentions is a direct consequence of the fact that in 1970 when Ehrenpreis [1970] appeared, the possibility of proving a Fundamental Principle for systems of convolution equations seemed extremely farfetched.

There are several ways to exploit this remark on exponential series, and we would like to begin by discussing the approach which was devised by Ehrenpreis (Ehrenpreis [1959; 1970]), and which, more than ten years later, has been revived by the authors.

The key point, both in Ehrenpreis's approach and ours, is based on the fact that a representation such as (4.4) is possible for a large class of spaces, and not just for the space of C^∞ functions. Therefore, one can think of what is usually known as the *comparison problem* (or also, with a more subdued emphasis, as the *general elliptic problem*). Consider two spaces (of functions, distributions, or other generalized functions, in the same number of variables though) and let T be an operator which acts on both of them (T might be a system of linear partial differential operators with constant coefficients, or even a system of convolution operators, or finally a system of linear partial differential operators with variable coefficients); the general problem consists of comparing the kernels of T in both spaces and to show, e.g., that they coincide or that one of them contains the other. As a concrete example, one may take Ω_1, Ω_2 to be two convex open sets in $\mathbb{R}^2$, $\Omega_1 \cap \Omega_2 \neq \emptyset$, and consider for T a system of partial differential operators acting on $\mathcal{E}(\Omega_1)$ and $\mathcal{E}(\Omega_2)$. One can therefore ask whether every solution $\vec{f} \in [\mathcal{E}(\Omega_1)]^r$ of the system

$$T(\vec{f}) = 0$$

extends to a solution of the same system in $[\mathcal{E}(\Omega_2)]^r$. The reader will notice that many of the classical problems in partial differential equations can be reduced to this general question, as the following examples show (partial differential equations are used in these examples, but convolution equations could replace them everywhere).

Example 4.1. Consider the spaces $\mathcal{E}(\mathbb{R}^n)$ and $\mathcal{D}'(\mathbb{R}^n)$. A system of differential equations is said to be *hypoelliptic* if its kernels in $\mathcal{D}'(\mathbb{R}^n)$ and $\mathcal{E}(\mathbb{R}^n)$ are the same (i.e., if every distribution solution is indeed a C^∞ solution).

Example 4.2. Fix an integer q, $1 \leq q \leq n$. A system of differential equations is said to be *hyperbolic* in $x_1, \ldots, x_q$ if there exists $C > 0$ such that every solution f of the system which is defined and indefinitely differentiable for $|x_1| < C, \ldots, |x_q| < C$, and for all values of $x_{q+1}, \ldots, x_n$, can be *uniquely* extended to a solution on all of $\mathbb{R}^n$. Again, this is a comparison property, since it is equivalent to the request of the equality of the kernel of the system in the spaces $\mathcal{E}(\mathbb{R}^n)$ and $\mathcal{E}(q, C)$, where this last symbol denotes the space of functions which are C^∞ for $|x_1| < C, \ldots, |x_q| < C$, and for all values of $x_{q+1}, \ldots, x_n$.

Example 4.3. Other examples can be easily constructed by the reader, either by eliminating the uniqueness condition in Example 4.2 or by totally changing attitude and by considering, e.g., spaces of holomorphic functions on convex subsets of $\mathbb{C}^n$, or spaces of ultradifferentiable functions or ultradistributions on convex subsets of $\mathbb{R}^n$.

The way to relate this general kind of comparison problem to our treatment of series of exponentials is easily understood once one realizes that a simple consequence of the Fundamental Principle enables us to give a quite general condition for the solvability of the comparison problem. Let us spend a few words on this, since the next theorem is the basis for most results in this chapter. As the reader has seen, all of the results discussed in the previous chapters are given for spaces X such that their strong dual X' is isomorphic (usually via Fourier transform, or some similar device) to a space of entire functions whose growth at infinity is bounded by a family $\mathcal{K} = \{k\}$ of positive weights, which satisfies suitable technical conditions: the so-called LAU-spaces of Ehrenpreis. (This statement, of course, is not completely precise, since it does not include the spaces considered by De Roever, Meril, and the authors—De Roever [1978]; Meril [1983a,b]; Berenstein and Struppa [1983]—for which, however, a completely similar argument works, as shown, e.g., in Meril [1983b].) One can then give the following result (which was proved in Ehrenpreis [1970] for the case of partial differential operators and, in its more general form, first appeared in Struppa [1983b]).

Theorem 4.1. *Let X, Y be two spaces as above and let T be a system of convolution equations defined on both of them. Denote by V the support of the multiplicity variety associated to T and let $\mathcal{K}_X$ and $\mathcal{K}_Y$ be the family of weights which describe the topology of X and Y, respectively. Then the kernel of T in X is contained in the kernel of T in Y if for each $k_X \in \mathcal{K}_X$, there exists $k_Y \in \mathcal{K}_Y$ and $A > 0$ such that*

$$k_Y(z) \leq A k_X(z) \quad for \quad z \in V. \tag{4.5}$$

In most concrete cases, condition (4.5) is also a necessary condition, since it is usually shown with use of *lacunary series* methods (see Ehrenpreis [1954–62; 1970]; Gårding [1951]; Gårding and Malgrange [1961]; Hörmander [1955a; 1971a]).

The road to generalization is now clear (as Ehrenpreis would say!). Indeed, we can now prove, for example, that a series of exponentials which converges in a given convex set Ω may *overconverge*, provided that the frequencies which appear in the exponentials (which are nothing but the connected components of V) satisfy a condition such as (4.5). The reader should note that the requirement that Ω be convex is essential, since no Fundamental Principle is known (nor is it clear which shape it could have) for solutions defined on nonconvex sets. In a similar way, one might be able to prove that a series of exponentials which converges to a holomorphic function with a given growth at infinity may actually satisfy a stronger growth condition (these are Hamburger-type theorems when the series is a Dirichlet series), provided again that the frequencies are suitably located.

Before giving the precise statements of our recent results in this area, we would like to mention the approach and the results of Ehrenpreis (Ehrenpreis [1970]) in this context.

In his philosophy, it was not possible to conceive the set S of frequencies of a series of exponentials as the variety associated to a system of convolution equations (since his methods would not have been powerful enough), but he could instead consider *infinitely many* differential operators and look at S as the union of the varieties associated to each one of them. In this framework, the kind of problem which Ehrenpreis is able to (partially) deal with has the following aspect.

Problem. Let X, Y, Z be three LAU-spaces such that (if we denote by $X \overset{a}{\cap} Y$ the largest LAU-space contained in $X \cap Y$)

$$X \overset{a}{\cap} Y \subset Z \subset X$$

(of course, every inclusion is a topological one, i.e., the smaller space has a stronger topology than the one induced by the larger space); let $\{T_j\}_{j=1}^{+\infty}$ be a sequence of systems of linear partial differential operators with constant coefficients, and let $\{f_j\}_{j=1}^{+\infty}$ be a sequence of elements of Z such that $T_j f_j = 0$ (for every $j = 1, \dots$) and such that the series $\sum f_j$ converges in the topology of X to some $f \in X \cap Y$. Then one wants to prove (under suitable conditions on the *sequence* $\{T_j\}$) that $f \in Z$ and that the series is convergent in the Z-topology.

As is obvious, such a problem is, in general, too ambitious, and indeed, most of Ehrenpreis's results are given for $f_j = e^{i\alpha_j x}$ and $T_j = \frac{d}{dx} - i\alpha_j$. In particular, Ehrenpreis can apply his methods to treat some questions on overconvergence (along the lines of the classical Fabry gap theorem) and some questions on Dirichlet series (along the lines of the classical Hamburger theorem); to convey the flavor of his approach, let us quote two of his main results (Ehrenpreis [1970]):

Theorem 4.2. *Let $\{a_j\}$ be a sequence of real numbers and let $\{a_{j_t}\}$ be a subsequence of it. Suppose that there exists an $\eta > 0$ such that for every t and every $\varepsilon > 0$, the number of a_j's lying in $|z - a_{j_t}| < \eta|a_{j_t}|$ is bounded by $\varepsilon|a_{j_t}|$, at least for $|a_{j_t}|$ large enough, and suppose that these a_j's satisfy $|a_j - a_k| \geq \ell > 0$ for some $\ell(j \neq k)$. Suppose further that the series*

$$f(z) = \sum c_j \exp(ia_j z), \quad z \in \mathbb{C},$$

converges uniformly and absolutely on compact subsets of an open strip Σ (which in this case is equivalent to convergence in substrips) in $\mathbb{C}$ and that f can be continued to be analytic in a neighborhood of a boundary point z_0 of Σ. Then the series

$$g(z) = \sum c_{j_t} \exp(ia_{j_t} z)$$

converges in an open strip, properly containing Σ and z_0.

Theorem 4.3. *Let $\{a_j\}$ be a sequence of real negative numbers such that (for $|a_j|$ large enough)*

(i) *$|a_j - a_k| \geq \exp(-2|a_j|)$ for $j \neq k$;*
(ii) *there exists $\delta > 0$ such that the number of a_j's in $(10a_k, 0)$ is less than or equal to $\exp(\delta|a_k|)$.*

Suppose that the series

$$f(z) = \sum c_j \exp(ia_j z), \quad z \in \mathbb{C},$$

is absolutely convergent in $\{z \in \mathbb{C} : \operatorname{Im} z < -1\}$, uniformly on compact subsets (or proper half-planes). Suppose, furthermore, that f can be continued to be an entire function which satisfies

$$|f(z)| = 0(\exp(\beta|z| \log |z|))$$

for some sufficiently small $\beta = \beta(\delta) > 0$. Then there exists $B > 0$ such that

$$c_j = 0(\exp(-\exp(B|a_j|))).$$

Let us now go back to the approach to overconvergence, which can be followed using Theorem 4.1. The first result we mention is a strengthened version of the Fabry gap theorem (Dienes [1957]), which has been proved by the authors in Gel'fand and Shilov [1968].

Theorem 4.4. *Let $\Gamma_\theta = \{z \in \mathbb{C} : \pi/2 - \theta < \arg z < \pi/2 + \theta\}$, for $\theta \in (0, \pi/2)$, and let $\mu \in H'(\mathbb{C})$ be a slowly decreasing analytic functional acting by convolution on $H(\Gamma_\theta)$. Suppose that μ satisfies the following extra condition: There exists $\varepsilon > 0$ such that (with at most a finite number of exceptions) all z which are in*

$$\{z \in \mathbb{C} : \operatorname{Im} z < 0 \quad and \quad \hat{\mu}(z) = 0\}$$

do not belong to

$$\{z \in \mathbb{C} : \pi + \theta - \varepsilon < \arg z < 2\pi - \theta + \varepsilon\}.$$

Let $f \in H(\Gamma_\theta)$ be a solution of $\mu * f = 0$. If there exists a neighborhood $\mathcal{U}$ of the origin in $\mathbb{C}$ to which f extends holomorphically, then we can find $\delta > 0$ such that f extends holomorphically to a function $\tilde{f}$, holomorphic in

$$\{z \in \mathbb{C} : \operatorname{Im} z < -\delta\}$$

and which there satisfies $\mu * \tilde{f} = 0$.

Several remarks can be made about this result: Its proof is just a combination of the Fundamental Principle for convolution equations in $H(\Gamma_\theta)$, due to Meril and Struppa [1987a], together with Theorem 4.1. This result, on the other hand, can be envisioned as a result concerning series of exponentials, in view of what we mentioned before, and one can therefore recognize its relation to the Fabry gap theorem.

Also, one may want to see how this theorem can be generalized to the case of $n > 1$ variables, or when the functions which are involved satisfy suitable growth conditions. As to the case of several variables, nothing new takes place, except for the fact that (in order to be dealing with series of exponentials) one needs to work with systems of convolution equations with discrete characteristic variety. The details of this extension are completely discussed in Berenstein and Struppa [1988b], where also some concrete cases with applications to microfunction theory (see Chap. 5) are discussed. Much more delicate is the situation which arises when growth conditions are considered. This is actually the central topic discussed in Berenstein and Struppa [1988b], where applications of this argument to Dirichlet series are given. Indeed, general Dirichlet series appear as series of the form

$$f(z) = \sum_{n=1}^{+\infty} c_n \exp(-\lambda_n z), \quad z \in \mathbb{C}, \tag{4.6}$$

where the c_n and λ_n can be chosen to be complex numbers. It is clear that a function such as the one in (4.6) is once again a solution of a convolution equation, and therefore its convergence and growth properties can be explored in terms of Theorem 4.1. As is well known (see, e.g., the classical treatises of V.I. Bernstein (Bernstein [1933]) and S. Mandelbrojt (Mandelbrojt [1972])), the general properties of $f(z)$ are more easily understood when the so-called density of the sequence $\{\lambda_n\}$ is *finite*; this corresponds to the case in which the λ_n are zeros of an entire function $g(z)$ and whose growth is bounded as follows:

$$|g(z)| \leq A_\varepsilon \exp[(D + \varepsilon)|z|], \tag{4.7}$$

for any $\varepsilon > 0$, and the smallest possible value of D which may appear in (4.7) is the density of $\{\lambda_n\}$ (density which is usually defined otherwise). Since the zeros of g are the frequencies of f, i.e., the characteristic variety of the convolutor μ for which f is μ-mean periodic, the restriction to Dirichlet series of finite density corresponds to considering solutions of equations such as

$$\mu * f = 0,$$

with $\mu \in H'(\mathbb{C})$, and carried by a ball of radius D. If $D = 0$ in (4.7), we have that g is of *infraexponential type*, and therefore it is the Fourier–Laplace transform of an analytic functional carried by the origin, i.e., $\mu*$ is an *infinite-order differential operator*. This case has been studied in detail by Kawai (Kawai [1984; 1987a]), and we shall come back to it later. However, many important problems (related e.g., to the Riemann ζ-function, for which $\lambda_n = \log n$) arise when the frequencies have infinite density, which corresponds to requiring that the growth of g must be much faster. The case in which (4.7) is replaced by something like

$$|g(z)| \leq A_\varepsilon \exp[(D + \varepsilon)|z|^\rho],$$

for $\rho > 1$, is now completely understood in Berenstein and Struppa [1988b], from which we quote two sample results.

Before we state them, we need some preliminary definitions for the spaces of the objects with which we will be dealing. For $p(z) = |z|^\rho$, with $\rho > 1$ being a real number, and for $\Gamma = \{z \in \mathbb{C} : -\psi < \arg z < \psi\}$, $\psi \in (0, \frac{\pi}{2}]$, we denote by $A_{p,0}(\Gamma)$ the space of all functions f which are holomorphic in Γ and such that for all $\varepsilon > 0$ and all cones $\Gamma' \subset\subset \Gamma$,

$$|f(z)| \leq C_{\varepsilon,\Gamma',f} \exp(\varepsilon p(z)).$$

$A_{p,0}$ is naturally endowed with a projective limit topology. A functional $\mu \in A_{p,0}(\Gamma)'$ is said to be (p, Γ)-*slowly decreasing* if its Fourier–Laplace transform $\hat{\mu}(\zeta) = \langle \mu, \exp(-z \cdot \zeta) \rangle$ satisfies a set of conditions similar to those described in Definition 1.6 (suitable modifications are necessary to deal with the different kinds of growth, see Berenstein and Struppa [1988b]). Finally, a sequence $\{\lambda_n\}$ of complex numbers is said to be *strictly (p, Γ)-slowly decreasing* (resp., *weakly (p, Γ)-slowly decreasing*) if there exists $\mu \in A_{0,p}(\Gamma)'$, with (p, Γ)-slowly decreasing, such that $\{\lambda_n\} = \{z \in \mathbb{C} : \hat{\mu}(z) = 0\}$ (resp., $\{\lambda_n\} \subset \{z \in \mathbb{C} : \hat{\mu}(z) = 0\}$). An analogous set of definitions can be easily given if the sequence $\{\lambda_n\}$ is replaced by a more general multiplicity variety $\{\lambda_n, m_n\}$, $n = 1, 2, \ldots$, and $m_n \in \mathbb{N}, \lambda_n \in \mathbb{C}$ (the meaning being that for each $\lambda_n \in \mathbb{C}$, an order of multiplicity m_n is specified). We therefore state our results.

Theorem 4.5. *Let* $\{\lambda_n, m_n\}$ *be a strictly (p, Γ)-slowly decreasing multiplicity variety, with (except possibly for a finite number indices n), $\arg(\lambda_n) \notin (-\theta + \psi - \frac{\pi}{2}, \theta - \psi + \frac{\pi}{2},)$, for some $\theta \in (0, \frac{\pi}{2})$; then any generalized Dirichlet series convergent in* $A_{p,0}(\Gamma)$,

$$\sum_{n=1}^{+\infty} \sum_{j=1}^{m_n} c_{n,j}(z) \exp(-\lambda_n z),$$

actually converges in $\mathbb{C}$ *to an entire function in* $A_{p,0}(\mathbb{C})$.

Theorem 4.6. *Let $\{\lambda_n\}$ be a weakly (p, π_+)-slowly decreasing sequence $(\pi_+ = \{z \in \mathbb{C} : \operatorname{Re} z > 0\})$, associated to some $\mu \in A_{p,0}(\pi_+)'$, such that $\{z \in \mathbb{C} : \hat{\mu}(z) = 0\}$ is a weak interpolating variety. Consider the series*

$$\sum_{n=1}^{+\infty} c_n \exp(-\lambda_n z),$$

and suppose it is convergent in π_+ to a holomorphic function f. Suppose, moreover, that f extends to an entire function $\tilde{f} \in A_{p,0}(\mathbb{C})$. Then

$$c_n = 0(\exp(-B|\lambda_n|^\sigma)),$$

for all $B > 0$, and σ such that $\rho^{-1} + \sigma^{-1} = 1$.

On the other hand, the case of even faster growth (such as would be given by a weight such as $\exp(\exp(|z|))$) is still open, and some work is in progress on this.

We wish to mention here the work of Kawai on the Fabry gap theorem, which stimulated our interest in this question. In Kawai [1987a], Kawai studied the Fabry theorem (or Fabry–Ehrenpreis theorem, as he calls it) in a way which is similar to ours, since he considered the exponential series as solutions of a very special class of convolution equations, namely the infinite-order differential equations (with linear and constant coefficients). He then proves an overconvergence theorem (in a sense similar to Theorem 4.4) in which the removability of the singularity, however, is *not* a consequence of any comparison result such as Theorem 4.1, but of a deep result of Kashiwara and Schapira (Kashiwara and Schapira [1979]), stemming from their work in algebraic analysis (see Chap. 6). Kawai's result states:

Theorem 4.7. *Let $\{a_j\}$ be a sequence of nonzero complex numbers satisfying the following conditions:*

(i) $\lim_{j \to +\infty} \frac{j}{|a_j|} = 0$;

(ii) *there exists $C > 0$ such that, for all i, j,*

$$|a_i - a_j| \geq C|i - j|;$$

(iii) *there exist finitely many unit vectors $e_k (k = 1, \ldots, t)$ in $S^1 \subseteq \mathbb{R}^2$ for which the following holds:*

     $\forall \varepsilon > 0$, and for each compact $K \subseteq S^1 \backslash \{e_1, \ldots, e_t\}$, there exists $n_0 = n_0(\varepsilon, K)$ such that

$$\inf_{\substack{n \geq n_0 \\ e \in K}} \left| \frac{(\operatorname{Re} a_n, \operatorname{Im} a_n)}{|a_n|} - e \right| > \varepsilon.$$

Suppose that there exists a holomorphic function $f(z)$, defined on $\Gamma_\theta = \{z \in \mathbb{C} : \frac{\pi}{2} - \theta < \arg z < \frac{\pi}{2} + \theta\}$, for some $\theta \in (0, \frac{\pi}{2})$, such that the series

$$\sum_{j=1}^{+\infty} c_j \exp(i\alpha_j z, \quad c_j \in \mathbb{C},$$

converges to $f(z)$ uniformly on compact subsets of Γ_0. If f can be analytically continued to $\Gamma_0 \cup U$, for some neighborhood U of the origin in $\mathbb{C}$, then there exists $\delta > 0$ such that f extends analytically to $\Gamma_0 - i\delta$, and the series above converges to f on $\Gamma_0 - i\delta$.

We now turn to a different application of series of exponentials, which as Mandelbrojt (Mandelbrojt [1935]) pointed out is ultimately related to the study of Fourier and Dirichlet series, that is, to the study of *quasianalytic classes* of functions, namely those spaces of functions for which certain vanishing conditions for a function in the space suffice to imply that the function is identically zero. Typical examples are the spaces $\mathcal{E}_B$ of ultradifferentiable functions of Beurling type (see Björck [1966], e.g., for their precise definitions), which, under some specific conditions, are such that any function $f \in \mathcal{E}_B$ which vanishes together with all of its derivatives at the origin is, indeed, identically zero. The usual idea, from a historical point of view, consists of looking at some *special* spaces whose functions already enjoy some natural *smallness property*. Among the many natural smallness properties, one might require the possibility of representing each function by a series of exponentials, which brings us back to our situation. To begin with, let us quote a result due to Mandelbrojt, whose book (Mandelbrojt [1935]), together with Beurling's lectures [1957; 1961], are the basic references on the subject.

Theorem 4.8. *Let $f \in L^1([0, 2\pi])$ be represented as*

$$f(x) = \sum_{j=1}^{+\infty} c_j \exp(ia_j x) \tag{4.8}$$

for some sequence $\{a_j\}$ of positive real numbers. Suppose

(i) *the sequence $\{a_j\}$ is lacunary, i.e., there exists $\sigma < 1$ such that, for any $\varepsilon > 0$,*

$$\sum a_j^{-\sigma-\varepsilon} < +\infty, \quad \text{while} \quad \sum a_j^{-\sigma+\varepsilon} = +\infty;$$

(ii) *there is a point $x_0 \in [0, 2\pi]$ which is a zero in the mean of exponential order ρ, i.e.,*

$$\rho = \varlimsup_{\alpha \to 0^+} \frac{\log(-\log \int_{x_0}^{x_0+\alpha} |f(x)| dx)}{-\log \alpha},$$

with $\rho > \sigma/(1-\sigma)$.

Then $f \equiv 0$.

The reader can see the similarity between this statement and our previous remarks: One has a function f which can be expanded as a series of exponentials, i.e., (from our point of view) a μ-mean periodic function, with respect to some compactly supported distribution (or analytic functional) μ; we have a geometrical condition on the set of frequencies of the expansion of f, namely on the characteristic variety of $\hat{\mu}$, and we have a vanishing condition on f; the conclusion is that $f \equiv 0$. Various generalizations of this kind of result are possible, either by employing Theorem 4.1 or again by following Kawai's treatment of infinite-order differential equations. Let us give three results in this direction, the first two due to Struppa (Struppa [1988a]) and the last one by Kawai (Kawai [1984]). Let us first introduce some notation. Take $B = \{b_j\}_{j=1}^{\infty}$ to be a convex sequence of positive numbers and set on $\mathbb{R}$,

$$\lambda(x) = \sum_{j=1}^{+\infty} \frac{|x|^j}{b_j}.$$

Define $\mathcal{E}_B = \mathcal{E}_B(\mathbb{R})$ to be the space of all C^{∞} functions f on $\mathbb{R}$ such that for every $\varepsilon > 0$ and every compact set K, there is a constant $C = C(\varepsilon, K)$ such that

$$\sup_{x \in K} |D^j f(x)| \leq C \cdot \varepsilon^j b_j$$

for all values of j.

Theorem 4.9. *Let $\tilde{\mathcal{E}}_B(\mathbb{R})$ be the subspace of $\mathcal{E}_B$ consisting of those functions which can be given a representation (4.8). Suppose*

$$\exists \delta > 0 \quad \text{such that} \quad \forall m > 0 \exists \ell > 0 \quad \text{such that for some} \quad A > 0,$$

$$\lambda(\ell|z|)(1 + |z|)^{\ell} \exp((\ell - \delta)|\operatorname{Im} z|) \geq A \exp(m|\operatorname{Re} z|).$$

Then, any function in $\tilde{\mathcal{E}}_B$ which vanishes at the origin together with all of its derivatives is identically zero.

Theorem 4.10. *Let g be a hyperfunction on $\mathbb{R}$, which we represent as the difference of two boundary values of holomorphic functions*

$$g = b_+(f^+) - b_-(f^-) = b(f^+, f^-),$$

with $f^+ \in H(\{z \in \mathbb{C} : \operatorname{Im} z > 0\})$, $f^- \in H(\{z \in \mathbb{C} : \operatorname{Im} z < 0\})$. Suppose f^+ admits an exponential series representation whose frequencies are the zeros of the Fourier–Borel transform of a slowly decreasing analytic functional μ which acts as a convolutor on $H(\{z \in \mathbb{C} : \operatorname{Im} z > 0\})$ and such that, for some $\theta \in (0, \pi/2)$ and $\varepsilon > 0$, if z belongs to

$$\{z \in \mathbb{C} : \operatorname{Im} z > 0 \quad \text{and} \quad \hat{\mu}(z) = 0\},$$

then it does not belong to

$$\{z \in \mathbb{C} : \pi + \theta - \varepsilon < \arg z < 2\pi - \theta + \varepsilon\}.$$

Then if g vanishes in a neighborhood of the origin, it vanishes identically.

The reader will not fail to notice the close relationship between Theorem 4.10 (which one might as well formulate in $\mathbb{R}^n$) and Theorem 4.4. One the other hand, the counterpart of Theorem 4.7 is Kawai's following result. Fix a sequence $\{a(\ell)\}$ of n-dimensional real vectors and denote by $\{a_j(\ell)\}$ its jth reduced sequence in the sense of Kawai [1984]. Set $|a(\ell)| = \sum_{j=1}^{n} |a(\ell)_j|$, where $a(\ell)_j$ is the jth component of $a(\ell)$.

Theorem 4.11. *Let $\{a(\ell)\}$ be as above such that $\{a_j(\ell)\}$ satisfies, for some $C > 0$, the following conditions:*

(i) $\lim_{m \to +\infty}(m/a_j(m)) = 0 \quad (j = 1, \dots, n)$;
(ii) $|a_j(m) - a_j(m')| \geq C|m - m'| \quad (j = 1, \dots, n)$.

Let $c(\ell)$ be a sequence of complex numbers of $a(\ell)$-infraexponential growth, i.e., such that for every $\varepsilon > 0$, there exists $C_\varepsilon > 0$ such that

$$|c(\ell)| \leq C_\varepsilon \exp(\varepsilon|a(\ell)|)$$

for all $\ell \in \mathbb{C}$. Let $f(x)$ be the hyperfunction on $\mathbb{R}^n$ given by

$$\sum_{\ell} c(\ell) \exp(ia(\ell) \cdot x).$$

Then, if f vanishes on an open neighborhood of the origin of $\mathbb{R}^n$, it vanishes identically.

As we are mentioning Kawai's work, we cannot fail to briefly recall the work of the Japanese school, among them Sato, Kashiwara, Kawai, and Aoki, who have applied microlocal analysis to the study of theta functions. This will probably be better understood after reading Chap. 6, but their results (Sato, Kawai and Kashiwara [1983, 1984]; Aoki, Kashiwara and Kawai [1986]) have once again very close connections with Hamburger's characterization of the Riemann ζ-function (Ehrenpreis [1970]; Hamburger [1921]). The starting point for most of their analysis, whose detailed description would take us too far away, is, however, what we would expect in this chapter. One considers the classical elliptic theta function

$$\theta(z, t) = \sum_{\nu \in \mathbb{Z}} \exp(\pi i \nu^2 t + 2\pi i \nu z),$$

which is, once again, a series of exponentials. Instead of thinking about it as a solution to the heat equation (t being the time), Sato and his coworkers show how to find θ as a solution of a system of (micro)-differential equations of infinite order, for which they prove the finite dimensionality of the space of solutions. We refer the reader to the original articles for further details on this topic.

Chapter 5
Residues and the Bernstein–Sato Polynomials

In this chapter, a new method to study mean-periodicity will be described briefly. We should like to point out that the results which we mention here are extremely new, most of them are not yet published, and therefore it is not yet completely understood to what extent this approach will succeed in the future. Its present applications, however, seem to indicate its strength.

To begin with, we recall that in the division and interpolation problems we described in Chap. 1 (and later applied in Chap. 2), one has p holomorphic functions $f_1, \ldots, f_p$ defining a (multiplicity) variety V; one then wants to find a way to write any holomorphic function f as

$$f = f_1 g_1 + \cdots + f_p g_p + h,$$

with g_i being holomorphic and so that the remainder h vanishes precisely when the holomorphic function f belongs to the ideal I generated by $f_1, \ldots f_p$; of course, one also wants to be able to keep track of the bounds in the way we have indicated in the previous chapters. For many practical purposes (related to some important questions in engineering, and for which we refer the reader to our survey, Berenstein and Struppa [1988a]), it is important, however, to provide explicit versions of the formula above, i.e., one wants to be able to explicitly produce $g_1, \ldots, g_p$ and h. The reader will immediately point out the possibility of using the explicit integral representations given by the Henkin–Ramirez kernels and the Andersson–Berndtsson kernels, which we briefly described in Chap. 1. Unfortunately, however, these descriptions *are not explicit enough*, in the sense that they require too much knowledge of the function f; what one is looking for is a way to construct $g_1, \ldots, g_p, h$ by the knowledge of f and its derivatives on $\bigcup_{i=1}^{p}\{z \in \mathbb{C}^n : f_i(z) = 0\}$. This can now be done by employing the notion of *residue current* introduced by Grothendieck, Dolbeault, Coleff, Herrera, and Lieberman (see Coleff and Herrera [1978] and Dolbeault [1990]) and recently made into a quite complete theory by Passare in his thesis (Passare [1988c]), and in a series of subsequent papers (Passare [1988a; 1987]). The use of these currents to solve division problems was already shown in Passare [1988c], but some problems were evident (such as their definition in terms of Hironaka's resolution of singularities (Hironaka [1964])), which Berenstein, Gay, and Yger (Berenstein, Gay and Yger [1989]) proved could be solved with the use of Bernstein–Sato polynomials. Before we get to the results in which we are interested, let us briefly describe this last idea. Consider a *real-valued nonnegative polynomial $P(x)$ defined on* $\mathbb{R}^n$; then for any $\lambda \in \mathbb{C}$ with $\operatorname{Re} \lambda > 0$, P^λ is a well-defined continuous function, which we want to think of as the following distribution: For any $f \in \mathcal{D}(\mathbb{R}^n)$, define

$$\Gamma_f(\lambda) = \int P^\lambda(x) f(x) dx.$$

It is easily seen that Γ_f is an analytic function of λ in $\{\lambda \in \mathbb{C} : \operatorname{Re} \lambda > 0\}$. A famous question posed by Gel'fand back in 1954 was whether Γ_f could be analytically continued to a meromorphic function defined on all of $\mathbb{C}$. For many concrete choices of f, meromorphic extensions of Γ_f were constructed over the years until—employing Hironaka's resolution of singularities—Atiyah (Atiyah [1970]) and Bernstein and Gel'fand (Bernstein and Gel'fand [1969]) gave a proof that such an extension always exists and that the poles of Γ_f are contained in a set of the form $\{-\frac{1}{N}, -\frac{2}{N}, \dots\}$, where N is a positive integer whose size is controlled by the degree of P and the number n of variables. But in 1972, a new and simple proof of this same fact was given by Bernstein (Bernstein [1972]), who proved the following theorem.

Theorem 5.1. *Given a polynomial $P(x)$, there exists a polynomial $b(\lambda) = \lambda^s + c_1 \lambda^{s-1} + \dots + c_s$ with complex coefficients and a finite set of differential operators $\{Q_j(x, \frac{\partial}{\partial x})\}$ with polynomial coefficients such that*

$$b(\lambda)\Gamma_f(\lambda) = \int P^{\lambda+1}(x)\left[\Sigma \lambda^j Q_j(x, \frac{\partial}{\partial x})(f(x))\right] dx$$

holds for any $f \in \mathcal{D}$.

The above identity has to be understood first for λ with $\operatorname{Re} \lambda > 0$. Then the meromorphic extension of Γ_f can be obtained by iterating this functional equation. The poles of Γ_f are contained in the set $\bigcup_{b(\alpha)=0}\{\alpha, \alpha - 1, \dots\}$. It is immediately seen that there is a unique polynomial $b(\lambda)$ of smallest possible degree, normalized so that its highest coefficient is 1, and which satisfies a functional equation as above; such a polynomial is called the *Bernstein–Sato polynomial of P*. It is a very deep result of Kashiwara that all the roots of the Bernstein–Sato polynomial are rational (Kashiwara [1976]).

Further progress in this direction can be done, and a *residue current* can be defined in this way for varieties which are *complete intersections*. In Berenstein, Gay and Yger [1989], it is shown that

Theorem 5.2. *Let $f_1, \dots, f_p$ be holomorphic functions in an open set $\Omega \subseteq \mathbb{C}^n$ such that $V = \bigcap_{j=1}^{p} f_j^{-1}(0)$ has dimension $n - p$. Let $t_1, \dots, t_p > 0$; then the $(0, p)$ current-valued holomorphic map $\lambda \mapsto S_\lambda^{(t)}$ defined in*

$$\left\{\lambda \in \mathbb{C} : \operatorname{Re} \lambda > \frac{1}{\min t_i}\right\}$$

by

$$\langle S_\lambda^{(t)}, \varphi \rangle = t_1 \cdot \dots \cdot t_p \int |f_1|^{2(t_1\lambda - 1)} \cdot \dots \cdot |f_p|^{2(t_p\lambda - 1)} \overline{\partial f_1} \wedge \dots \wedge \overline{\partial f_p} \wedge \varphi,$$

for $\varphi \in \mathcal{D}_{n,n-p}(\Omega)$, can be analytically continued to a meromorphic map on the whole complex plane with values in $\mathcal{D}'_{(n,n-p)}(\Omega)$. Furthermore, the limit *in $\mathcal{D}'_{n,n-p}(\Omega)$ of $\lambda^p S_\lambda^{(t)}$ when $\lambda \to 0$ exists, is independent of t, and it is orthogonal to $f \cdot \mathcal{D}_{n,n-p}(\Omega)$.*

Let us proceed to see how this method relates to the classical residue calculus. If $f = (f_1, \ldots, f_n)$ is a holomorphic mapping of a neighborhood Ω of 0 in $\mathbb{C}^n$ to $\mathbb{C}^n$ such that $f^{-1}(0) = \{0\}$, then one can define for every $h \in H(\Omega)$, its *residue (Grothendieck residue) at $\{0\}$ with respect to f* by

$$\text{Res}_f(h) = \left(\frac{1}{2\pi i}\right)^n \int_\Gamma h \frac{dz_1 \wedge \cdots \wedge dz_n}{f_1 \cdot \cdots \cdot f_n},$$

where $\Gamma = \{z \in \Omega : |f_1(z)| = \cdots = |f_n(z)| = \varepsilon\}$ and ε is small enough (the result is, actually, independent on ε). This was the starting point for the research of Dolbeault and of Coleff and Herrera, who came up with a definition where $h dz_1 \wedge \cdots \wedge dz_n$ was replaced by $\omega \in \mathcal{D}_{n,n-p}(\Omega)$ (with $\bar{\partial}\omega$ vanishing near V in the Dolbeault case, and not necessarily so in the Coleff–Herrera case). In the case of a *complete intersection*, Coleff and Herrera have shown (Coleff and Herrera [1978]) that Res_f is indeed a current which annihilates $f \cdot \mathcal{D}_{n,n-p}(\Omega)$; finally, Passare (Passare [1988c]) has shown that this current coincides (up to a multiplicative constant) with the one defined in Theorem 5.2. One shall then write

$$\text{Res}_f(\omega) = \langle \bar{\partial}\frac{1}{f_1} \wedge \cdots \wedge \bar{\partial}\frac{1}{f_p}, \omega \rangle, \quad \omega \in \mathcal{D}_{n,n-p}(\Omega)$$

(or, more briefly, $\langle \bar{\partial}\frac{1}{f}, \omega \rangle$). In this framework, one has

Theorem 5.3 (Berenstein, Gay and Yger [1989]). *Let $f_1, \ldots, f_p$ be holomorphic functions in an open connected subset Ω of $\mathbb{C}^n$, defining a complete intersection variety V. Let $\varphi \in \mathcal{D}_{n,n-p}(\Omega)$ be $\bar{\partial}$-closed in a neighborhood of V. The function defined by*

$$\lambda \mapsto \frac{(-1)^{p(p-1)/2}}{(2\pi i)^p} \int \frac{|f_1 \cdot \cdots \cdot f_p|^{2p\lambda}}{|f_1|^2 + \cdots + |f_p|^2} \overline{\partial f_1} \wedge \cdots \wedge \overline{\partial f_p} \wedge \varphi$$

is meromorphic in the complex plane. Furthermore, the pole at $\lambda = 0$ is simple, and its residue has the value

$$\frac{1}{p \cdot p!} \langle \bar{\partial}\frac{1}{f_1} \wedge \cdots \wedge \bar{\partial}\frac{1}{f_p}, \varphi \rangle.$$

In the case $p = n$, one can also show (Berenstein, Gay and Yger [1989]) that the condition of φ being $\bar{\partial}$-closed in a neighborhood of V can be removed. It seems that this is still true for the other values of p.

These results immediately yield important applications (from our point of view), since they can be used (Berndtsson and Passare [1989]; Berenstein and Yger [1991b]) to explicitly construct the Noetherian operators of Ehrenpreis and Palamadov, which we described in Chap. 1 and used in Chap. 2.

In Berenstein, Gay and Yger [1989], these results are used to provide new division and interpolation formulas by introducing a complex parameter in the integral representation formulas of Andersson and Berndtsson (Andersson and Berndtsson [1982]; Berndtsson [1980]), and they conclude that " ... the method of analytic continuation of currents and computation of their residues is a more systematic approach to interpolation problems ... ."

To concretely show the applications of this method, we quote the following result, first proved in Passare's thesis (Passare [1988c]) and later given a different proof in Berenstein, Gay and Yger [1989].

Theorem 5.4. *Let Ω be a pseudoconvex domain in $\mathbb{C}^n$ and let $f_1, \dots, f_p \in H(\Omega)$ define a complete intersection variety. The following two properties are equivalent for $h \in H(\Omega)$:*

(i) $h \in I(f_1, \dots, f_p)$.
(ii) $h \cdot \bar{\partial} \frac{1}{f} = 0$.

Let us now follow Passare [1988b] to show how these division formulas can be employed toward an *explicit version* of Ehrenpreis's Fundamental Principle. Let P be a polynomial in $\mathbb{C}^n$. Then the residue current $R[\frac{1}{P}]$ associated to it can be defined as we saw before or, equivalently and more simply, as

$$R\left[\frac{1}{P}\right] = \frac{1}{2\pi i} \lim_{\varepsilon \to 0} \frac{\bar{\partial}\chi_\varepsilon}{P},$$

where $\chi_\varepsilon = \chi(|P|/\varepsilon)$ and χ is a smooth approximation of the characteristic function of the interval $[1, +\infty)$. Let us see how a representation for a function

$$f \in \mathrm{Exp}(\mathbb{C}^n) = A_p(\mathbb{C}^n), \quad p(z) = |z|,$$

satisfying

$$P(D)f = 0,$$

can be derived. One first starts with the well-known formula

$$f(w) = \int_{\mathbb{C}^n} f(z) \exp(\bar{z} \cdot (w - z))\omega_n(z),$$

where $\omega_n(z) = (2\pi i)^{-n}(\bar{\partial}\partial|z|^2)^n/n!$. By interchanging z and $\bar{z}$, we get

$$f(w) = \int_{\mathbb{C}^n} f(\bar{z}) \exp(z \cdot (w - \bar{z}))\omega_n(z), \tag{5.1}$$

and it follows that if $P(D)f = 0$, then for all $w \in \mathbb{C}^n$,

$$\int_{\mathbb{C}^n} P(z)f(\bar{z}) \exp(z \cdot (w - \bar{z}))\omega_n(z) = 0. \tag{5.2}$$

Now one uses the division formula, via residue currents as in Passare [1988b], to get (for $[\frac{1}{P}] = \lim_{\varepsilon \to 0} \frac{\chi_\varepsilon}{P}$, the *principal value current*):

$$\exp(z \cdot w) = P(z) \left[\frac{1}{P(\zeta)} \right] (\exp[(z \cdot w) + (\bar{\zeta} \cdot (z - \zeta))]\omega_n(\zeta))$$

$$+ R \left[\frac{1}{P(\zeta)} \right] (p(\zeta, z) \wedge \exp[(\zeta \cdot w) + (\bar{\zeta} \cdot (z - \zeta))]\omega_{n-1}(\zeta)).$$

$$(5.3)$$

In (5.3), $(P_1, \ldots, P_n)$ is a *Hefer-vector* for the polynomial P, i.e.,

$$P(z) - P(\zeta) = \sum_{j=1}^{n} P_j(\zeta, z)(z_j - \zeta_j),$$

and $p(\zeta, z) = \sum_{j=1}^{n} P_j(\zeta, z)d\zeta$. By inserting (5.3) into (5.1) and reversing the order of integration, one gets (Passare [1988b]):

Theorem 5.5. *Suppose that* $f \in \mathrm{Exp}(\mathbb{C}^n)$ *and that* $P(D)f = 0$. *Then*

$$f(z) = R \left[\frac{1}{P(\zeta)} \right] (p(\zeta, D)f(\bar{\zeta}) \wedge \exp(\zeta \cdot (z - \zeta))\omega_{n-1}(\zeta)).$$

This representation result is obtained in much greater generality in Berndtsson and Passare [1989] (see also Yger [1987]). To state their result, we need a preliminary division theorem, very similar to others we have already stated.

Theorem 5.6. *Let* $P : \mathbb{C}^n \to \mathbb{C}^m$ *be a complete intersection polynomial mapping and let* $\Omega \subseteq \mathbb{R}^n$ *be a bounded strictly convex domain given as* $\Omega = \{x \in \mathbb{R}^n : \rho(x) < 1\}$, *where* ρ *is a function which is smooth except at the origin and which satisfies* $\rho(\lambda t) = \lambda \rho(t)$ *for all* $\lambda \in \mathbb{R}^+$. *Then one can find linear operators*

$$\hat{S}_1, \ldots, \hat{S}_m, \hat{T} : \hat{\mathcal{D}}(\bar{\Omega}) \to \hat{\mathcal{E}}'(\bar{\Omega})$$

such that

$$h(z) = \sum (\hat{S}_j h)(z)P_j(z) + (\hat{T}h)(z)$$

for all $h \in \hat{\mathcal{D}}(\bar{\Omega})$. *Moreover,* $\hat{T}$ *may be explicitly defined by*

$$(\hat{T}h)(z) = (2\pi i)^{-n} \langle \bar{\partial}\frac{1}{P}(\zeta), g(\zeta, z) \wedge h(\zeta) \exp[-i(\varphi'(\zeta) \cdot (z - \zeta))](2\bar{\partial}\partial\varphi(\zeta))^{n-m} \rangle,$$

where φ *is the supporting function for* Ω, φ' *is its gradient, and* g *is the Hefer matrix associated to* P. *It follows that* $\hat{T}h = 0$ *precisely when* h *is in the ideal generated by* $P_1, \ldots, P_m$.

As for the Hefer matrix mentioned in Theorem 5.6, we recall that to every polynomial mapping $z \mapsto (P_1(z), \ldots, P_m(z))$, we can associate a matrix (g_{jk}) consisting of polynomials $g_{jk}(\zeta, z)$ such that

$$P_j(z) - P_j(\zeta) = \sum_{k=1}^{n} g_{jk}(\zeta, z)(z_k - \zeta_k), \quad j = 1, \ldots, m,$$

then define

$$g_j(\zeta, z) = \sum_{k=1}^{n} g_{jk}(\zeta, z) d\zeta_k,$$

and finally

$$g(\zeta, z) = g_m(\zeta, z) \wedge \cdots \wedge g_1(\zeta, z),$$

which is the object that appears in the statement of Theorem 5.6.

Let us now consider a polynomial mapping $P : C^n \to C^m$ and the associated system of differential equations

$$P_j(D)f(x) = 0, \quad j = 1, \ldots, m; \quad x \in \Omega \tag{5.4}$$

for $f \in \mathcal{E}(\bar{\Omega})$. In Berndtsson and Passare [1989], Theorem 5.6 is the key tool for the proof of

Theorem 5.7. *Let $P : \mathbb{C}^n \to \mathbb{C}^m$ be a complete intersection hypoelliptic polynomial mapping. Then the operator $\hat{T}$ described in Theorem 5.6 induces an operator*

$$T : \mathcal{E}'(\Omega) \to \mathcal{E}'(\bar{\Omega}),$$

whose transpose

$$^tT : \mathcal{E}(\bar{\Omega}) \to \mathcal{E}(\Omega)$$

is a projection operator on the space of solutions to (5.4). In fact, if $f \in \mathcal{E}(\bar{\Omega})$ satisfies (5.4), the natural representation

$$f(x) = (^tTf)(x), \quad x \in \Omega$$

may be written explicitly as

$$f(x) = (2\pi i)^{-n} \langle \bar{\partial} \left[\frac{1}{P} \right] (\zeta), g(\zeta, D)f(\varphi'(\zeta)) \tag{5.5}$$

$$\wedge \exp[-i(\zeta \cdot (t - \varphi'(\zeta)))](2\bar{\partial}\partial\varphi(\zeta))^{n-m} \rangle,$$

where $g(\zeta, D)f(\varphi')$ is the form obtained from $g(\zeta, z)$ by considering its coefficients as polynomials in z and by replacing z^k by $D^k f|_{t=\varphi'}$.

We refer the reader to Berndtsson and Passare [1989] for many illuminating concrete examples of such an explicit result. We content ourselves with a few remarks: In general, it may be quite hard to construct the residue current (and one should use here the approach proposed in Berenstein, Gay and Yger [1989] instead of the resolution of singularities), but in many cases (e.g., when the variety $\bigcap_{j=1}^{m} P_j^{-1}(0)$ is nonsingular), the construction can actually be done by hand; an important feature of the formula (5.5) is clearly the fact that the function f itself appears in the representation; so if $n = 1$ and $P(z) = z - \alpha$, formula (5.5) gives the general solution as

$$f(w) = \exp(-i\alpha(t - w))$$

instead of

$$\text{const.} \exp(-i\alpha t).$$

Also, if the supporting function φ, which appears in (5.5), is suitably modified, one can find that in many instances only the boundary values of f, and of some of its derivatives, appear in the representation. The last remark concerns the conditions which are imposed on the polynomial mapping P; one is the *complete intersection* condition with which we have been confronted since Chap. 1: This method does not allow to eliminate it (notice, however that the Ehrenpreis–Palamodov method works without it); the second condition is the *hypoellipticity*, which, on the other hand, is not necessary and can be dispensed with, at the cost of complicating the formulas (see Berenstein and Yger [1991b]).

We now conclude this chapter by going back to Theorem 5.5 and to one possible application: In Passare [1988b], Passare deals with some classical problems posed by D. J. Newman and H. S. Shapiro (Newman and Shapiro [1968]) back in the sixties (these problems are quoted in Ehrenpreis [1970], in the chapter on the Cauchy problem), and which recently reappeared in Barth, Brannan and Hayman [1984] and in the survey of Shapiro [1988]. Let us briefly recall them:

(a) *Let (P,Q) be a pair of polynomials in $\mathbb{C}[z_1,\dots,z_n]$ and consider the following property:*
    $(\mathcal{P})$ the map $T : H(\mathbb{C}^n) \to H(\mathbb{C}^n)$ defined by $T(f) : P(D)(Qf)$ is a bijection.
    Is it true that if (P,Q) satisfies $(\mathcal{P})$, then also (Q,P) satisfies $(\mathcal{P})$?
(b) *Let P be a polynomial of degree m, with a nonzero coefficient for z_1^m, and let $Q = z_1^m$. Does (P,Q) satisfy $(\mathcal{P})$? Does (Q,P) satisfy $(\mathcal{P})$?*
(c) *Let P be a polynomial and let $P^*(z) = \overline{P(\bar{z})}$. Does (P,P^*) satisfy $(\mathcal{P})$?*

Problems (a) and (b) have been completely solved by Meril and Struppa in Meril and Struppa [1985]. Indeed, they show that (P,Q) satisfies $(\mathcal{P})$ if and only if every function $g \in H(V)$, with V being the multiplicity variety associated to the polynomial Q, extends uniquely to an entire function, which is a solution of $P(D)f = 0$. Therefore $(\mathcal{P})$ is equivalent to the well-posedness of a Cauchy problem in $H(\mathbb{C}^n)$. On the other hand, a simple duality argument shows that (Q,P) satisfies $(\mathcal{P})$ if and only if the *same* Cauchy problem is well posed in the space $\mathrm{Exp}(\mathbb{C}^n)$. It is then easy to show that (a) can be answered affirmatively if $n = 1$, and negatively if $n \geq 2$. As to problem (b), one can use the same kind of argument to show that the answer is positive, and it is also possible to characterize all pairs (P,Q) satisfying $(\mathcal{P})$ in the case in which $P(z) = z_1^m$ (Meril and Struppa [1985]).

Then it is clear that the explicit representations proved by Berenstein, Gay, Yger, and Passare could be used toward this same purpose; the first attempt in this direction was done by Passare in Passare [1988b] with respect to question (b). This seems quite interesting, especially in view of the fact that no definite answer is known, as of now, for problem (c), and it is therefore possible to hope that these new methods may shed some light on it. For more details on this matter, we refer to Shapiro [1988] and to Meril and Yger [1992].

Chapter 6
Algebraic Analysis

We have decided to end our survey by mentioning one of the most exciting new developments in modern analysis, namely the impressive building of *algebraic analysis*, which has mainly developed out of the work of M. Sato and his Kyoto school.

Algebraic analysis is too big a subject to be given a complete description in a short chapter. What we aim to do, therefore, is to give a reasonable overview of the results which this new branch of mathematics has been able to produce, especially with respect to the problems raised in this portion of the book and with special attention to the case of operators with variable coefficients. For most definitions and/or notations, the reader is referred to the two basic works, Sato, Kawai and Kashiwara [1973] and Kashiwara, Kawai and Kimura [1986].

Among all of the topics which we touched upon in Chap. 2 and which form the basis of most results in Chaps. 3, 4, and 5, one can certainly single out the Fundamental Principle of Ehrenpreis and Palamodov, which is limited, however, to the case of differential operators with *constant coefficients*. This limitation, of course, has its excellent reasons: On one hand, exponential polynomials do not play any particular role as solutions of differential equations with variable coefficients; on the other hand, one of the pillars of the theory of linear differential equations with constant coefficients on $\mathcal{D}'$ (or $\mathcal{E}$) is the surjectivity of such operators (the so-called Ehrenpreis–Malgrange lemma, if one wants to reduce things to their essentials); however, in 1957, H. Lewy found (Lewy [1957]) that there exist very simple linear partial differential equations with (*variable coefficients*) without local solutions. Since then, the question of solvability of differential equations with variable coefficients has played a central role, as well as the attempt to understand to what extent the Fundamental Principle could be true for these equations. It now becomes clear that what the Fundamental Principle really points out is the role of the characteristic variety, which is maintained even in the variable coefficient case. Here and in the following, we shall *always* refer to *real analytic* coefficients, which are those that algebraic analysis is able to deal with, and which Ehrenpreis himself considered as the first possible extension of this theory (see Ehrenpreis [1970, Chap. XI]); as Kashiwara, Kawai, and Kimura mention in Kashiwara, Kawai and Kimura [1986]: " ... This central problem has long been recognized in the case of equations with constant coefficients (i.e., Ehrenpreis's Fundamental Principle). But one had to wait until the advent of microfunction theory to consider the above for the case of variable coefficients ... ." In this chapter, we shall try to explain this sentence, as well as to show how algebraic analysis can contribute to a deeper understanding of Lewy's example.

In order to do so, one needs to formulate the Fundamental Theorem of Sato, for which some preparation is necessary.

Let M be a real-analytic manifold (the reader may think of it as $\mathbb{R}^n$, to fix the ideas) and denote by $\mathcal{B} = \mathcal{B}_M$ the sheaf of hyperfunctions on M. The abstract definition of $\mathcal{B}_M$ is quite subtle: If X is a compactification of M (i.e., $X = \mathbb{C}^n$ if $M = \mathbb{R}^n$) and if $\mathcal{O}_X$ and ω_M are the sheaf of germs of holomorphic functions on X, and the sheaf of orientations on M, respectively ($\omega_M = \mathbb{Z}$, the constant sheaf, when $M = \mathbb{R}^n$), then one defines

$$\mathcal{B}_M := \mathcal{H}_M^n(\mathcal{O}_X) \bigotimes_{\mathbb{Z}} \omega_M,$$

where $\mathcal{H}_M^n(\mathcal{O}_X)$ denotes the nth derived sheaf of $\mathcal{O}_X$ with support in M. Hyperfunctions can be more easily understood in terms of boundary values of holomorphic functions, subject to some natural equivalence relation (for example, if $M = \mathbb{R}$, then for any open set $\Omega \subset \mathbb{R}$, the set $\mathcal{B}(\Omega)$ is defined by

$$\frac{\mathcal{O}(V \backslash \Omega)}{\mathcal{O}(V)},$$

where V is any complex neighborhood of Ω in $\mathbb{C}$). The sheaf $\mathcal{B}$ of hyperfunctions is a sheaf of generalized functions in the sense that both the sheaf $\mathcal{A}$ of real-analytic functions and the sheaf $\mathcal{D}'$ of distributions are subsheaves of $\mathcal{B}$. From the point of view of the analyst, the great advantage of $\mathcal{B}$ is its flabbiness, which makes possible the extension of any hyperfunction defined, say, on an open subset Ω of $\mathbb{R}^n$ to all of $\mathbb{R}^n$. It also allows the construction of flabby resolutions of the sheaf of solutions of a system of partial differential equations. The main object in algebraic analysis, however, is not so much the sheaf $\mathcal{B}$, but the sheaf $\mathcal{C}$ of *microfunctions*. We refer to Kashiwara, Kawai and Kimura [1986] for its formal definition, but let us try to give here an idea of its meaning. Suppose $M = \mathbb{R}$ and let f be a hyperfunction defined in a neighborhood of a point $x_0 \in \mathbb{R}$. If f is not real-analytic on x_0, one may ask *to which extent f fails to be real-analytic* and how this can be expressed (for the C^∞ category, this same idea was introduced in Hörmander [1971a,b] and in that case is known as the *analytic wave front set*). It turns out that the proper context for such a problem is the *spheric cotangent bundle* iS^*M of M. For $M = \mathbb{R}$, the points of iS^*M can be written as $(x, +i\infty)$ or $(x, -i\infty)$, where $\pm i\infty$ indicate the *duals* to the two possible directions approaching $x \in \mathbb{R}$ when embedded in $\mathbb{C}$.

Definition 6.1. A hyperfunction f defined in a neighborhood of $x_0 \in \mathbb{R}$ is said to be *microanalytic* at $(x_0, +i\infty)$ (or at $(x_0, -i\infty)$, respectively) if it is the boundary value *from below* of a function holomorphic in the lower half-plane (or is the boundary value *from above* of a function holomorphic in the upper half-plane, respectively).

A hyperfunction f is real-analytic at x_0 if and only if it is microanalytic both at $(x_0, +i\infty)$ and at $(x_0, -i\infty)$, i.e., the notion of microanalyticity is the right notion to explain *in which way f fails to be analytic*. Also, Definition 6.1 can be extended to $\mathbb{R}^n$ as well as to any real-analytic manifold, where everything becomes more complicated, however, in view of the greater complexity of the notion of boundary value. The reader is referred to Kashiwara, Kawai and Kimura [1986] for details.

Definition 6.2. Let f be a hyperfunction defined on a real-analytic manifold M. We define its *singularity spectrum* to be the subset of iS^*M defined by

$$\text{S.S.}(f) := \{(x_0, i\xi_0) \in iS^*M : f \text{ is not microanalytic here}\}.$$

By employing such a notion, one can define a new object which describes just the *singularities* of the sheaf $\mathcal{B}$: Consider the presheaf which associates to each open set $\Omega \subseteq iS^*M$ the set

$$\frac{\mathcal{B}(M)}{\{f \in \mathcal{B}(M) : \text{S.S.}(f) \cap \Omega = \emptyset\}}.$$

Definition 6.3. The sheaf associated to this presheaf (which is not complete for $n > 1$) is said to be the sheaf $\mathcal{C}$ of *microfunctions*.

Such an object (roughly speaking, the sheaf of singularities of hyperfunctions) is a flabby sheaf, and if $\pi : iS^*M \to M$ is the canonical projection, one can characterize $\mathcal{C}$ with the following exact sequence of sheaves on M:

$$0 \to \mathcal{A} \hookrightarrow \mathcal{B} \to \pi_*\mathcal{C} \to 0.$$

There is a natural map (the *spectrum* map)

$$\text{sp} : \mathcal{B} \to \mathcal{C},$$

which associates to every hyperfunction its equivalence class as a microfunction.

As we pointed out before, one aims to study the surjectivity of differential operators with variable (real-analytic) coefficients. The best way to prove such a surjectivity is to concretely produce an inverse to the operator itself. This idea is quite old, but, of course, when we try to invert a differential operator such as $\frac{\partial}{\partial x}$, we are compelled to leave the space of differential operators and to enter a larger space in which integral operators have to be allowed; for this reason, the Sato school looked for a suitable space of differential operators where you could always construct an inverse (Sato, Kawai and Kashiwara [1973]). Let us briefly describe how such a space can be constructed. For a real-analytic manifold M, let Δ_M be its diagonal

$$\Delta_M = \{(x, y) \in M \times M : x = y\},$$

and define

$$\Delta^a_{iS^*M} = \{(x, y; i(\xi, \eta)\infty) : x = y, \xi = -\eta\}.$$

78   C.A. Berenstein, D.C. Struppa

Now it can be shown that if $K(x,y)$ is a hyperfunction on $M \times M$, with $\mathrm{supp}(K) \subseteq \Delta_M$ and $\mathrm{S.S.}(K) \subseteq \Delta^a_{iS^*M}$, then the integral operator $\mathcal{K}$ defined by

$$\mathcal{K}(f)(x) = \int K(x,y)f(y)dy$$

is a sheaf homomorphism from $\mathcal{B}_M$ to itself, and it is easily seen that such an operator is a natural generalization of differential operators of finite order (indeed, by using John's plane waves expansion (John [1955]) of Dirac's δ-function, it is easy to explicitly compute the kernel $K(x,y)dy$ corresponding to a differential operator such as $\sum_{|\alpha| \le m} a_\alpha(x)\frac{\partial^{|\alpha|}}{\partial x^\alpha}$). A similar construction (though much more complicated from the technical point of view) can be carried out to find sheaf homomorphisms for the sheaf $\mathcal{C}_M = \mathcal{C}$. In particular, if v_M denotes the sheaf of volume elements of M, and if $K(x,y)dy$ belongs to the somewhat mysterious sheaf

$$\mathcal{H}^0_{iS^*_M(M \times M)}(\mathcal{C}_{M \times M} \otimes v_M),$$

then the integral operator

$$\mathcal{K}(f)(x) = \int K(x,y)f(y)dy$$

defines a sheaf homomorphism from $\mathcal{C}_M$ to itself.

Definition 6.4. The sheaf

$$\mathcal{L}_M := \mathcal{H}^0_{iS^*_M(M \times M)}(\mathcal{C}_{M \times M} \otimes v_M)$$

is said to be the *sheaf of microlocal operators*.

Among the crucial properties of the sheaf $\mathcal{L}_M$, we may just mention the fact that $\mathcal{L}_M$ is a ring whose identity $\delta(x - y)dy$ acts on $\mathcal{C}_M$ as an identity map. Consider now a linear differential operator of order m,

$$P(x, D_x) = \sum_{|\alpha| \le m} a_\alpha(x)D_x^\alpha$$

(with obvious meaning of the symbols); its *principal symbol* is defined as

$$\sigma(P)(x, \xi) = \sum_{|\alpha|=m} a_\alpha(x)\xi^\alpha$$

and belongs to $\mathcal{O}_{T^*M}$; it is then clear that $P(x, D_x)$ is a microlocal operator with kernel function

$$P(x, D_x)\delta(x - y)dy.$$

We now have all the necessary terminology to state the Fundamental Theorem of Sato (Sato, Kawai and Kashiwara [1973]).

Theorem 6.1. *A linear differential operator of finite order $P(x, D_x)$ is left- and right-invertible on the ring $\mathcal{L}_M$ over $\{(x, i\xi\infty) \in iS^*M : \sigma(P)(x, i\xi) \neq 0\}$.*

The proof of Theorem 6.1, although microlocal in technique, is quite classical in spirit, since it can be considered as a *microlocalization* of John's construction (John [1950; 1951]) of a fundamental solution for an elliptic differential equation. Indeed, one can prove Theorem 6.1 as follows. Consider the statement (local in nature) in a neighborhood of $x = 0$. Therefore, let $\sigma(P)(0, i\xi_0) \neq 0$; by the Cauchy–Kovalevsky theorem, there is a real-analytic function $u(x, \xi, p)$, defined in a neighborhood of $(x, \xi, p) = (0, \xi_0, 0)$ in $S^*M \times \mathbb{R}$, such that

$$P(x, D_x)u(x, \xi, p) = 1 \quad \text{and} \quad u \equiv 0 \mod (x \cdot \xi - p)^m;$$

such a solution $u(x, \xi, p)$ is what is usually called a *unitary solution*; by using such a solution and employing the classical Duhamel principle (namely, a superposition principle), one can finally construct (this is the most technical part of the proof) a *fundamental solution* $E(x, y)$ such that

$$P(x, D_x)E(x, y) = \delta(x - y).$$

One can use this fundamental solution E to finish the proof of the theorem.

The reader will certainly notice that Theorem 6.1 points out the role (which we mentioned at the beginning of this chapter) of the characteristic variety, since what it really says is that outside

$$\{(x, i\xi\infty) \in iS^*M : \sigma(P)(x, i\xi) = 0\},$$

every partial differential operator is invertible. How this relates to the previously discussed work can be made even more evident by considering the following simple (but important) consequence of Theorem 6.1.

Theorem 6.2. *Let u, f be two hyperfunctions which satisfy*

$$P(x, D_x)u(x) = f(x);$$

then

$$\text{S.S.}(u) \subseteq \text{S.S.}(f) \cup \{(x, i\xi\infty) \in iS^*M : \sigma(P)(x, i\xi) = 0\}.$$

The proof of this result is immediate from Theorem 6.1. In fact, from $P(x, D_x)u(x) = f(x)$, one immediately deduces that

$$P(x, D_x)\text{sp}(u(x)) = \text{sp}(f(x)).$$

By Theorem 6.1, if $\sigma(P)(x, i\xi) \neq 0$, one can find a microlocal operator E such that $EP = 1$; therefore, one has

$$\text{sp}(u) = (EP)\text{sp}(u) = E(\text{sp}(f)),$$

which yields the desired conclusion.

From Theorem 6.2, one concludes that a solution $u(x)$ of the homogeneous equation

$$P(x, D_x)u(x) = 0$$

has its singularity spectrum contained in

$$\{(x, i\xi\infty) \in iS^*M : \sigma(P)(x, i\xi) = 0\}.$$

Therefore, the study of the structure of the space of solutions of $P(x, D_x)u = 0$ on the characteristic variety is a central theme in the theory of linear differential equations.

To get a deeper understanding of this role and to link it even more to the surjectivity problem, one needs to proceed further into the field of algebraic analysis, to the notion of *quantized contact transformation*. The first step is to understand that, despite the many consequences of Theorem 6.1, the sheaf $\mathcal{L}_M$ is *too big* for practical purposes. It is therefore useful to introduce a new class of operators, the so-called microdifferential operators, which lies between the class of microlocal operators and the class of differential operators. Actually, such operators arise quite naturally already in Theorem 6.1, since it is possible to show that the inverse of a differential operator (which microlocally exists in view of Theorem 6.1, at least outside the characteristic variety) is indeed a very special object in $\mathcal{L}_M$, namely a microdifferential operator. The idea for the construction (Kashiwara, Kawai and Kimura [1986]) of the sheaf $\mathcal{E}$ of microdifferential operators is quite simple (the use of the standard symbol $\mathcal{E}$ for this sheaf will not induce any confusion with C^∞-functions, since everything in this chapter is about *real-analytic* objects). Namely, if $P(x, D_x) = \sum_{|\alpha| \le m} a_\alpha(x) D_x^\alpha$ is a differential operator of order m, then the integral expression for $P(x, D_x)\delta(x - y)$ is given by

$$\left(\frac{-1}{2\pi i}\right)^n \int \sum_j \frac{(-1)^j (j + n - 1)! p_j(x, \xi)}{[(x - y) \cdot \xi + i0]^{n+j}} \omega(\xi),$$

where $\omega(\xi) = \sum_{i=1}^n (-1)^{i-1} \xi_i d\xi_1 \wedge \cdots \wedge d\xi_{i-1} \wedge d\xi_{i+1} \wedge \cdots \wedge d\xi_n$ and $p_j(x, \xi) = \sum_{|\alpha|=j} a_\alpha(x)\xi^\alpha$. Now a microdifferential operator corresponds to a generalization of this formula in which p_j is no longer a homogeneous polynomial in ξ of degree j but just a holomorphic function (of course, some technical conditions have to be imposed to ensure that we have a sheaf of operators acting on microfunctions). It is interesting to notice that the fundamental theorem of Sato (Theorem 6.1) can be extended to sheaf $\mathcal{E}$:

Theorem 6.3. *Let P be a microdifferential operator of finite order m. If its principal symbol $\sigma_m(P)(x, i\xi)$ does not vanish in an open set $\Omega \subseteq iS^*M$, then there exists a microdifferential operator E, defined on Ω, such that*

$$PE = EP = 1.$$

This result shows that, for a microdifferential operator P, the structure of the space of solutions of $P(D)u = 0$ is trivial outside its characteristic variety

$$V = \{(x, i\xi\infty) \in iS^*M : \sigma_m(P)(x, i\xi) = 0\}.$$

The crucial point in all of algebraic analysis is the fact that the analysis of the space of solutions of $P(D)u = 0$ can be beautifully described in terms of the geometry of V. In fact, there exist some classes of canonical transformations (the so-called quantized contact transformations, see Kashiwara, Kawai and Kimura [1986]) which turn any (overdetermined) system into a *canonical form*, where its properties can be easily recognized, since these transformations do not modify the structure of the solution sheaf. It is with the use of these incredibly powerful ideas that the following kind of result came to be proved (in the statement, $\{\cdot, \cdot\}$ denotes the *Poisson bracket*).

Theorem 6.4. *Let $P(x, D_x)$ be a microdifferential operator of order m, which is defined in a neighborhood of $(x_0, i\xi_0\infty) \in iS^*M$, and let $f(x, \xi) = \sigma_m(P)(x, i\xi)$. Then the following statements hold:*

(i) *if $f(x_0, \xi_0) = 0$ and $\{f, \bar{f}\}(x_0, \xi_0) > 0$ hold, then $P(x, D_x)$ is surjective in a neighborhood of $(x_0, i\xi_0\infty)$, and $\mathrm{Ker}(P(x, D_x))$ is equal to the image of a microlocal operator $\mathcal{K}$;*

(ii) *if $f(x_0, \xi_0) = 0$ and $\{f, \bar{f}\}(x_0, \xi_0) < 0$ hold, then $P(x, D_x)$ is injective but not surjective in a neighborhood of $(x_0, i\xi_0\infty)$, and its image equals the kernel of a microlocal operator $\mathcal{K}$.*

Thus, at least for a *generic* equation, the question of solvability is perfectly understood: Microlocal analysis shows its worth with these kinds of results. A similar theorem could be proved for operators of infinite order. The importance of the study of operators of infinite order is now commonly accepted, even when studying the structure of microdifferential equations of finite order (see, e.g., Kawai [1986a; 1987b]). Important developments, for example, have recently taken place in the case of *constant coefficients*, but *infinite-order* linear differential operators, i.e., vanishing of cohomology, as well as division theorems, have been proved (see mainly Kawai and Struppa [1990], or the lengthy report Struppa [1988]), and a Fundamental Principle for some special classes of overdetermined systems of such operators is now essentially known (Kawai and Struppa [1990]).

Among the other important applications of algebraic analysis to our topic, we should mention the treatment of the theta zero-values (which we briefly mentioned in Chap. 4), or the deep results by Kashiwara (Kashiwara [1976]) on the Bernstein–Sato polynomials (see Chap. 5). This by no means exhausts the list of the possible applications of algebraic analysis. For the beautiful theory of $\mathcal{D}$-modules and $\mathcal{E}$-modules that has been developed in recent years, we refer the reader to Kashiwara [1986], Borel [1987], Schapira [1985], Kashiwara and Kawai [1983], as well as to Kashiwara and Schapira [1979]. Let us finally end this part by pointing out the relations between algebraic analysis and the construction of the Henkin–Ramirez reproducing kernels. As mentioned in Chaps. 1 and 5, we saw that Henkin and Ramirez constructed their kernels by using the Leray formula (also Cauchy–Leray–Koppelman–Fantappiè ... formula); on the other hand, in Sato, Kawai and Kashiwara [1973], the authors defined a remarkable generalization of the plane wave decomposition of the δ-function, which they called *curvilinear wave expansion*, actually a form of the spherical Radon transform which, as Bony noticed in Bony [1976], is constructed by taking the *boundary value* of the Leray kernel. An analogous approach is taken by Tajima (see Tajima [1982; 1985]), who studies the boundary values of the Henkin–Ramirez kernel from the microlocal point of view and who shows that *such a kernel is actually a holomorphic reproducing kernel for microfunction solutions of the tangential Cauchy–Riemann equations.*

References[1]

Abramczuk, W.
[1984]   A class of surjective convolution operators. Pac. J. Math. *110*, 1–7. Zbl.509.46029
[1985]   On continuation of quasi-analytic solutions of partial differential equations to compact convex sets. J. Aust. Math. Soc., Ser. A *39*, 306–316. Zbl.595.35015

Agranovich, M.S.
[1961]   On partial differential equations with constant coefficients. Usp. Mat. Nauk *16*, 27–93. English transl.: Russ. Math. Surv. *16*, No. 2, 23–90 (1961). Zbl.101,74

Aizenberg, L.A., Dautov, Sh.A.
[1975]   Differential Forms Orthogonal to Holomorphic Functions or Forms and Their Properties. Novosibirsk: Nauk. English transl.: Providence, R.I.: Am. Math. Soc., 1983. Zbl.323.32002

Aizenberg, L.A., Yuzhakov, A.P.
[1979]   Integral Representations and Residues in Multidimensional Complex Analysis. English trans.: Providence, R.I.: Am. Math. Soc., 1983. Zbl.445.32002; Zbl.537.32002

Amar, E.
[1980]   Extension de formes $\bar{\partial}_b$ fermées et solutions de l'équation $\bar{\partial}_b u = f$. Ann. Sc. Norm. Super. Pisa, Cl. Sci., IV. Ser. *7*, 155–179. Zbl.456.32012
[1983]   Extension de fonctions holomorphes et courant. Bull. Sci. Math., II. Ser. *107*, 25–48. Zbl.543.32007
[1984]   Cohomologie complexe et applications. J. Lond. Math. Soc., II. Ser. *29*, 127–140. Zbl.583.32033
[1985]   Non division dans $A^{\infty}(\Omega)$. Math. Z. *188*, 493–511. Zbl.547.32010

---

[1] For the convenience of the reader, references to reviews in Zentralblatt für Mathematik (Zbl.), compiled using the MATH database, and Jahrbuch über die Fortschritte der Mathematik (Jbuch) have, as far as possible, been included in this bibliography.

[1987]   Estimations, par noyaux, des solutions de l'équation $\bar{\partial}u = f$. Springer Lecture Notes in Mathematics *1276*, 12–24. Zbl.627.32015

Amar, E., Bonami, A.
[1979]   Mésures de Carleson d'ordre α et solutions au bord de l'équation $\bar{\partial}$. Bull. Soc. Math. Fr. *107*, 23–48. Zbl.409.46035

Andersson, M., Berndtsson, B.
[1982]   Henkin–Ramirez kernels with weight factors. Ann. Inst. Fourier *32*, 91–110. Zbl.466.32001

Andersson, M., Passare, M.
[1988]   A shortcut to weighted representation formulas for holomorphic functions. Ark. Mat. *26*, 1–12. Zbl.659.32006

Andreotti, A.
[1975]   Complexes of Partial Differential Operators. New Haven, London: Yale Univ. Press. Zbl.309.58020

Andreotti, A., Vesentini, E.
[1965]   Carleman estimates for the Laplace–Beltrami equation on complex manifolds. Inst. Hautes Etud. Sci., Publ. Math. *25*, 313–362. Zbl.138,66

Aoki, T.
[1982]   Invertibility of microdifferential operators of infinite order. Publ. Res. Inst. Math. Sci. *18*, 421–449. Zbl.512.35077

[1984]   The exponential calculus of microdifferential operators of infinite order V. Proc. Japan Acad., Ser. A *60*, 8–9. Zbl.599.35138

[1988]   Existence and continuation of holomorphic solutions of differential equations of infinite order. Adv. Math. *72*, 261–283. Zbl.702.35253

Aoki, T., Kashiwara, M., Kawai, T.
[1986]   On a class of linear differential operators of infinite order with finite index. Adv. Math. *62*, 155–168. Zbl.628.35003

Atiyah, M.F.
[1970]   Resolution of singularities and division of distributions. Commun. Pure Appl. Math. *23*, 145–150. Zbl.188,194

Avanissian, V., Gay, R.
[1974a]   Sur une transformation des fonctionnelles analytiques portables par des convexes compactes de $\mathbb{C}^d$, et la convolution d'Hadamard. C. R. Acad. Sci., Paris, Ser. A *279*, 133–136. Zbl.291.32006

[1974b]   Sur les fonctions entières arithmétiques de type exponentiel et le quotient d'exponentielle-polynômes de plusieures variables. C. R. Acad. Sci., Paris, Ser. A *279*, 161–164. Zbl.291.32007

Bagchi, S.C., Sitaram, A.
[1979]   Spherical mean-periodic functions on semi-simple Lie groups. Pac. J. Math. *84*, 241–250. Zbl.442.43017

Bang, T.
[1946]   Om Quasi-analytiske Funktioner (thesis), Copenhagen. Zbl.60,151

Barlet, D.
[1985]   Prolongement analytique de $|f|^{2\alpha}$ et connexion de Gauss–Manin. Astérisque *130*, 48–55. Zbl.564.32007

Barth, K.F., Brannan, D.A., Hayman, W.K.
[1984]   Research problems in complex analysis. Bull. Lond. Math. Soc. *16*, 490–517. Zbl.593.30001

Bartolomeis, P.
[1984]   Fibrati positivi e teorie L^2 per l'operatore $\bar{\partial}$. Racc. Sem. Dip. di Mat. Univ. della Calabria *4*: Atti del Convegno "Geometria Analitica ed Analisi Complessa."

Bartolomeis, P., Tomassini, G.
[1981]   Idéaux de type fini dans $A^\infty(D)$. C. R. Acad. Sci., Paris, Ser. I *293*, 133–134. Zbl.477.32016
[1982]   Finitely generated ideals in $A^\infty(D)$. Adv. Math. *46*, 162–170. Zbl.499.32012

Beckenbach, E.F., Reade, M.
[1943]   Mean values and harmonic polynomials. Trans. Am. Math. Soc. *53*, 230–238.

Bellman, R., Cooke, K.
[1963]   Differential-Difference Equations. New York: Academic Press. Zbl.105,64

Berenstein, C.A.
[1970]   Convolution operators and related quasi-analytic classes (thesis). New York.
[1976]   On the converse to Pompeiu's problem. Notas e Comunicacoes de Matematica: Univ. Fed. de Pernambuco *73*.
[1978]   Interpolation and Fourier analysis in several complex variables. In: Harmonic Analysis and Several Complex Variables: Proc. Int. Conf. Cortona/Italy 1976–1977, 1–13. Zbl.421.32018
[1980]   An inverse spectral theorem and its relation to the Pompeiu problem. J. Anal. Math. *37*, 128–144. Zbl.449.35024
[1987]   Spectral synthesis on symmetric spaces. Contemp. Math. *63*, 1–25. Zbl.613.53030
[1988]   On an overdetermined Neumann problem. Sem. Geom. Univ. Bologna 1986, 21–26. Zbl.676.35070

Berenstein, C.A., Dostal, M.A.
[1972]   Analytically uniform spaces and their applications to convolution equations. Springer Lecture Notes in Mathematics *256*. Zbl.237.47025
[1973a] Some remarks on convolution equations. Ann. Inst. Fourier *23*, 55–74. Zbl.241.46039
[1973b] On convolution equations I. Springer Lecture Notes in Mathematics *336*, 79–94. Zbl.259.46040
[1974a] The Ritt theorem in several variables. Ark. Mat. *12*, 267–280. Zbl.293.33001
[1974b] A lower estimate for exponential sums. Bull. Am. Math. Soc. *80*, 687–691. Zbl.304.33001
[1975]   On a property of indicators of smooth convex bodies. Mich. Math. J. *22*, 237–246. Zbl.305.42023

Berenstein, C.A., Gay, R.
[1986a] A local version of the two-circles theorem. Isr. J. Math. *55*, 267–288. Zbl.624.31002
[1986b] Sur la synthèse spectrale dans les espaces symétriques. J. Math. Pures Appl., IX. Ser. *65*, 323–333. Zbl.635.46049
[1989]   Le problème de Pompeiu local. J. Anal. Math. *52*, 133–166. Zbl.668.30037

Berenstein, C.A., Gay, R., Yger, A.
[1989]   Analytic continuation of currents and division problems. Forum Math. *1*, 15–51. Zbl.651.32005
[1990]   The three squares theorem, a local version. Lect. Notes Pure Appl. Math. *122*, 35–50. Zbl.728.44005

Berenstein, C.A., Krishnaprasad, P.S., Taylor, B.A.
[1984]   Deconvolution methods for multi-sensors. DTIC #ADA 152351.

Berenstein, C.A., Lesmes, J.
[1979] The Cauchy problem for convolution operators. Uniqueness. Mich. Math. J. *26*, 333-349. Zbl.403.45004

Berenstein, C.A., Shahshahani, M.
[1983] Harmonic Analysis and the Pompeiu problem. Am. J. Math. *105*, 1217-1229. Zbl.522.43006

Berenstein, C.A., Struppa, D.C.
[1983] Interpolation problems in cones I, II. Atti Accad. Naz. Lincei, VIII. Ser., Rend., Cl. Sci. Fis. Mat. Nat. *74*, 267-273; 331-335. Zbl.578.42014

[1984] On explicit solutions to the Bezout equation. Syst. Control Lett. *4*, 33-39. Zbl.538.15005

[1987a] Solutions of convolution equations in convex sets. Am. J. Math. *109*, 521-543. Zbl.628.46036

[1987b] A remark on "Convolutors in space of holomorphic functions." Springer Lecture Notes in Mathematics *1276*, 276-280. Zbl.628.42009

[1987c] On the Fabry-Ehrenpreis-Kawai theorem. Publ. Res. Inst. Math. Sci. *23*, 565-574. Zbl.628.30003

[1988a] Small degree solutions for the polynomial Bezout equation. Linear Algebra Appl. *98*, 41-55. Zbl.643.13006

[1988b] Dirichlet series and convolution equations. Publ. Res. Inst. Math. Sci. *24*, 783-810. Zbl.668.30005

Berenstein, C.A., Taylor, B.A.
[1977] The "three-squares" theorem for continuous functions. Arch. Ration. Mech. Anal. *63*, 253-259. Zbl.353.46027

[1979] A new look at interpolation theory for entire functions of one variable. Adv. Math. *33*, 109-143. Zbl.432.30028

[1980a] Mean-periodic functions. Int. J. Math. Sci. *3*, 199-236. Zbl.438.42012

[1980b] Interpolation problems in $\mathbb{C}^n$ with applications to harmonic analysis. J. Anal. Math. *38*, 188-254. Zbl.464.42003

[1982] On the geometry of interpolating varieties. Springer Lecture Notes in Mathematics *919*, 1-25. Zbl.484.32004

Berenstein, C.A., Taylor, B.A., Yger, A.
[1983a] Sur les systèmes d'équations différence-différentielles. Ann. Inst. Fourier *33*, 109-130. Zbl.493.34052

[1983b] Sur quelques formules explicites de déconvolution. J. Optics *14*, 75-82.

Berenstein, C.A., Yang, P.
[1982] An overdetermined Neumann problem in the unit disk. Adv. Math. *44*, 1-17. Zbl.488.35063

[1987] An inverse Neumann problem. J. Reine Angew. Math. *382*, 1-21. Zbl.623.35078

Berenstein, C.A., Yger, A.
[1983] Le problème de la déconvolution. J. Funct. Anal. *54*, 113-160. Zbl.526.46042

[1986a] Ideals generated by exponential polynomials. Adv. Math. *60*, 1-80. Zbl.586.32019

[1986b] On Lojasiewicz-type inequalities for exponential polynomials. J. Math. Anal. Appl. *129*, 166-195. Zbl.638.30003

[1989] Analytic Bezout identies. Adv. Appl. Math. *10*, 51-74. Zbl.673.32011

[1990] Bounds for the degrees in the division problem. Mich. Math. J. *37*, 25-43. Zbl.702.32006

[1991a] Effective Bezout identities in $\mathbb{Q}[z_1, \ldots, z_n]$. Acta. Math. *166*, 69-120. Zbl.724.32002

[1991b] About the Ehrenpreis Principle. In: Geom. Algebraic Aspects in Several Complex Variables. EditEl, 47–61.

Berenstein, C.A., Zalcman, L.

[1976] Pompeiu's problem on spaces of constant curvature. J. Anal. Math. *30*, 113–130. Zbl.332.35033

[1980] Pompeiu's problem on symmetric spaces. Comment. Math. Helv. *55*, 593–621. Zbl.452.43012

Berndtsson, B.

[1980] Integral formulas for the $\partial\bar{\partial}$-equation and zeros of bounded holomorphic functions in the unit ball. Math. Ann. *249*, 163–176. Zbl.414.31007

[1983] A formula for interpolation and division in $\mathbb{C}^n$. Math. Ann. *263*, 399–418. Zbl.499.32013

Berndtsson, B., Passare, M.

[1989] Integral formulas and an explicit version of the Fundamental Principle. J. Funct. Anal. *84*, 358–372. Zbl.686.46031

Bernstein, I.N.

[1971] Modules over the ring of differential operators. A study of the fundamental solutions of equations with constant coefficients. Funkts. Anal. Prislozh. *5*, No. 2, 1–16. English transl.: Funct. Anal. Appl. *5*, 89–101 (1971). Zbl.233.47031

[1972] The analytic continuation of generalized functions with respect to a parameter. Funkts. Anal. Prilozh. *6*, No. 4, 26–40. English transl.: Funct. Anal. Appl. *6*, 273–285. Zbl.282.46038

Bernstein, I.N., Gel'fand, S.I.

[1969] Meromorphy of the function P^λ. Funkts. Anal. Prilozh. *3*, No. 1, 84–85. English transl.: Funct. Anal. Appl. *3*, 68–69 (1969). Zbl.208,152

Bernstein, V.I.

[1933] Lecons sur les progrès recents de la théorie des séries de Dirichlet. Paris: Gauthier-Villars. Zbl.8,115

Beurling, A.

[1957] Lectures on Quasi-analytic Functions. Princeton: The Institute for Advanced Studies, 1956–57.

[1961] Quasi-Analyticity and General Distributions. Stanford: AMS Summer Inst.

[1963] Notes on Dirichlet series. J. Indiana Math. Soc. *27*, 19–26. Zbl.115,288

Björck, G.

[1966] Linear partial differential operators and generalized distributions. Ark. Mat. *6*, 351–407. Zbl.166,365

[1971] Beurling Distributions and Linear Partial Differential Equations, 1st ed. Roma: Alta Matematica.

Björk, J.E.

[1979] Rings of Differential Operators. Amsterdam: North-Holland. Zbl.499.13001

Bochner, S.

[1943] Analytic and meromorphic continuation by means of Green's formula. Ann. Math., II. Ser. *44*, 652–673. Zbl.60,242

[1952] Partial Differential Equations and Analytic Continuation. Proc. Natl. Acad. Sci. U.S.A. *38*, 227–230. Zbl.46,99

Bony, J.M.

[1976] Propagation des singularités différentiables pour une classe d'opérateurs différentiels à coefficients analytiques. Astérisque *34–35*, 43–91. Zbl.344.35075

Bony, J.M., Schapira, P.

[1972] Existence et prolongement des solutions holomorphes des équations aux dérivées partielles. Invent. Math. *17*, 95–105. Zbl.225.35008

Borel, A.

[1987] Algebraic *D*-Modules. Boston: Academic Press. Zbl.642.32001

Borisevich, A.I., Lapin, G.P.

[1968] On the interpolation of entire functions. Sib. Mat. Zh. *9*, 522–529. English transl.: Sib. Math. J. *9*, 394–399 (1968). Zbl.183,340

Bose, N.K.

[1985] Multidimensional Systems Theory. Boston: Reidel. Zbl.562.00017

Brown, L., Kahane, J.P.

[1982] A note on the Pompeiu problem for convex domains. Math. Ann. *259*, 107–110. Zbl.464.30035

Brown, L., Schnitzer, F., Shields, A.L.

[1968] A note on a problem of *D*. Pompeiu. Math. Z. *105*, 59–61. Zbl.161,247

Brown, L., Schreiber, B.M., Taylor, B.A.

[1973] Spectral synthesis and the Pompeiu problem. Ann. Inst. Fourier *23*, 125–154. Zbl.265.46044

Brownawell, W.D.

[1987] Bounds for the degrees in the Nullstellensatz. Ann. Math., II. Ser. *126*, 577–591. Zbl.641.14001

Bruna, J.

[1987] Boundary absolute continuity of functions in the ball algebra. Springer Lecture Notes in Mathematics *1276*, 55–59. Zbl.623.32010

Bruna, J., Ortega, J.M.

[1984a] Idéaux fermes de type fini dans les algèbres $A^m(D)$. Springer Lecture Notes in Mathematics *1094*, 20–28. Zbl.582.32021

[1984b] Closed finitely generated ideals in algebras of holomorphic functions and smooth to the boundary in strictly pseudoconvex domains. Math. Ann. *268*, 137–157. Zbl.533.32002

Buchsbaum, D.

[1964] A generalized Koszul complex I. Trans. Am. Math. Soc. *111*, 183–196. Zbl.131,278

Carleman, T.

[1926] Les fonctions quasi-analytiques. Paris: Gauthier-Villars. Jbuch52,255

[1944] L'intégrale de Fourier et questions qui s'y rattachent. Uppsala: Almqvist and Wiksells. Zbl.60,255

Cartan, H.

[1950] Idéaux et modules de fonctions analytiques de variable complexes. Bull. Soc. Math. Fr. *78*, 29–64. Zbl.38,237

Charpentier, P.

[1980] Formules explicites pour les solutions minimales de l'équation $\bar{\partial}u = f$ dans la boule et le polydisque. Ann. Inst. Fourier *30*, 121–154. Zbl.425.32009

Choquet, G., Deny, J.

[1944] Sur quelques propriétés de moyenne caractéristiques des fonctions harmoniques et polyharmoniques. Bull. Soc. Math. Fr. *72*, 118–140.

Chou, C.C.

[1973] La transformation de Fourier complexe et l'équation de convolution. Springer Lecture Notes in Mathematics *325*. Zbl.257.46037

Čirka, E.M., Khenkin, G.M.
[1975] Boundary properties of holomorphic functions of several complex variables. Itogi Nauki Tekh., Ser. Sovrem. Probl. Mat. 4, 12–142. English transl.: J. Sov. Math. 5, 612–687 (1976). Zbl.335.32001

Coleff, N.R., Herrera, M.E.
[1978] Les courants résiduels associés à une forme méromorphe. Springer Lecture Notes in Mathematics 633. Zbl.371.32007

Cormack, A.M., Quino, E.T.
[1980] A Radon transform on spheres through the origin in $\mathbb{R}^n$ and applications to the Darboux equation. Trans. Am. Math. Soc. 260, 575–581. Zbl.444.44003

Curley, W.
[1972] On Beurling spaces (thesis). Stevens Inst. of Technology.

Cohoon, D.K.
[1972] Nonexistence of a continuous right inverse for linear partial differential operators with constant coefficients. Math. Scand. 29, 337–342. Zbl.235.35018

Delsarte, J.
[1935] Les fonctions moyenne-périodiques. J. Math. Pures Appl., IX. Ser. 14, 403–453. Zbl.13,254
[1958] Note sur une propriété nouvelle des fonctions harmoniques. C. R. Acad. Sci., Paris 246, 1358–1360. Zbl.84,94
[1960] Théories des fonctions moyenne-périodiques de deux variables. Ann. Math., II. Ser. 72, 121–178. Zbl.104,339
[1961] Lectures on Topics in Mean Periodic Functions and the Two-Radius Theorem. Bombay: Tata Institute of Fundamental Research.

Delsarte, J., Lions, J.L.
[1959] Moyennes généralisées. Comment. Math. Helv. 33, 59–69. Zbl.84,94

Demailly, J.P.
[1979] Fonctions holomorphes à croissance polynomiale sur la surface d'équation $e^x + e^y = 1$. Bull. Sci. Math., II. Ser. 103, 179–191. Zbl.412.32007
[1982a] Relations entre les différentes notions de fibres et de courants positifs. Springer Lecture Notes in Mathematics 919, 56–76. Zbl.481.32010
[1982b] Scindage holomorphe d'un morphisme de fibres vectorials semi-positifs avec estimations L^2. Springer Lecture Notes in Mathematics 919, 77–107. Zbl.481.32011

De Roever, J.W.
[1978] Complex Fourier Transformation and Analytic Functionals with Unbounded Carriers. Math. Centre Tracts 89, Amsterdam. Zbl.406.46032

Dickson, D.G.
[1957] Expansions in Series of Solutions of Linear Difference-Differential Equations and Infinite Order Differential Equations with Constant Coefficients. Mem. Am. Math. Soc. 23. Zbl.79,113
[1964] Analytic mean periodic functions. Trans. Am. Math. Soc. 110, 361–374. Zbl.124,288

Dienes, P.
[1957] Taylor Series. New York: Dover. Zbl.78,59

Dieudonné, J., Schwartz, L.
[1950] La dualité dans les espaces $(\mathcal{F})$ et $(\mathcal{LF})$. Ann. Inst. Fourier 1, 61–101. Zbl.35,355

Djakov, P.B., Mityagin, B.A.
[1980] The structure of polynomial ideals in the algebra of entire functions. Stud. Math. 68, 85–104. Zbl.434.46034

Dolbeault, P.
[1990]   General theory of multidimensional residues. Encyclopaedia of Mathematical Sciences, Vol. 7 (A.G. Vitushkin, ed.), Several complex variables I. Berlin, Heidelberg: Springer-Verlag.

Duistermaat, J.J., Hörmander, L.
[1972]   Fourier integral operators II. Acta Math. *128*, 183–269. Zbl.232.47055

Egorov, Yu.V.
[1969]   On canonical transformations of pseudo-differential operators. Usp. Mat. Nauk *24*, 235–236. Zbl.191,438

Eguchi, M., Hashizume, M., Okamoto, K.
[1973]   The Paley–Wiener theorem for distributions on symmetric spaces. Hiroshima Math. J. *3*, 109–120. Zbl.271.22007

Ehrenpreis, L.
[1954]   Solutions of some problems of division I. Am. J. Math. *76*, 883–903. Zbl.56,106
[1955a] Solutions of some problems of division II. Am. J. Math. *77*, 286–292. Zbl.64,115
[1955b] Mean Periodic Functions I. Am. J. Math. *77*, 293-328, 731–733. Zbl.68,317; Zbl.68,318
[1956]   Solutions of some problems of division III. Am. J. Math. *78*, 685–715. Zbl.72,328
[1957]   Sheaves and differential equations. Proc. Am. Math. Soc. *7*, 1131–1138. Zbl.77,92
[1958]   Some applications of the theory of distributions to severeal complex variables. Semin. Analytic Functions *1*, 65–79. Zbl.95,62
[1959]   Theory of infinite derivatives. Am. J. Math. *81*, 799–845. Zbl.103,330
[1960]   Solutions of some problems of division IV. Am. J. Math. *82*, 522–588. Zbl.98,84
[1961a] A fundamental principle for systems of linear partial differential equations with constant coefficients, and some of its applications. Proc. Intern. Symp. Linear Spaces, Jerusalem 1960, 161–174. Zbl.117,334
[1961b] A new proof and an extension of Hartogs' theorem. Bull. Am. Math. Soc. *67*, 507–509. Zbl.99,78
[1961c] Analytically uniform spaces and some applications. Trans. Am. Math. Soc. *101*, 52–74. Zbl.101,91
[1962]   Solutions of some problems of division V. Am. J. Math. *84*, 324–348. Zbl.107,328
[1967]   Complex Fourier transform technique in variable coefficient partial differential equations. J. Anal. Math. *19*, 75–95. Zbl.152,295
[1970]   Fourier Analysis in Several Complex Variables. New York: Wiley Interscience. Zbl.195,104
[1980]   Reflection, removable singularities, and approximation for partial differential equations I. Ann. Math., II. Ser. *112*, 1–20. Zbl.463.35005
[1986]   Lewy's operator and its ramifications. J. Funct. Anal. *68*, 329–365. Zbl.638.35002
[1987]   Reflection, removable singularities, and approximation for partial differential equations II. Trans. Am. Math. Soc. *302*, 1–45. Zbl.667.35005
[1992]   The Radon transform, book in preparation.

Ehrenpreis, L., Kawai, T.
[1982]   Poisson's summation formula and Hamburger's theorem. Publ. Res. Inst. Math. Sci. *18*, 833–846. Zbl.499.10042

Ehrenpreis, L., Malliavin, P.
[1974] Invertible operators and interpolation in AU-spaces. J. Math. Pures
Appl., IX. Ser. 53, 165–182. Zbl.331.46031

Ehrenpreis, L., Mautner, F.I.
[1955] Some properites of the Fourier transform on semi-simple Lie groups I.
Ann. Math., II. Ser. 61, 406–439. Zbl.66,357
[1957] Some properties of the Fourier transform on semi-simple Lie groups II.
Trans. Am. Math. Soc. 84, 1–55. Zbl.79,132
[1957] Some properties of the Fourier transform on semi-simple Lie groups III.
Trans. Am. Math. Soc. 90, 431–484. Zbl.86,99

El Mir, H.
[1984] Sur le prolongement des courants positifs fermés. Acta Math. 153, 1–45.
Zbl.557.32003

Ephron, M.
[1983] Indicators of products and surjective convolution maps. J. Math. Anal.
Appl. 91, 287–299. Zbl.538.30028

Epifanov, O.V.
[1969] On the character of the solution of a differential equation of infinite order.
Rostov: Compendium of Mathematical Analysis and its Applications 1,
72–82.
[1974] Solvability of a convoluton equation in convex domains. Mat. Zametki 15,
787-796. English transl.: Math. Notes 15, 472-477 (1974). Zbl.314.46050

Eskin, G.I., Vishik, M.I.
[1965] Convolution equations in a bounded domain. Usp. Mat. Nauk 20, No.
3, 89–152. English transl.: Russ. Math. Surv. 20, No. 3, 86–151 (1965).
Zbl.152,342
[1966] Convolution equations in a bounded domain in spaces with weighted
norms. Mat. Sb., Nov. Ser. 69 (111), 65–110. English transl.: Am. Math.
Soc., Transl., II. Ser. 67, 33–82 (1968). Zbl.152,343
[1967] Elliptic convolution equations in a bounded domain and their applications. Usp. Mat. Nauk 22, No. 1 (133), 15–76. English transl.: Russ.
Math. Surv. 22, 13–75 (1967). Zbl.167,448

Euler, L.
[1743] De integratione aequationum differentialium altiorum graduum. Misc.
Berol. 7, 193–242.

Fantappiè, L.
[1943] L'indictrice proiettiva dei funzionali lineari e i prodotti funzionali
proiettivi. Ann. Mat. Pura Appl., IV. Ser. 22, 181–289. Zbl.60,278

Flatto, L.
[1961] Functions with a mean value property. J. Math. Mech. 10, 11–18.
Zbl.97,306
[1963] Functions with a mean value property II. Am. J. Math. 85, 248–270.
Zbl.145,371
[1965] The converse of Gauss' theorem for harmonic functions. J. Differ. Equations 1, 483–490. Zbl.132,78

Flensted-Jensen, M.
[1972] Paley–Wiener-type theorems for a differential operator connected with
symmetric spaces. Ark. Mat. 10, 143–162. Zbl.233.42012

Forster, O., Ohsawa, T.
[1987] Complete intersections with growth conditions. Adv. Stud. Pure Math.
10, 91–104. Zbl.675.14022

Gårding, L.
[1950] Linear hyperbolic partial differential equations with constant coefficients.
        Acta Math. *85*, 1–62. Zbl.45,202
Gårding, L., Malgrange, B.
[1961] Opérateurs différentiels partiellement hypoelliptiques et partiellement
        elleptiques. Math. Scand. *9*, 5–21. Zbl.108,101
Gay, R.
[1976a] Division des fonctionelles analytiques et fonctions entières de type expo-
        nentiel de plusieurs variables (thesis). Strasbourg.
[1979b] Sur une problème de division des fonctionnelles analytiques. Application
        aux fonctions moyenne-périodiques. C. R. Acad. Sci., Paris, Ser. A *283*,
        835–838. Zbl.343.32011
[1980] Division des fonctionnelles analytiques. Application aux fonctions en-
        tières de type exponentiel moyenne-périodiques. Springer Lecture Notes
        in Mathematics *822*, 77–89. Zbl.449.35015
Gay, R., Sebbar, A.
[1985] Division et extension dans l'algèbre $A^\infty(\Omega)$ d'un ouvert pseudo-convexe à
        bord lisse de $\mathbb{C}^M$. Math. Z. *189*, 421–447. Zbl.547.32009
Gel'fand, S.I., Shilov, G.E.
[1964] Generalized Functions I. New York: Academic Press. Translation from the
        Russian original (1959). Zbl.91,111
[1968] Generalized Functions II. New York: Academic Press. Translation from
        the Russian original (1958). Zbl.91,111.
Gel'fond, A.O.
[1951] Linear Differential Equations of Infinite Order with Constant Coefficients
        and Asymptotic Periods of Entire Functions. Tr. Mat. Inst. Steklova *38*,
        42–67. English transl.: Am. Math. Soc., Transl. *84* (1953). Zbl.49,342
[1963] Calcul des différences finies. Paris: Hermann. Translation from the Rus-
        sian original (1952). Zbl.47,332
Gentili, G., Struppa, D.C.
[1987] Corona problems in spaces of entire functions with growth conditions.
        Complex Variables *9*, 41–48. Zbl.589.32006
Gleason, A.
[1963] The Cauchy–Weil Theorem. J. Math. Mech. *12*, 429–444. Zbl.122,320
Globevnik, J.
[1983] Analyticity on rotation invariant families of curves. Trans. Am. Math.
        Soc. *280*, 247–254. Zbl.575.30033
[1990] Zero integrals on circles and characterizations of harmonic and analytic
        functions. Trans. Am. Math. Soc. *317*, 313–330. Zbl.696.31001
Globevnik, J., Rudin, W.
[1988] A characterization of harmonic functions. Indagationes Math. *50*, 419–
        426. Zbl.705,31001
Goldschmidt, H.
[1985] The Radon transform for forms of type (1,1) on complex projective space.
        Math. Ann. *271*, 485–492. Zbl.542.53038
Golovin, V.D.
[1963] Mean periodic functions. Dokl. Akad. Nauk SSSR *150*, 17–20. English
        transl.: Sov. Math., Dokl. *4*, 569–572 (1963). Zbl.161,341
Grauert, H., Lieb, I.
[1970] Das Ramirezsche Integral und die Lösung der Gleichung $\bar{\partial}f = \alpha$ im
        Bereich der beschränkten Formen. Rice Univ. Stud. *56*, No. 2, 29–50.
        Zbl.217,392

Grinberg, E.

[1985]	On images of Radon transforms. Duke J. Math. *52*, 939–972.
	Zbl.623.44005

[1987]	Euclidean Radon transforms: ranges and restrictions. Integral Geometry, Proc. Conf., Brunswick/Maine 1984, Cont. Math. *63*, 109–133.
	Zbl.626.47050

Grudzinski, O.V.

[1976]	Einige elementare Ungleichungen für Exponentialpolynome. Math. Ann. *221*, 9–34. Zbl.318.33002

Gruman, L.

[1984]	Solutions of difference equations with nonconstant coefficients. Springer Lecture Notes in Mathematics *1094*, 84–138. Zbl.551.39001

Gruman, L., Lelong, P.

[1986]	Entire Functions of Several Complex Variables. New York: Springer-Verlag. Zbl.583.32001

Gurevich, D.I.

[1971]	On a problem of L. Schwartz. Funkts. Anal. Prilozh. *5*, No. 1, 76–77. English transl.: Funct. Anal. Appl. *5*, 63–64 (1971). Zbl.245.46050

[1972]	Generalized bases in some rings of holomorphic functions. Izv. Akad. Nauk SSSR, Ser. Mat. *36*, 568–582. English transl.: Math. USSR, Izv. *6*, 564–578 (1973). Zbl.264.46023

[1974a]	Closed ideals with exponential-polynomial generators in rings of entire functions of two variables. Izv. Akad. Nauk Armjan. SSR, Mat. *9*, 459–472. Zbl.304.46032

[1974b]	Closed ideals with the zero-dimensional root set in certain rings of holomorphic functions. Zap. Nauchn. Semin. Leningr. Otd. Mat. Inst. Steklova *47*, 55–66. English transl.: J. Sov. Math. *9*, 172–182 (1978). Zbl.372.46029

[1975]	Counterexamples to a problem of L. Schwartz. Funkts. Anal. Prilozh. *9*, No. 2, 29–35. English transl.: Funct. Anal. Appl. *9*, 116–120 (1975). Zbl.326.46020

Hadamard, J.

[1923]	Lectures on Cauchy's Problem in Linear Partial Differential Equations. New Haven: Yale University Press. Jbuch49,725

Hamburger, H.

[1921]	Über die Riemannsche Funktionalgleichung der ζ-Funktion. Math. Z. *10*, 240–254. Jbuch48,1210

Hansen, S.

[1982]	On the Fundamental Principle of L. Ehrenpreis (Habilitationsschrift). Paderborn. See also: Banach Cent. Publ. *10*, 185–201 (1983). Zbl.555.35009

[1981]	Localizable Analytically Uniform Spaces and the Fundamental Principle. Trans. Am. Math. Soc. *264*, 235–250. Zbl.482.46023

Hartogs, F.

[1906]	Einige Folgerungen aus der Cauchyschen Integralformel bei Funktionen mehrerer Veränderlicher. Münchener Sitz. Ber. *36*, 223–242. Jbuch37,443

Haslinger, F.

[1986]	Weighted spaces of entire functions. Indiana Univ. Math. J. *35*, 193–208. Zbl.559.46012

Haslinger, F., Smejkal, M.
[1987] Representation and duality in weighted Fréchet spaces of entire functions.
Springer Lecture Notes in Mathematics *1275*, 168–196.
Zbl.648.46025

Helgason, S.
[1962] Differential Geometry and Symmetric Spaces. New York: Academic Press.
Zbl.111,181
[1965] The Radon transform on Euclidean spaces, compact two-point homogeneous spaces and Grassmann manifolds. Acta Math. *113*, 153–180.
Zbl.163,166
[1970] A duality for symmetric spaces with applications to group representations.
Adv. Math. *5*, 1–154. Zbl.209,254
[1973] The surjectivity of invariant differential operators on symmetric spaces I.
Ann. Math., II. Ser. *98*, 451–479. Zbl.274.43013
[1980] The Radon Transform. Boston: Birkhäuser. Zbl.453.43011
[1981] The X-ray transform on a symmetric space. Springer Lecture Notes in Mathematics *838*, 145–148. Zbl.437.53036
[1984] Groups and Geometric Analysis. Integral geometry, invariant differential operators, and spherical functions. New York: Academic Press.
Zbl.543.58001

Henkin, G.M. (= Khenkin), Leiterer, J.
[1981] Global integral formulas for solving the $\bar{\partial}$-equation on Stein manifolds.
Ann. Pol. Math. *39*, 93–116. Zbl.477.32020
[1984] Theory of Functions on Complex Manifolds. Basel: Birkhäuser.
Zbl.573.32001

Hermann, G.
[1929] Die Frage der endlich vielen Schritte in der Theorie der Polynomideale.
Math. Ann. *95*, 736–788. Jbuch52,127

Hertle, A.
[1984] On the range of the Radon transform and its dual. Math. Ann. *267*, 91–99. Zbl.554.46016

Herrera, M.E., Lieberman, D.I.
[1971] Residues and principal values on complex spaces. Math. Ann. *194*, 259–294. Zbl.224.32012

Hickel, M.
[1986] Quelques résultats de division dans $A^{\infty}(\Omega)$. Preprint. Bordeaux. Appeared in: Ann. Sc. Norm. Super. Pisa, Cl. Sci., IV. Ser. *15*, No. 1, 35–63 (1988). Zbl.683.46040
[1988] Sur une problème de division dans l'algèbre $A^{\infty}(\Omega)$ d'un ouvert faiblement pseudo-convexe de $\mathbb{C}^2$. Math. Z. *197*, 201–218. Zbl.617.32024

Hironaka, H.
[1964] Resolution of singularities of an algebraic variety I, II. Ann. Math., II.
Ser. *79*, 109–326. Zbl.122,386

Holmgren, E.
[1901] Über Systeme von linearen partiellen Differentialgleichungen. Öfversigt af Vetenskaps Akad. Förh. *58*, 91–105. Jbuch32,357

Hörmander, L.
[1955a] On the theory of general partial differential operators. Acta Math. *94*, 161–248. Zbl.67,322
[1955b] La transformation de Legendre et la théorème de Paley–Wiener. C. R.
Acad. Sci., Paris *240*, 392–395. Zbl.64,103
[1958] On the division of distributions by polynomials. Ark. Mat. *3*, 555–568.
Zbl.131,119

[1961] Hypoelliptic differential operators. Ann. Inst. Fourier *11*, 477–492.
        Zbl.99,301

[1962] On the range of convolution opertors. Ann. Math., II. Ser. *76*, 148–170.
        Zbl.109,85

[1963a] Linear Partial Differential Operators. Berlin: Springer-Verlag. Zbl.108,93

[1963b] Supports and singular supports of convolutions. Acta Math. *110*, 279–302.
        Zbl.188,194

[1965] L^2-estimates and existence theorems for the $\bar\partial$-operator. Acta Math. *113*,
        89–152. Zbl.158,110

[1967] Generators for some rings of analytic functions. Bull. Am. Math. Soc. *73*,
        943–949. Zbl.172,417

[1968] Convolution equations in convex domains. Invent. Math. *4*, 306–317.
        Zbl.229.44008

[1971a] Uniqueness theorems and wave front sets for solutions of linear differential
        equations with analytic coefficients. Commun. Pure Appl. Math. *24*, 671–
        704. Zbl.226.35019

[1971b] Fourier integral operators I. Acta Math. *127*, 79–183. Zbl.212,466

[1973] An Introduction to Complex Analysis in Several Variables. 2nd ed. New
        York: Elsevier. Zbl.271.32001; Zbl.138,62

[1983] The Analysis of Linear Partial Differential Operators I, II. Berlin:
        Springer-Verlag. Zbl.521.35001; Zbl.521.35002

[1985] The Analysis of Linear Partial Differential Operators III, IV. Berlin:
        Springer-Verlag. Zbl.601.35001; Zbl.612.35001

Horváth, J.
[1966] Topological Vector Spaces and Distributions I. New York: Addison-
        Wesley. Zbl.143,151

Ishimura, R.
[1985] Existence locale de solutions holomorphes pour les équations différen-
        tielles d'ordre infini. Ann. Inst. Fourier *35*, No. 3, 49–57. Zbl.544.35085

Jacobi, K.G.J.
[1835] Theoremata nova algebraica circa systema duarum aequationum inter
        duas variables propositorum. Crelle J. *14*, 281–288.

Jennane, B.
[1978] Extension d'une fonction défini sur une sous-variété avec contrôle de
        la croissance. Springer Lecture Notes in Mathematics *694*, 126–133.
        Zbl.403.32008

John, F.
[1949] On linear partial differential equations with analytic coefficients. Com-
        mun. Pure Appl. Math. *2*, 209–253. Zbl.35,346

[1950] The fundamental solution of linear elliptic differential equations with ana-
        lytic coefficients. Commun. Pure Appl. Math. *3*, 273–304. Zbl.41,62

[1951] General properties of solutions of linear elliptic partial differential equa-
        tions. Proc. Symp. Spectral Theor. Differ. Probl., Stillwater, 113–175.

[1955] Plane Waves and Spherical Means Applied to Partial Differential Equa-
        tions. New York: Interscience. Zbl.67,321

Kahane, J.P.
[1957a] Sur les fonctions moyenne-périodiques bornées. Ann. Inst. Fourier *7*, 293–
        314. Zbl.83,344

[1957b] Lectures on Mean Periodic Functions. Bombay: Tata Institute of Funda-
        mental Research. Zbl.99,323

Kataoka, K.
[1981] On the theory of Radon transformations of hyperfunctions. J. Fac. Sci.,
        Univ. Tokyo, Sect. I A *28*, 331–413. Zbl.576.32008

Kaneko, A.
[1970] On continuation of regular solutions of partial differential equations to compact convex sets. J. Fac. Sci., Univ. Tokyo, Sect. I A *17*, 567–580. Zbl.218.35022
[1972] On continuation of regular solutions of partial differential equations to compact convex sets II. J. Fac. Sci., Univ. Tokyo *18*, 415–433. Zbl.235.35001
[1973] Fundamental Principle and extension of solutions of linear differential equations with constant coefficients. Springer Lecture Notes in Mathematics *287*, 122–134. Zbl.259.35017
Kashiwara, M.
[1971] Algebraic study of systems of partial differential equations (Master's thesis). Tokyo: University of Tokyo.
[1976] *B*-functions and holonomic systems. Invent. Math. *38*, 33–53. Zbl.354.35082
[1986] Introduction to microlocal analysis. Enseign. Math., II. Ser. *32*, 227–259. Zbl.632.58030
Kashiwara, M., Oshima, T.
[1977] Systems of differential equations with regular singularities and their boundary value problems. Ann. Math., II. Ser. *106*, 145–200. Zbl.358.35073
Kashiwara, M., Schapira, P.
[1979] Micro-hyperbolic systems. Acta Math. *142*, 1–55.
Kashiwara, M., Kawai, T.
[1983] Microlocal Analysis. Publ. Res. Inst. Math. Sci. *19*, 1003–1032. Zbl.536.58030
Kashiwara, M., Kawai, T., Kimura, T.
[1986] Foundations of Algebraic Analysis. Princeton: Princeton University Press. Zbl.605.35001
Kawai, T.
[1970] On the Fourier hyperfunctions and its applications to partial differential equations with constant coefficients. J. Fac. Sci., Univ. Tokyo, Sect. I A *17*, 467–517. Zbl.212,461
[1971] Construction of local elementary solutions for linear partial differential operators with real analytic coefficients I. Publ. Res. Inst. Math. Sci. *7*, 363–397. Zbl.216,123
[1972] On the global existence of real analytic solutions of linear differential equations I. J. Math. Soc. Japan *24*, 481–517. Zbl.234.35012
[1983] An example of a complex of linear differential operators of infinite order. Proc. Japan Acad., Ser. A *59*, 113–115. Zbl.525.35082
[1984] The Fabry–Ehrenpreis gap theorem for hyperfunctions. Proc. Japan Acad., Ser. A *60*, 276–278. Zbl.554.46017
[1986a] Systems of microdifferential equations of infinite order. Hyperbolic equations and related topics, Proc. Taniguchi: Int. Symp. Katata and Kyoto 1984, 143–154.
[1986b] On the global existence of real analytic solutions and hyperfunction solutions of linear differential equations. Proc. Japan Acad., Ser. A *62*, 77–79. Zbl.618.35002
[1987a] The Fabry–Ehrenpreis gap theorem and linear differential equations of infinite order. Am. J. Math. *109*, 57–64. Zbl.618.30003
[1987b] Systems of linear differential equations of infinite order: an aspect of infinite analysis. Prepubl. RIMS *599*. See also Proc. Symp. Pure Math. *49*, I, 3–17 (1989). Zbl.694.35225

Kawai, T., Struppa, D.C.
[1990] On the existence of holomorphic solutions of systems of linear differential
        equations of infinite order with constant coefficients. Int. J. Math. *1*, 63-
        82. Zbl.704.35075
Kawai, T., Takei, Y.
[1986] On a closed range property of a linear differential operator. Proc. Japan
        Acad., Ser. A *62*, 386–388. Zbl.607.35033
Kelleher, J.J.
[1966] An application of the corona theorem to rings of entire functions. AMS
        Summer Institute, San Diego.
Kelleher, J.J., Taylor, B.A.
[1967] An application of the corona theorem to some rings of entire functions.
        Bull. Am. Math. Soc. *73*, 246–249. Zbl.154,150
[1971a] Finitely generated ideals in rings of analytic functions. Math. Ann. *193*,
        225–237. Zbl.207,129
[1971b] On finitely generated modules over some rings of analytic functions.
        Preprint.
[1972b] Closed ideals in locally convex algebras of analytic functions. J. Reine
        Angew. Math. *255*, 190–209. Zbl.237.46052
Kellogg, O.D.
[1934] Converses of Gauss' theorem on the arithmetic mean. Trans. Am. Math.
        Soc. *36*, 227–242. Zbl.9,112
Khenkin, G.M. (= Henkin)
[1970] Integral representation of functions in strongly pseudoconvex domains and
        applications to the $\bar{\partial}$-problem. Mat. Sb., Nov. Ser. *82*, 300–308. English
        transl.: Math. USSR Sb. *11*, 273–281 (1970). Zbl.206,91
[1990] The method of integral representations in complex analysis, Encyclopae-
        dia of Mathematical Sciences, Vol. 7 (A.G. Vitushkin, ed.), Several com-
        plex variables I. Berlin, Heidelberg: Springer-Verlag, 19–113.
Khenkin, G.M., Mityagin, B.S.
[1971] Linear problems of complex analysis. Usp. Mat. Nauk *26*, No. 4, 93–152.
        English transl.: Russ. Math. Surv. *26*, No. 4, 99–164 (1972). Zbl.245,46027
Kiselman, C.O.
[1967] On entire functions of exponential type and indicators of analytic func-
        tionals. Acta Math. *117*, 1–35. Zbl.152,76
[1969] Prolongement des solutions d'une équation aux dérivées partielles à coeffi-
        cients constants. Bull. Soc. Math. Fr. *97*, 329–356. Zbl.189,405
Kohn, J.J.
[1961] Solution of the $\bar{\partial}$ anti-Neumann problem on strongly pseudoconvex mani-
        folds. Proc. Natl. Acad. Sci. U.S.A. *47*, 1198–1202. Zbl.123,78
Kohn, J.J., Nirenberg, L.
[1973] A pseudoconvex domain not admitting a holomorphic support function.
        Math. Ann. *201*, 265–268. Zbl.248.32013
Kóllar, J.
[1988] Sharp effective Nullstellensatz. J. Am. Math. Soc. *1*, 963–975.
        Zbl.682.14001
Komatsu, H.
[1967] Projective and injective limits of weakly compact sequences of locally con-
        vex spaces. J. Math. Soc. Japan *19*, 366–383. Zbl.168,106
[1968] Resolution by hyperfunctions of sheaves of solutions of differential equa-
        tions with constant coefficients. Math. Ann. *176*, 77–86. Zbl.161,298
[1973a] Relative cohomology of sheaves of solutions of differential equations.
        Springer Lecture Notes in Mathematics *287*, 192–261. Zbl.278.58010

[1973b] Ultradistributions I: Structure theorems and a characterization. J. Fac. Sci., Univ. Tokyo, Sect. I A *20*, 25–105. Zbl.258.46039

[1977] Ultradistributions II: The kernel theorem and ultradistributions with support in a manifold. J. Fac. Sci., Univ. Tokyo, Sect. I A *24*, 607–628. Zbl.385.46027

Koppelman, W.

1967] The Cauchy integral for differential forms. Bull. Am. Math. Soc. *73*, 554–556. Zbl.186,138

Krasichkov-Ternovskij, I.F.

[1966] Closed ideals in a locally convex algebra of entire functions with arbitrary growth majorant. Dokl. Akad. Nauk SSSR *170*, 1018–1019. English transl.: Sov. Math., Dokl. *7*, 1324–1325 (1966). Zbl.169,178

[1967] Closed ideals in locally convex algebras of entire functions I. Izv. Akad. Nauk SSSR, Ser. Mat. *31*, 37–60. English transl.: Math. USSR, Izv. *1*, 35–55 (1967). Zbl.152,132

[1968a] Closed ideals in locally convex algebras of entire functions II. Izv. Akad. Nauk SSSR, Ser. Mat. *32*, 1024–1032. English transl.: Math. USSR, Izv. *2*, 979–986 (1968). Zbl.169,178

[1968b] Closed ideals in locally convex algebras of entire functions. Minimal-type algebras. Sib. Mat. Zh. *9*, 77–96. English transl.: Sib. Math. J. *9*, 59–71 (1968). Zbl.161,108

[1972a] Invariant subspaces of analytic functions I: Spectral analysis on convex domains. Mat. Sb. *87*, 459–489. English transl.: Math. USSR, Sb., Nov. Ser. *16*, 471–500 (1973). Zbl.253.46041

[1972b] Invariant subspaces of analytic functions II: Spectral synthesis on convex domains. Mat. Sb., Nov. Ser. *88*, 3–30. English transl.: Math. USSR, Sb. *17*, 1–29 (1973). Zbl.253.46042

[1972c] Invariant subspaces of analytic functions III: Extension of spectral synthesis. Mat. Sb., Nov. Ser. *88*, 331–352. English transl.: Math. USSR, Sb. *17*, 327–348 (1973). Zbl.253.46043

[1973a] Invariant subspaces of analytic functions: Analytic continuation. Izv. Akad. Nauk SSSR, Ser. Mat. *37*, 931–945. English transl.: Math. USSR, Izv. *7*, 933–947 (1974). Zbl.343.46019

[1973b] Invariant subspaces of analytic functions: Dirichlet coefficients. Funkts. Anal. Prilozh. *7*, No. 4, 38–43. English transl.: Funct. Anal. Appl. *7*, 285–289 (1973). Zbl.343.46020

Kuchment, P.A.

[1981] Representations of solutions of invariant differential equations on certain symmetric spaces. Dokl. Akad. Nauk SSSR *259*, 532–535. English transl.: Sov. Math., Dokl. *24*, 104–106 (1981). Zbl.497.58025

[1982] Functions on symmetric spaces which are periodic in the mean. Funkts. Anal. Prilozh. *16*, No. 3, 68–69. English transl.: Funct. Anal. Appl. *16*, 213–214 (1983). Zbl.509.43007

[1985a] Spherical representation of solutions of invariant differential equations on a Riemannian symmetric space of noncompact type. Izv. Akad. Nauk SSSR, Ser. Mat. *49*, 1260–1273. English transl.: Math. USSR, Izv. *27*, 535–548 (1986). Zbl.593.58044

[1985b] Spectral synthesis in spaces of solutions of differential equations invariant with respect to groups of transformations. Primen. Topol. Sovrem. Anal., Nov. Global'nom Anal. 1985, 87–105. English transl.: Springer Lecture Notes in Mathematics *1214*, 85–100 (1986). Zbl.574.58029

Lazard, D.

[1977] Algèbre linéaire sur $k[x_1, \ldots, x_n]$ et élimination. Bull. Soc. Math. Fr. *105*, 165–190. Zbl.447.13008

Lelong, P.

[1964]   Les fonctions plurisousharmoniques. Ann. Sci. Ec. Norm. Supér., III. Ser.
         62, 301–338. Zbl.61,232

[1964]   Fonctions entières (n-variables) et fonctions plurisousharmoniques d'ordre
         fini dans $\mathbb{C}^n$. J. Anal. Math. 12, 365–407. Zbl.126,296

[1967]   Plurisubharmonic Functions and Positive Differential Forms. New York:
         Gordon and Breach. Zbl.192,201; Zbl.195,116

[1968]   Fonctionnelles Analytiques et Fonctions Entières (n Variables). Montreal:
         Presses de l'Université de Montréal. Zbl.194,388

Leont'ev, A.F.

[1949]   Differential-difference equations. Mat. Sb., Nov. Ser. 24, 347-374.
         Zbl.41,423

[1951]   Series of Dirichlet polynomials and their generalizations. Tr. Mat. Inst.
         Steklova 39, 214pp. Zbl.45,351.

[1964]   Differential equations of infinite order and their applications. Tr. IV
         Vsesoyuz. Mat. S'ezda 2, 648–660. Zbl.217,400

[1966a] Mean-periodic functions. 3rd Summer Math. School, Constructive Func-
         tion Theory, Kiev, 84–134.

[1966b] Representation of arbitrary analytical functions by Dirichlet series and
         more general series. Sovrem. Probl. Teor. Anal. Funkts., Konf. Teor.,
         Erevan 1965, 209–216. Zbl.172,94

[1969]   Representation of functions by generalized Dirichlet series. Usp. Mat.
         Nauk 24, No. 2, 97–164. English transl.: Russ. Math. Surv. 24, No. 2,
         101–178 (1969). Zbl.176,24

[1970]   Representation of analytic functions in an open region by Dirichlet se-
         ries. Mat. Sb., Nov. Ser. 81, 552–579. English transl.: Math. USSR, Sb.
         10, 503–530 (1970). Zbl.194,104

[1971]   Summation of a Dirichlet series with complex exponents and its applica-
         tions. Tr. Mat. Inst. Steklova 112, 300–326. English transl.: Proc. Steklov
         Inst. Math. 112, 310–338 (1973). Zbl.226.30006

[1972]   On conditions of expandibility of analytic functions in Dirichlet series.
         Izv. Akad. Nauk SSSR, Ser. Mat. 36, 1282–1295. English transl.: Math.
         USSR Izv. 6, 1265–1277 (1972). Zbl.244.30002

Leray, J.

1956]   Fonctions des variables complexes: sa représentation comme somme de
         puissances négatives de fonctions linéaires. Rend. Acc. Naz. Lincei 20,
         589–590. Zbl.71,296

[1959]   Le calcul différentiel et intégral sur une variété analytique complexe. Bull.
         Soc. Math. Fr. 87, 81–180. Zbl.199,412

[1961–62] Séminaire sur les équations aux dérivées partielles. Collège de France.

Levin, B.Ya.

[1964]   Distribution of Zeros of Entire Functions. Providence, R.I.: Am. Math.
         Soc. Zbl.152,67

Levinson, N.

[1940]   Gap and Density Theorems. New York: Am. Math. Soc. Zbl.26,216

Lewy, H.

[1957]   An example of a smooth linear partial differential operator without solu-
         tions. Ann. Math., II. Ser. 66, 155–158. Zbl.78,81

Lieb, I.
[1969]   Ein Approximationssatz auf streng pseudokonvexen Gebieten. Math. Ann.
          184, 56–60. Zbl.181,89
[1970–71] Die Cauchy–Riemannschen Differentialgleichungen auf streng pseu-
          dokonvexen Gebieten. Math. Ann. *190*, 6–44. Zbl.199,427
[1972]   Die Cauchy–Riemannschen Differentialgleichungen auf streng pseu-
          dokonvexen Gebieten: Stetige Randwerte. Math. Ann. *199*, 241–256.
          Zbl.231.35055
Lieb, I., Range, M.
[1980]   Lösungsoperatoren für den Cauchy–Riemann Komplex mit C^k-Ab-
          schätzungen. Math. Ann. *253*, 145–164. Zbl.441.32007
Liess, O.
[1972]   On the Fundamental Principle of Ehrenpreis–Palamodov. Manuscript.
Lojasiewicz, S.
[1959]   Sur le problème de division. Stud. Math. *18*, 87–136. Zbl.115,102
Ludwig, D.
[1966]   The Radon transform on Euclidean space. Commun. Pure Appl. Math.
          19, 49–81. Zbl.134,113
Malgrange, B.
[1956]   Existence et approximation des solutions des équations aux dérivées par-
          tielles et des équations de convolution. Ann. Inst. Fourier *6*, 271–355.
          Zbl.71,90
[1961]   Sur les équations de convolution. Rend. Semin. Mat. Univ. Politec Torino
          19, 19–27. Zbl.96,321
[1964]   Systèmes differentiels à coefficients constants. Sémin. Bourbaki *15*, Exp.
          No. 246. Zbl.141,273
[1961–62] Sur les systèmes différentiels à coefficients constants. In: Séminaire sur
          les Equations aux Dérivées Partielles. Collège de France.
[1974a]   Sur les polynômes de Bernstein. Usp. Mat. Nauk *29*, No. 4, 81–88. English
          transl.: Russ. Math. Surv. *29*, No. 4, 81–88 (1974). Zbl.308.32008
[1974b]   Sur les points singuliers des équations différentiels. Enseign. Math., II.
          Ser. *20*, 147–176. Zbl.299.34011
[1976]   Le polynôme de Bernstein d'une singularité isolée. Springer Lecture Notes
          in Mathematics *459*, 98–119. Zbl.308.32007
[1978]   Algebraic aspects of the theory of partial differential equations. Enseign.
          Math. *24*, 179–188. Zbl.395.35082
Mandelbrojt, S.
1935]   Séries de Fourier et classes quasianalytiques de fonctions. Paris:
          Gauthier-Villars. Zbl.13,110
[1936]   Séries lacunaires. Paris: Hermann. Zbl.13,270
[1952]   Séries adhérents, régularization des suites, applications. Paris: Hermann.
          Zbl.48,52
[1972]   Dirichlet Series: Principles and Methods. Dordrecht: Reidel.
          Zbl.241.30010
Martineau, A.
[1961]   Les hyperfonctions de M. Sato. Sémin. Bourbaki 13 *214*, 1–13.
[1963]   Sur les fonctionnelles analytiques et la transformation de Fourier–Borel.
          J. Anal. Math. *11*, 1–164. Zbl.124,318
[1964a]   Distributions et valeurs au bord des fonctions holomorphes. Theory of
          Distributions, Proc. Int. Summer School, Lisbon, 195–326.
[1964b]   Equations différentielles d'ordre infini. Deuxième Colloque Anal. Fonct.,
          37–47. Zbl.138,380

[1968]   Unicité du support d'une fonctionnelle analytique: un théorème de C. O. Kiselman. Bull. Sci. Math., II. Ser. *91*, 131–141. Zbl.189,366

[1970]   Les supports des fonctionnelles analytiques. Springer Lecture Notes in Mathematics *116*, 175–195. Zbl.198,482

Martinelli, E.

[1938]   Alcuni teoremi integrali per le funzioni analitiche di più variabili complesse. Mem. Accad. Ital. *9*, 269–283. Zbl.22,240

Maslov, V.P.

[1972]   Théorie des Perturbations et Méthodes Asymptotiques. Paris: Dunod. Zbl.247.47010

Masser, D.W.

[1981]   On polynomials and exponential polynomials in several complex variables. Invent. Math. *63*, 81–95. Zbl.436.32005

Masser, D.W., Wüstholz, G.

[1983]   Fields of large transcendence degree generated by values of elliptic functions. Invent. Math. *72*, 407–464. Zbl.516.10027

May, D.C.

[1941]   An integral formula for analytic functions of n variables with some applications (thesis). Princeton University.

Mebkhout, Z.

[1979]   Cohomologie locale des espaces analytiques complexes. Point de vue des $\mathcal{D}_x$-modules (thesis). Paris. Zbl.455.32006

Meise, R.

[1985]   Sequence space representations for (DFN)-algebras of entire functions modulo closed ideals. J. Reine Angew. Math. *363*, 59–95. Zbl.574.46043

Meise, R., Momm, S., Taylor, B.A.

[1987]   Splitting of slowly decreasing ideals in weighted algebras of entire functions. Springer Lecture Notes in Mathematics *1276*, 229–252. Zbl.636.46021

Meise, R., Schwerdtfeger, K., Taylor, B.A.

[1986]   Kernels of slowly decreasing convolution operators. Doga Tr. J. Math. *10*, 176–197.

Meise, R., Taylor, B.A.

[1987]   Splitting of closed ideals in (DFN)-algebras of entire functions and the property (DN). Trans. Amer. Math. Soc. *302*, 341–370. Zbl.621.32022

[1988]   Each nonzero convolution operator on the entire function admits a continuous linear right inverse. Math. Z. *197*, 139–152. Zbl.618.32014

[1989]   Linear extension operators for ultradifferentiable functions of Beurling type on compact sets. Am. J. Math. *111*, 309–337. Zbl.696.46001

Meise, R., Taylor, B.A., Vogt, D.

[1987a]   Equivalence of slowly decreasing conditions and local Fourier expansions. Indiana Univ. Math. J. *36*, 729–756. Zbl.637.46037

[1987b]   Characterization of convolution operators on spaces of C^∞-functions which admit a continuous linear right inverse. Math. Ann. *279*, 141–155. Zbl.607.42011

[1988]   Caractérization des opérateurs linéaires aux dérivées partielles avec coefficients constants sur $\mathcal{E}(\mathbb{R}^N)$ admettant un inverse à droite qui est linéaire et continu. C. R. Acad. Sci., Paris, Ser. I *307*, 239–242. Zbl.649.46031

Meril, A.

[1983a]   Analytic functionals with unbounded carriers and mean periodic functions. Trans. Am. Math. Soc. *278*, 115–136. Zbl.519.46048

[1983b]   Problèmes d'interpolation dans quelques espaces de fonctions non entières. Bull. Soc. Math. Fr. *111*, 251–285. Zbl.535.46020

[1983c] Fonctionnelles analytiques à porteur non borné sur $\mathbb{C}$. Tokyo J. Math. *6*,
        447–472. Zbl.535.46021
Meril, A., Struppa, D.C.
[1985] Equivalence of Cauchy problems for entire and exponential-type functions.
        Bull. Lond. Math. Soc. *17*, 469–473. Zbl.561.35007
[1987a] Convolutors in spaces of holomorphic functions. Springer Lecture Notes in
        Mathematics *1276*, 253–275. Zbl.628.42008
[1987b] Phénomène de Hartogs et équations de convolution. Springer Lecture
        Notes in Mathematics *1295*, 146–156. Zbl.631.32012
Meril, A., Yger, A.
[1992] Problèmes de Cauchy globaux. Bull. Soc. Math. Fr. *120*, 87–111.
Meyer, Y.
[1976] Remarques sur une théorème de J. Delsarte. Ann. Inst. Fourier *26*, 133–
        152. Zbl.318.42028
Mitrinovič, D.S., Kečkic, J.D.
[1984] The Cauchy Method of Residues. Dordrecht: Reidel. Zbl.405.30030;
        Zbl.546.30004
Morimoto, M.
[1978] Analytic functionals with noncompact carrier. Tokyo J. Math. *1*, 77–103.
        Zbl.384.46027
Morzhakov, V.V.
[1974] Convolution equations in spaces of functions holomorphic in convex do-
        mains and on convex compacta in $\mathbb{C}^n$. Mat. Zametki *16*, 431–440. English
        transl.: Math. Notes *16*, 846–851 (1975). Zbl.313.45009
Muñoz Diaz, J.
[1972] Caracterizatión de las álgebras diferenciabiles y síntesis espectral para
        modulos sobre tales algebras. Collect. Math. *23*, 17–83. Zbl.274.46037
Myshkis, A.D.
[1949] General theory of differential equations with a retarded argument. Usp.
        Mat. Nauk *4*, 99–141. Zbl.35,178
Naftalevich, A.
[1962] Application of the iteration method for the solution of a difference equa-
        tion. Mat. Sb., Nov. Ser. *57*, 151–178. Zbl.123,273
Nagel, A.
[1974] On algebras of holomorphic functions with C^∞-boundary values. Duke
        Math. J. *41*, 527–535. Zbl.291.32023
[1976] Smooth zero sets and interpolation sets for some algebras of holomorphic
        functions on strictly pseudoconvex domains. Duke Math. J. *43*, 323–348.
        Zbl.343.32016
Nagel, A., Rudin, W.
[1978] Local boundary behavior of bouded holomorphic functions. Can. J. Math.
        30, 583–592. Zbl.427.32006
Nakano, S.
[1955] On complex analytic vector bundles. J. Math. Soc. Japan *7*, 1–12.
        Zbl.68,344
[1986] Extensions of holomorphic functions with growth conditions. Publ. Res.
        Inst. Math. Sci. *22*, 247–258. Zbl.599.32011
Napalkov, V.V.
[1973] Subspaces of entire functions of exponential type, invariant under shifts.
        Sib. Mat. Zh. *14*, 427–436. English transl.: Sib. Math. J. *14*, 294–300
        (1973). Zbl.256.46042
[1974] On one class of inhomogeneous convolution-type equations. Usp. Math.
        Nauk *29*, 217–218. Zbl.318.45006

[1979a] Convolution equations in multidimensional spaces. Mat. Zametki *25*, 761–774. English transl.: Math. Notes *25*, 393–400 (1979). Zbl.403.45005

[1979b] Systems of infinite-order nonhomogeneous partial differential equations. Mat. Zametki *26*, 217–226. English transl.: Math. Notes *26* (1979), 600–605 (1980). Zbl.425.35030

Narasimhan, R.

[1970] Cohomology with bounds in complex spaces. Lect. Notes Math. *155*, 141–150. Zbl.207,380

Natterer, F.

[1986] The Mathematics of Computerized Tomography. Stuttgart: B. G. Teubner; Chichester: John Wiley & Sons. Zbl.617.92001

Newman, D.J., Shapiro, H.S.

[1968] Fischer spaces of entire functions. Proc. Symp. Pure Math. *11*, 360–369. Zbl.191,415

Nikol'skij, N.K.

[1968] Closed ideals in certain algebras of entire functions. Sib. Mat. Zh. *9*, 211–215. English transl.: Sib. Math. J. *9*, 160–162 (1968). Zbl.161,108

[1974] Invariant subspaces in the theory of operators and theory of functions. Itogi Nauki Tekh., Ser. Mat. Anal. *12*, 199–412. English transl.: J. Sov. Math. *5*, 129–249 (1976). Zbl.333.47003

Nishimura, Y.

[1980] Problème d'extension dans la théorie des fonctions entières d'ordre fini. J. Math. Kyoto Univ. *20*, 635–650.

Ohsawa, T.

[1983] Cohomology vanishing theorems on weakly 1-complete manifolds. Publ. Res. Inst. Math. Sci. *19*, 1181–1201. Zbl.537.32014

Ohsawa, T., Takegoshi, K.

[1987] On the extension of L^2 holomorphic functions. Math. Z. *195*, 197–204. Zbl.625.32011

Oka, K.

[1936] Sur les fonctions analytiques de plusieurs variable I. J. Sci. Hiroshima Univ. A *6*, 245–255. Zbl.15,309

[1937] Sur les fonctions analytiques de plusieurs variable II. J. Sci. Hiroshima Univ. *7*, 115–130. Zbl.17,122

Ostrowski, A.

[1926] On representation of analytic functions by power series. J. Lond. Math. Soc. *1*, 251–263. Jbuch52,292

Ovrelid, N.

[1971a] Integral representation formulas and L^p estimates for the $\bar{\partial}$-equation. Math. Scand. *29*, 137–160. Zbl.227.35069

[1971b] Generators of the maximal ideals of $A(\bar{D})$. Pac. J. Math. *39*, 219–223. Zbl.231.46090

Palamodov, V.P.

[1968] Systems of linear differential equations. Itogi Nauki, Mat. Anal. 1968, 5–37. English transl.: Progr. Math. *10*, 1–35 (1971). Zbl.197,364

[1970] Linear Differential Operators with Constant Coefficients. Berlin: Springer-Verlag. Translation from the Russian original (1967). Zbl.191,434

Paley, R.E.A.C., Wiener, N.

[1934] Fourier Transforms in the Complex Domain. New York: Am. Math. Soc. Zbl.11,16

Pallu de la Barrièrre, P.
[1976]   Existence et prolongement des solutions holomophes des équations aux
         dérivées partielles. J. Math. Pures Appl., IX. Ser. 55, 21–46.
         Zbl.293.35018
Passare, M.
[1987]   Courants méromorphes et égalité de la valeur principale et de la partie
         finie. Springer Lecture Notes in Mathematics 1295, 157–166.
         Zbl.634.32009
[1988a] A calculus for meromorphic currents. J. Reine Angew. Math. 392, 37–56.
         Zbl.645.32007
[1988b] Residue solutions to holomophic Cauchy problems. Seminar in Complex
         Analysis and Geometry 1987. Dept. Math. Univ. Calabria, Conf. 1, 99–
         105. Zbl.655.32006
[1988c] Residues, currents and their relation to ideals of holomorphic functions.
         Math. Scand. 62, 75–152. Zbl.633.32005
Petrowski, I.G.
[1939]   Sur l'analyticité des solutions des systèmes d'équations différentielles.
         Mat. Sb., Nov. Ser. 5, 3–68. Zbl.22,226
Philippon, P.
[1988]   A propos du texte de W. D. Brownawell, "Bounds for the degrees in the
         Nullstellensatz." Ann. Math., II. Ser. 127, 367–371. Zbl.641.14002
Polyakov, P.L.
[1971]   The Cauchy–Weil formula for differential forms. Mat. Sb., Nov. Ser. 85,
         388–402. English transl.: Math. USSR, Sb. 14, 383–398 (1972). Zbl.217,204
Pompeiu, D.
[1929a] Sur une propriété intégrale des fonctions de deux variables réelles. Bull.
         Acad. Bruxelles 15, 265–269. Jbuch55,139
[1929b] Sur une propriété des fonctions continue dépendent de plusieurs variables.
         Bull. Sci. Math. 53, 328–332. Jbuch55,138
[1929c] Sur certains systèmes d'équations linéaires et sur une propriété intégrale
         des fonctions de plusieurs variables. C. R. Acad. Sci., Paris 188, 1138–
         1139. Jbuch55,139
Radon, J.
[1917]   Über die Bestimmung von Funktionen durch ihre Integralwerte längs
         gewisser Mannigfaltigkeiten. Ber. Math. Phys. Kl. Sächs. Ges. Wiss.
         Leipzig 69, 262–277. Jbuch46,436
Ramirez de Arellano, E.
[1970]   Ein Divisionsproblem und Randintegraldarstellungen in der Komplexen
         Analysis. Math. Ann. 184, 172–187. Zbl.189,97
Range, R.M.
[1986]   Holomorphic Functions and Integral Representations in Several Complex
         Variables. New York: Springer-Verlag. Zbl.591.32002
Range, R.M., Siu, Y.T.
[1973]   Uniform estimates for the $\bar{\partial}$-equation on domains with piecewise smooth
         strictly pseudoconvex boundaries. Math. Ann. 206, 325–354.
         Zbl.248.32015
Rashevskij, P.K.
[1965]   Closed ideals in a countable normed algebra of entire analytic functions.
         Dokl. Akad. Nauk SSSR 162, 513–515. English transl.: Sov. Math., Dokl.
         6, 717–719 (1965). Zbl.139,304

Ritt, J.F.
[1929] On the zeros of exponential polynomials. Trans. Am. Math. Soc. *31*, 680–686. Jbuch55,212

Roth, B.
[1970] Finitely generated ideals of differentiable functions. Trans. Am. Math. Soc. *150*, 213–225. Zbl.199,463

Roumieu, C.
[1960] Sur quelques extensions de la notion de distribution. Ann. Sci. Ec. Norm. Supér., III. Ser. *77*, 47–121. Zbl.104,334

Rubel, L.A., Taylor, B.A.
[1968] A Fourier series method for meromorphic and entire functions. Bull. Soc. Math. Fr. *96*, 53–96. Zbl.157,396

Saburi, Y.
[1978] Vanishing theorems of cohomology groups with coefficients in sheaves of holomophic functions with bounds. Proc. Japan Acad., Ser. A *54*, 274–278. Zbl.445.32024

Saerens, R.
[1987] Interpolation theory in $\mathbb{C}^n$: A survey. Springer Lecture Notes in Mathematics *1268*, 158–188. Zbl.646.32011

Sato, M.
[1959] Theory of hyperfunctions I. J. Fac. Sci. Univ. Tokyo, Sect. I *8*, 139–193. Zbl.87,314

[1960] Theory of hyperfunctions II. J. Fac. Sci. Univ. Tokyo, Sect. I *8*, 387–437. Zbl.97,314

[1973] Pseudo-differential equations and theta functions. Astérisque *2–3*, 286–291. Zbl.288.35045

Sato, M., Kawai, T., Kashiwara, M.
[1973] Microfunctions and pseudodifferential equations. Springer Lecture Notes in Mathematics *287*, 263–529. Zbl.277.46039

[1983] Linear differential equations of infinite order and theta functions. Adv. Math. *47*, 300–325. Zbl.546.35047

[1984] Microlocal analysis of theta functions. Adv. Stud. Pure Math. *4*, 267–289. Zbl.603.10023

Schneider, R.
[1969] Functions on a sphere with vanishing integrals over certain subspheres. J. Math. Anal. Appl. *26*, 381–384. Zbl.162,355

Schwartz, L.
[1943a] Approximation d'une fonction quelquonque par des sommes d'exponentielles imaginaires. Ann. Fac. Sci. Univ. Toulouse, IV. Ser. *6*, 111–176. Zbl.61,136

[1943b] Etude des sommes d'exponentielles réelles. Paris: Hermann. Zbl.61,136

[1947] Théorie générale des fonctions moyenne périodiques. Ann. Math., II. Ser. *48*, 857–929. Zbl.30,150

[1955] Division par une fonction holomorphe sur une variété analytique complexe. Summa Brasiliensis Math. *3*, 181–209.

[1950–51] Théorie des distributions I, II. Paris: Hermann. Zbl.37,73; Zbl.42,114

Sebbar, A.
[1980] Prolongement des solutions holomorphes de certains opérateurs différentiels d'ordre infini à coefficients constants. Springer Lecture Notes in Mathematics *822*, 199–220. Zbl.451.32009

[1985] Espaces fibres A^∞ et théorème de Grauert (Thesis Doct. d'Etat). Bordeaux.

Sedletskij, A.M.
[1970]   Mean-periodic functions. Izv. Akad. Nauk SSSR, Ser. Mat. *34*, 1391–1415.
         English transl.: Math. USSR, Izv. *4*, 1406–1428 (1972).
         Zbl.218.42016
Seidenberg, A.
[1974]   Constructions in algebra. Trans. Am. Math. Soc. *197*, 273–313.
         Zbl.356.13007
Shahshahani, M., Sitaram, A.
[1987]   The Pompeiu problem in exterior domains in symmetric spaces. Contemp.
         Math. *63*, 267–277. Zbl.615.53060
Shambayati, R., Zielezny, R.
[1985]   Convolution equations in spaces of distributions with one-sided bounded
         support. Trans. Am. Math. Soc. *289*, 707–713. Zbl.612.46037
Shamoyan, F.A.
[1969]   Closed ideals in an algebra of rapidly increasing analytic functions. Izv.
         Akad. Nauk Arm. SSR, Mat. *4*, 267–277. Zbl.187,386
[1970]   Description of closed ideals and certain factorization problems in algebras
         of increasing functions in the circle. Izv. Akad. Nauk Arm. SSR, Mat. *5*,
         419–433. Zbl.219.30039
Schapira, P.
[1970]   Théorie des hyperfonctions. Springer Lecture Notes in Mathematics *126*.
         Zbl.192,473
[1985]   Microdifferential Systems in the Complex Domain. New York: Springer-
         Verlag. Zbl.554.32022
Shapiro, H.S.
[1958]   The expansion of mean-periodic functions in series of exponentials. Com-
         mun. Pure Appl. Math. *11*, 1–21. Zbl.168,318
[1988]   Fischer's decomposition revisited. Royal Institute of Technology.
Shapiro, V.L.
[1969]   Sets of uniqueness for the vibrating string problem. Trans. Am. Math.
         Soc. *141*, 127–146. Zbl.216,128
Shephard, G.C., Todd, J.A.
[1954]   Finite unitary reflection groups. Can. J. Math. *6*, 274–304. Zbl.55,143
Shepp, L.A., Kruskal, J.B.
[1978]   Computerized tomography: The new medical X-ray technology. Am.
         Math. Mon. *85*, 420–439. Zbl.381.68079
Shields, A.
[1963]   On quotients of exponential polynomials. Commun. Pure Appl. Math. *16*,
         27–31. Zbl.113,54
Skoda, H.
[1972a] Sous-ensembles analytiques d'ordre fini ou infini dans $\mathbb{C}^n$. Bull. Soc.
         Math. Fr. *100*, 353–408. Zbl.246.32009
[1972b] Applications des techniques L^2 à la théorie des idéaux d'une algèbre de
         fonctions holomorphes avec poids. Ann. Sci. Ec. Norm. Supér., IV. Ser. *5*,
         545–579. Zbl.254.32017
[1973]   Formulation hilbertienne du Nullstellensatz dans les algèbres de fonc-
         tions holomorphes. Springer Lecture Notes in Mathematics *336*, 95–103.
         Zbl.259.32004
Smith, K.T., Solmon, D.C., Wagner, S.L.
[1977]   Practical and mathematical aspects of the problem of reconstructing ob-
         jects from radiographs. Bull. Am. Math. Soc. *83*, 1227–1270.
         Zbl.521.65090

Solmon, D.C.
[1976] The X-ray transform. J. Math. Anal. Appl. *56*, 61–83. Zbl.334.44007

Strichartz, R.S.
1971] The stationary observer problem for $\Box u = Mu$ and related questions. J. Differ. Equations *9*, 205–223. Zbl.216,128

Struppa, D.C.
[1983a] The Fundamental Principle for systems of convolution equations. Mem. Am. Math. Soc. *273*. Zbl.503.46027

[1983b] Comparison for convolution operators. Rend., Sci. Mat. Appl., A *117*, 135–143. Zbl.598.47035

[1986] Convolution equations and spaces of ultradifferentiable functions. Isr. J. Math. *54*, 60–70. Zbl.605.32001

[1988a] The first eighty years of Hartogs' theorem. Semin. Geom., Univ. Studi Bologna 1987–1988, 127–211. Zbl.657.35018

[1988b] Hartogs-type theorems for solutions of convolution equations. Rend. Semin. Mat. Brescia *10*, 51–64. Zbl.675.32011

[1991a] An extension of Fantappiè's theory of analytic functionals. Lect. Notes Pure Appl. Math. *132*, 329–356.

[1991b] A new look at the theory of quasianalytic classes. In: Analyse Complexe Multivariable. EditEl, Rende.

Tacklind, S.
[1936] Sur les classes quasianalytiques des solutions des équations aux dérivées partielles du type parabolique. Nova Acta Soc. Sci. Uppsal., IV. Ser. *10*, No. 3, 1–57. Zbl.14,22

Tajima, S.
[1982] Analyse microlocale sur les variétés de Cauchy–Riemann et problème du prolongement des solutions holomorphes des équations aux dérivées partielles. Publ. Res. Inst. Math. Sci. *18*, 911–945. Zbl.553.58028

[1985] CR-microfunctions and the Henkin–Ramirez reproducing kernel. Proc. Japan Acad., Ser. A *61*, 137–139. Zbl.609.32007

Taylor, B.A.
[1966] Some locally convex spaces of entire functions. Proc. Symp. Pure Math. *11*, 431–467. Zbl.181,133

[1982] Linear extension operators for entire functions. Mich. Math. J. *29*, 185–197. Zbl.471.30014

Tougeron, J.C.
[1972] Idéaux de fonctions différentiables. Berlin: Springer-Verlag. Zbl.251.58001

Treves, F.
1967] Topologival Vector Spaces, Distributions and Kernels. New York: Academic Press. Zbl.171,104

[1975] Basic Linear Partial Differential Equations. New York: Academic Press. Zbl.305.35001

Trimèche, K.
1988] Transmutation Operators and Mean-Periodic Functions Associated with Differential Operators. London–Paris–New York–Melbourne: Harwood Academic Publishers.

Tsuno, Y.
[1974] On the prolongation of local holomorphic solutions of partial differential equations. J. Math. Soc. Japan *26*, 523–548. Zbl.279.35012

Turán, P.
[1953] Eine neue Methode in der Analysis und deren Anwendungen. Budapest: Akadémiai Kiadó. Zbl.52,46

Vogt, D.
1977]   Charakterisierung der Unterräume von s. Math. Z. *155*, 109–117.
         Zbl.337.46015
[1983]   Frécheträume, zwischen denen jede stetige lineare Abbildung be-
         schränkt ist. J. Reine Angew. Math. *345*, 182–200. Zbl.514.46003
Vogt, D., Wagner, M.J.
[1980]   Charakterisierung der Quotientenräume von s und eine Vermutung von
         Martineau. Stud. Math. *67*, 225–240. Zbl.464.46010
Waldschmidt, M.
[1974]   Nombres transcendants. Springer Lecture Notes in Mathematics *402*.
         Zbl.302.10030
[1981   Transcendance et exponentielles en plusieurs variables. Invent. Math. *63*,
         97–127. Zbl.454.10020
Watson, G.N.
[1962]   A Treatise on the Theory of Bessel Functions. 2nd ed. Cambridge: Cam-
         bridge University Press. Zbl.174,362
Wawrzynczyk, A.
[1985]   Spectral analysis and mean periodic functions on rank-one symmetric
         spaces. Bol. Soc. Mat. Mex., II. Ser. *30*, No.2, 15–29. Zbl.652.22012
[1987]   Spectral analysis and systhesis on symmetric spaces. J. Math. Anal. Appl.
         127, 1–17. Zbl.647.43004
Weil, A.
[1935]   L'intégrale de Cauchy et les fonctions de plusieurs variables. Math. Ann.
         111, 178–182. Zbl.11,123
Weit, Y.
[1980]   On closed ideals in the motion group algebra. Math. Ann. *248*, 279–283.
         Zbl.425.43009
[1981]   On spectral analysis in locally compact motion groups. J. Funct. Anal. *40*,
         45–53. Zbl.461.43013
[1982]   On spectral synthesis in a non-commutative locally compact group. Am.
         J. Math. *104*, 779–793. Zbl.506.43003
Wiegerinck, J.
[1985]   A support theorem for Radon transforms on $\mathbb{R}^n$. Indagationes Math. *47*,
         87–93. Zbl.567.44004
[1987]   Convergence of formal power series and analytic extension. Springer Lec-
         ture Notes in Mathematics *1276*, 313–320. Zbl.623.32001
Williams, S.A.
[1976]   A partial solution of the Pompeiu problem. Math. Ann. *223*, 183–190.
         Zbl.329.35045
Yang, P.
[1977]   Curvature of complex submanifolds of $\mathbb{C}^n$. J. Differ. Geom. *12*, 499–511.
         Zbl.409.53043
Yger, A.
[1977]   Une généralisation d'un théorème de J. Delsarte. Publ. Math. Orsay 77,
         Sémin. Anal. Harmon. 1976/1977, 53-75. Zbl.373.42008
[1979]   Une généralisation d'une théorème de J. Delsarte. C. R. Acad. Sci., Paris,
         Ser. A *288*, 497–499. Zbl.402.46027
[1980]   Propriétés de certains systèmes d'équations de convolutions dans $\mathbb{R}^2$.
         C. R. Acad. Sci., Paris, Ser. A *289*, 169–171. Zbl.415.46036
[1981a] Systèmes de distributions à support fini et à spectre discret. C. R. Acad.
         Sci., Paris, Ser. I *292*, 543–546. Zbl.462.46027
[1981b] Fonctions définies dans le plan et vérifiant certaines propriétés de
         moyenne. Ann. Inst. Fourier *31*, No. 3, 115–146. Zbl.438.43010

[1982]   Fonctions moyenne-périodiques de deux variables et systèmes d'équations différentielles (thesis). Orsay.

[1987]   Formules de division et prolongement méromorphe. Springer Lecture Notes in Mathematics *1295*, 226–283. Zbl.632.32010

[1988]   Inégalités de Lojasiewicz ou de Malgrange pour certaines familles d'éléments de $\mathcal{E}'(\mathbb{R}^n)$. Preprint.

Yoshioka, T.

[1982]   Cohomologie à estimation L^2 avec poids plurisousharmoniques et extension des fonctions holomorphes avec contrôle de la croissance. Osaka J. Math. *19*, 787–813. Zbl.521.32015

Yuzhakov, A.P.

[1984]   On the computation of the complete sum of residues relative to a polynomial mapping in $\mathbb{C}^n$. Dokl. Akad. Nauk SSSR *275*, 817–820. English transl.: Sov. Math. Dokl. *29*, 321–324 (1984). Zbl.584.32002

Zalcman, L.

1972]   Analyticity and the Pompeiu problem. Arch. Ration. Mech. Anal. *47*, 237–254. Zbl.251.30047

[1973]   Mean values and differential equations. Isr. J. Math. *14*, 339–352. Zbl.263.35013

1980]   Offbeat integral geometry. Amer. Math. Mon. *87*, 161–175. Zbl.433.53048

Zharinov, V.V.

[1977]   Laplace transformation of Fourier hyperfunctions and related classes of analytic functionals I. Teor. Mat. Fiz. *33*, 291–309. English transl.: Theor. Math. Phys. *33*, 1027–1039 (1978). Zbl.364.46026

[1978]   Laplace transformation of Fourier hyperfunctions and related classes of analytic functionals II. Teor. Mat. Fiz. *37*, 12–29. English transl.: Theor. Math. Phys. *37*, 843–855 (1979). Zbl.389.46030

II. The Yang–Mills Fields, the Radon–Penrose Transform, and the Cauchy–Riemann Equations

G.M. Khenkin, R.G. Novikov

Translated from the Russian
by G.M. Khenkin

Contents

Introduction

In this part, we consider a number of problems of complex analysis and mathematical physics connected with the theory of Yang–Mills gauge fields on the one hand and the theory of Cauchy–Riemann equations on the other.

The main topics of this part are as follows:

(1) The nonabelian Radon transform for gauge fields. The Radon–Penrose transform.

(2) The connection between the Radon–Penrose transform and the Faddeev-type scattering data.

(3) The Yang–Mills equations as compatibility conditions of a linear system of differential equations with a complex spectral parameter.

(4) The Yang–Mills equations in terms of the $\bar{\partial}$-equation for scattering data.

(5) The interpretation of solutions of the Yang–Mills–Higgs–Dirac equations in terms of holomorphic vector bundles over a manifold of complex light rays.

In the following, we briefly discuss the contents of this part topic by topic. The first topic is mainly described in Chaps. 4 and 6.

We consider the gauge field $a = \sum_{j=0}^{3} a_j(x)\, dx^j$ in the open convex domain $\mathcal{D} \subset \mathbb{C}^4$, where $\{a_j(x)\}$ are holomorphic functions with values in $\mathrm{gl}(n, \mathbb{C})$. The field a is considered up to the gauge transform $a \sim g^{-1}ag + g^{-1}dg$, where the g are holomorphic functions with values in $\mathrm{GL}(n, \mathbb{C})$.

We consider the manifold $P(\mathcal{D})$ of all complex straight lines in $\mathbb{C}^4$ crossing $\mathcal{D}$.

The nonabelian Radon transform introduced in Chap. 4 converts the field a into the form $\theta = \sum_{j=0}^{3} \theta_j(x, m)\, dm_j$, $x \in \mathcal{D}$, $m \in \mathbb{C}^4 \setminus \{0\}$ with the following properties:

The functions $\theta_j(x, m)$ are holomorphic with respect to $x \in \mathcal{D}$;

$$\frac{\partial \theta_k(x, m)}{\partial \bar{m}_j} - \frac{\partial \theta_j(x, m)}{\partial \bar{m}_k} + [\theta_k, \theta_j] = 0;$$

$$\sum_{j=0}^{3} m_j \frac{\partial}{\partial x^j} \theta_k(x, m) = 0; \qquad (0.1)$$

$$\bar{c}\theta_j(x, c \cdot m) = \theta_j(x, m), \quad c \in \mathbb{C}^1 \setminus \{0\};$$

the form θ is considered up to the $\bar{\partial}$-gauge equivalence $\theta \sim H^{-1}\bar{\partial}H + H^{-1}\theta H$, where H is a holomorphic function with respect to x with values in $\mathrm{GL}(n, \mathbb{C})$ satisfying the equation

$$\left(m \cdot \frac{\partial}{\partial x}\right) H = 0; \qquad \bar{\partial}H = \sum \frac{\partial H}{\partial \bar{m}_j} d\bar{m}_j.$$

The form θ with such properties is the $(0,1)$-form on the manifold $P(\mathcal{D})$ and defines on $P(\mathcal{D})$ the holomorphic vector bundle E_θ, which is a deformation of the trivial n-dimensional bundle E. The holomorphic sections of the bundle E_θ are smooth sections h of the bundle E which satisfy the Cauchy–Riemann equation of the form $\bar{\partial}h + \theta h = 0$.

The inversion of the nonabelian Radon transform is reduced to the solution of a linear $\bar{\partial}$-equation.

We provide the space $\mathbb{C}^4$ with the complex metric $dx^2 = \sum_{j=0}^{3} dx_j^2$.

The restriction of the form θ with the properties (0.1) to the manifold $L(\mathcal{D})$—all complex light rays (zero lines: $x = ms + x^0, s \in \mathbb{C}, m^2 = 0$), crossing $\mathcal{D}$, is naturally called the Radon–Penrose transform, since the fundamental idea to represent physical fields by holomorphic objects of the space of the light rays belongs to R. Penrose (see Penrose [1969; 1976]; Penrose and Rindler [1986]). In the works of Ward [1977], Witten [1978], and Isenberg, Yasskin and Green [1978], the gauge fields in domain $\mathcal{D}$ were first interpreted as the holomorphic vector bundles over the manifold $L(\mathcal{D})$. The Radon–Penrose transform in a similar but more complicated form than the one mentioned earlier was introduced in Khenkin [1980].

The complex dimension of the manifold $P(\mathcal{D})$ is equal to six, the one of the manifold $L(\mathcal{D})$ equals five, when the gauge field is given in the four-dimensional domain $\mathcal{D}$. Thus, the full nonabelian Radon transform θ and the Radon–Penrose transform are overdetermined. So, it is natural and useful to consider the Radon–Penrose transform along the part of the light rays that is non-overdetermined. This is carried out in Chap. 4, where inversion formulae and the characterization and interpretation in terms of holomorphic bundles for such partial Radon–Penrose transforms are obtained.

Let us note that in this part the restrictions of the nonabelian Radon transform θ only to such submanifolds in $P(\mathcal{D})$ are studied that naturally appear in the context of the Yang–Mills equations (see Chap. 1).

If the domain $\mathcal{D}$ is real, for example, $\mathcal{D} \subset \mathcal{E}_0$ or $\mathcal{D} \subset \mathcal{M}_0$, where $\mathcal{E}_0 = \{x \in \mathbb{C}^4 : \mathrm{Im}\, x = 0\}$ and

$$\mathcal{M}_0 = \{x \in \mathbb{C}^4 : \mathrm{Im}\, x_0 = \mathrm{Re}\, x_1 = \mathrm{Re}\, x_2 = \mathrm{Re}\, x_3 = 0\}$$

are the Euclidean and Minkowski spaces, then $P(\mathcal{D})$ and $L(\mathcal{D})$ are the Cauchy–Riemann manifolds (CR-manifolds) (see Chaps. 3 and 4). The formulae of Chap. 4 make it possible to define the nonabelian Radon transform θ of (smooth) gauge fields a given in the domain $\mathcal{D}$ and for this case. In addition, the Radon transform of gauge fields may be interpreted as CR-bundles over $P(\mathcal{D})$, $L(\mathcal{D})$, and so on.

If the complex Minkowski space $\mathbb{C}\mathcal{M}_0 \simeq \mathbb{C}^4$ is provided with the (spinor) coordinates $x = \{x_{AB'}\}$, $A = 0, 1$; $A' = 0', 1'$, and with the metric $\det|dx_{AB'}|$, then null-planes in this metric are given by the equations

$$\zeta_{B'} = x_{AB'}\zeta^A \qquad \text{or} \qquad \eta_A = x_{AB'}\eta^{B'}, \tag{0.2}$$

where $\zeta = (\zeta^A, \zeta_{B'})$ and $\eta = (\eta_A, \eta^{B'})$ are points of the three-dimensional projective spaces $\mathbb{C}P^3_+$ and $\mathbb{C}P^3_-$. The point ζ gives the so-called α-plane, and η is the β-plane in $\mathbb{C}M_0$. Together the equations (0.2) give a null-line in $\mathbb{C}M_0$. For a fixed point $x \in \mathbb{C}M_0$, the equations (0.2) considered as equations for ζ and for η generate the two-dimensional subspaces $S_\pm(x)$ in $\mathbb{C}^4_\pm$ and the compact submanifolds $\mathcal{L}_\pm(x) \approx \mathbb{C}P^1$ in $\mathbb{C}P^3_\pm$, respectively. The manifolds of all α- or β-planes crossing the domain $\mathcal{D} \subset \mathbb{C}M_0$ have the form

$$L_\pm(\mathcal{D}) = \bigcup_{x \in \mathcal{D}} \mathcal{L}_\pm(x),$$

respectively.

The manifold of the null-lines crossing $\mathcal{D}$ may be represented in the form

$$L(\mathcal{D}) = \{(\zeta, \eta) \in L_+(\mathcal{D}) \times L_-(\mathcal{D}) : \langle \zeta \cdot \eta \rangle = 0\}.$$

The so-called scalar and spinor fields $\varphi_\pm$, ψ_A, $\psi_{A'}$ in the domain $\mathcal{D} \subset \mathbb{C}M_0$ may be considered as sections over $\mathcal{D}$ of the trivial n-dimensional and $2n$-dimensional bundles E and $E \otimes S_\pm$, respectively.

In Chap. 6, we review the results and the constructions of a number of papers (Atiyah [1979]; Eastwood, Penrose and Wells [1981]; Gindikin and Khenkin [1981]; Khenkin [1984; 1980]; Khenkin and Manin [1982; 1981]; Hitchin [1980]; Manin [1981]; Woodhouse [1985]) according to the representation by the Penrose-type transform of different fields in the domain $\mathcal{D}$ through $\bar{\partial}$-cohomologies of the manifold $L(\mathcal{D})$. In particular, Theorem 6.1 gives explicit formulae for the Penrose-type transform that establish the isomorphism between the spaces of holomorphic (respectively, smooth) sections over $\mathcal{D} \subset \mathbb{C}M_0$ (respectively, $\mathcal{M}_0$) of the bundles E or $E \otimes S_\pm$ and spaces of one-dimensional $\bar{\partial}$-cohomologies $L(\mathcal{D})$ with coefficients in $E_\theta(-2, 0)$, $E_\theta(0, -2)$, or $E_\theta(-1, 0)$, $E_\theta(0, -1)$, where $E_\theta(k, \ell) = E_\theta \otimes \mathcal{O}(k, \ell)$ (see Chap. 3). Theorem 6.1 essentially also gives the interpretation of all the spaces of the cohomologies $H^j(L(\mathcal{D}), E_\theta(k, \ell))$, $k \leq 0, \ell \leq 0$ in terms of the fields on $\mathcal{D}$. Here (see Chap. 3) these spaces should be understood as factor spaces of $(\bar{\partial} + \theta)$-closed $(0, j)$-forms on $L(\mathcal{D})$ with coefficients that are smooth sections of $E_\theta(k, \ell)$ by the subspace of the $(\bar{\partial} + \theta)$-exact forms of the same type. The Penrose-type transforms of these cohomologies on $L(\mathcal{D})$ into the fields on $\mathcal{D}$ consist of integrating the corresponding forms along the compact submanifolds

$$\mathcal{L}(x) = \mathcal{L}_+(x) \times \mathcal{L}_-(x) \subset L(\mathcal{D}).$$

In connection with the classical Yang–Mills equation (see Chap. 1), the Radon-type transform of the fields in the four-dimensional space is of the greatest interest. However, for other problems (see, for example, Khenkin and Novikov [1987; 1988]; Khenkin and Polyakov [1990]; Ward [1984]; Witten [1986]), there naturally appear the Radon-type transforms (both abelian and nonabelian) of different fields given in the spaces of the dimension different from 4. Let us note that the nonabelian Radon transform considered here may be defined by $\theta = \sum_{j=0}^{n-1} \theta_j(x,m)\, d\bar{m}_j$ and characterized in the same way as in Chap. 4 for the fields $(a = \sum_{j=0}^{n-1} a_j(x)\, dx^j)$ on the space of any dimension $n \geq 2$.

Topic (2) is considered in Chap. 5.

In investigating the multidimensional inverse scattering problem for the Schrödinger equation

$$-\Delta\psi(x,k) + v(x) \cdot \psi(x,k) = k^2\psi(x,k), \quad x \in R^n,$$

L.D. Faddeev in 1966 introduced (on the basis of a suitable choice of the Green function and the Lippman–Shwinger-type equation) the generalized scattering data $h(k,\ell)$, where $k,\ell \in \mathbb{C}^n$, $k^2 = \ell^2$, $\operatorname{Im} k = \operatorname{Im} \ell$ (see Khenkin and Novikov [1987]; Faddeev [1974]). The function h is a very useful (nonanalytical) extension of the classical scattering amplitude $f(k,\ell)$, where $k,\ell \in R^n$, $k^2 = \ell^2$. When describing the properties of the function h, it is convenient to carry out a change of variables $k = k$, $p = k - \ell$ and to consider the function $H(k,p) = h(k,k-p)$, where $k \in \mathbb{C}^n$, $p \in R^n$, $p^2 = 2k \cdot p$. The most essential (characteristic) properties of the function $H(k,p)$ are the following (see Ablowitz and Nachman [1986]; Khenkin and Novikov [1987]; Newton [1989]).

$$\left. \begin{aligned} h(k+i0k, \ell+i0k) &= f(k,\ell) \quad \text{if} \quad k,\ell \in R^n, \quad k^2 = \ell^2 \\ \frac{\partial}{\partial \bar{k}_j} H(k,p) &= -2\pi \int_{\zeta \in R^n} \zeta_j H(k,-\zeta) H(k+\zeta, p+\zeta)\delta(\zeta^2 + 2k\cdot\zeta)\, d\zeta \end{aligned} \right\}$$

$$(0.3)$$

$$\lim_{|k|\to\infty} H(k,p) = \left(\frac{1}{2\pi}\right)^n \int_{x \in R^n} v(x)e^{ipx}\, dx = \hat{v}(p), \quad k \in \mathbb{C}^n, \quad p \in R^n.$$

$$(0.4)$$

The generalized scattering data may be introduced, and for the Schrödinger equation for the magnetic field (Khenkin and Novikov [1988; 1987]),

$$\sum_{j=0}^{n-1}\left(i\frac{d}{dx^j} + a_j(x)\right)^2 \psi + v(x)\psi = k^2\psi.$$

In this case, the equalities (0.3) are also valid. The equality (0.4) is changing for the following equality

$$H(k,p) = H_0(m,p) \cdot |k| + H_1(m,p) + o(1),$$

where

$$H_0(e^{i\varphi}m,p) = e^{i\varphi}H_0(m,p), \quad m = \frac{k}{|k|}, \quad m \cdot p = 0.$$

The following formula connects the function $H_0(m, p)$ with the nonabelian Radon transform of the field a given on R^n

$$\theta_j(x, m) = -2\pi \int_{\zeta \in R^n} H_0(m, -\zeta) e^{i\zeta x} \delta(2\zeta \cdot m) \, d\zeta. \tag{0.5}$$

It is important to note that all formulae, deduced in Khenkin and Novikov [1988; 1987] for the case where a, v, h, ψ are scalar functions, remain almost without changes (for the right order of the factors) and for the case where a, v, h take values in $\mathrm{gl}(n, \mathbb{C})$, and ψ takes values in $\mathrm{GL}(n, \mathbb{C})$. In Chap. 5, it is shown, in particular, that for the case $n = 4$, the scattering data $H(k, p)$, restricted to the fixed level of energy $k^2 = E$, go over to the Radon–Penrose transform for $k \to \infty$ according to formula (0.5).

The connection of the Faddeev-type scattering data with the nonabelian Radon transform is very interesting and useful at least for two reasons. First, it is important in solving the inverse scattering problem for the Schrödinger operator in the gauge (magnetic) field (Khenkin and Novikov [1988; 1987]). Second, the Radon–Penrose transform considered as the limit case of the Faddeev-type scattering data is one of the most important examples of the spectral transform connected with the Schrödinger equation (or Klein–Gordon) that was applied to the investigation of nonlinear equations. For the other application of the Faddeev-type scattering data to nonlinear equations, see Novikov [1989].

The third topic is stated in Chap. 1.2.

As is known, the representation of a nonlinear equation in the form of the compatibility condition of a linear system is one of the bases of the inverse problem method. As a rule, such a representation allows us to rewrite an original nonlinear equation in terms of scattering data where the equation is usually greatly simplified. The inverse problem method turned out to be very effective in the investigation of many important nonlinear equations (see Ablowitz and Nachman [1986]; Beals and Coifman [1986]; Belavin and Zakharov [1978]; Forgacs and Horvath [1981]; Manakov, Novikov, Pitaevski and Zakharov [1980]; Manakov and Zakharov [1981]).

In Chap. 1, the classical facts about fundamental equations of mathematical physics—the Yang–Mills equations—are stated.

The representation of the Yang–Mills equation in the form of the compatibility condition of a linear system was first obtained for self-dual and for anti-self-dual cases in the work Belavin and Zakharov [1978].

Such a representation is equivalent to the fact discovered in Ward [1977] and Yang [1977] that those and only those gauge fields in $\mathbb{C}^4$ whose curvature vanishes on one of two connected components of the space of two-dimensional complex null planes (in the complex metric $(y - z)^2 = \sum_{j=0}^{3}(y_j - z_j)^2$) satisfy the Yang–Mills self-dual or anti-self-dual equation. The linear system has the form

$$(\alpha_\nu \cdot \nabla_y)\eta(y, \lambda_1) = 0 \quad \text{for the self-dual field } a(y),$$
$$(\beta_\nu \cdot \nabla_z)\eta(z, \lambda_2) = 0 \quad \text{for the anti-self-dual field } a(z), \tag{0.6}$$

where

$$\nu = 1,2, \quad \lambda_1 \in \mathbb{C}P^1, \quad \lambda_2 \in \mathbb{C}P^1, \quad y \in \mathcal{D}, \quad z \in \mathcal{D},$$
$$\alpha_1 = (1, -i, -\lambda_1, -i\lambda_1); \qquad \beta_1 = (1, -i, -\lambda_2, i\lambda_2);$$
$$\alpha_2 = (\lambda_1, i\lambda_1, 1, -i); \qquad \beta_2 = (\lambda_2, i\lambda_2, 1, i);$$
$$\nabla_{y^j} = \frac{\partial}{\partial y^j} + a_j(y); \qquad \nabla_{z^j} = \frac{\partial}{\partial z^j} + a_j(z).$$

From the papers Witten [1978], Isenberg and Yasskin [1979], Isenberg, Yasskin and Green [1978], and Forgacs, Horvath and Palla [1982], it follows (Theorem 2.1) that the full Yang–Mills homogeneous equation may be considered as the compatibility conditions for the linear system of the form

$$\begin{aligned}
(\alpha_\nu \cdot \nabla_y)\eta(y, z, \lambda_1, \lambda_2) &= O((y - z)^k), \\
(\beta_\nu \cdot \nabla_z)\eta(y, z, \lambda_1, \lambda_2) &= O((y - z)^k),
\end{aligned} \tag{0.7}$$

where

$$\nabla_{y^j} = \frac{\partial}{\partial y^j} + a_{y^j}(y, z), \qquad \nabla_{z^j} = \frac{\partial}{\partial z^j} + a_{z^j}(y, z), \quad \nu = 1, 2; \quad k = 3.$$

In Chap. 2, these results are stated together with a number of new results concerning equations obtained in the form of the compatibility condition of the system (0.7).

Theorem 2.2, for example, determines that the Yang–Mills equation for the gauge field $a(x)$, supplemented with the commutation condition between self-dual and anti-self-dual parts of the curvature $f = da + a \wedge a$, is exactly the compatibility condition of the system (0.7) for $k = 4$.

Further (for the fields $a(x)$ with values in $\mathrm{gl}(2, \mathbb{C})$), Theorem 2.3 completely describes all the Yang–Mills fields satisfying the conditions of Theorem 2.2.

The linear system (0.7) has two spectral parameters, λ_1 and λ_2. Developing the ideas of the paper Forgacs, Horvath and Palla [1982] in Chap. 2, several useful substitutions (for the interpretation of the Yang–Mills equations in terms of scattering data) are given (for example, $\lambda_1 = \lambda^2 + \lambda$, $\lambda_2 = \lambda^2 - \lambda$) that reduce the system (0.7) to the system with one spectral parameter but with the same compatibility conditions.

The central topic (4) is described in Chaps. 7 and 8.

The Radon–Penrose transform (after the corresponding calibration) may be chosen in the form of $\theta_1(y, \lambda_1)\, d\bar{\lambda}_1$ for the self-dual field and in the form of $\theta_2(z, \lambda_2)\, d\bar{\lambda}_2$ for the anti-self-dual field. In addition, the following formulae are valid

$$\theta_1(y, \lambda_1) = \eta^{-1}(y, \lambda_1)\frac{\partial}{\partial \bar{\lambda}_1}\eta(y, \lambda_1),$$

$$\theta_2(z, \lambda_2) = \eta^{-1}(z, \lambda_2)\frac{\partial}{\partial \bar{\lambda}_2}\eta(z, \lambda_2),$$

where $\eta(y, \lambda_1)$ and $\eta(z, \lambda_2)$ are solutions of the system (0.6).

116 G.M. Khenkin, R.G. Novikov

In terms of such scattering data, the Yang–Mills self-dual and anti-self-dual equations, respectively, take the form

$$\left(\alpha_\nu \cdot \frac{\partial}{\partial y}\right)\theta_1(y,\lambda_1) = 0; \qquad \left(\beta_\nu \cdot \frac{\partial}{\partial z}\right)\theta_2(z,\lambda_2) = 0;$$

$$\nu = 1,2, \quad \lambda_1 \in \mathbb{C}P^1, \quad \lambda_2 \in \mathbb{C}P^1, \quad y \in \mathcal{D}, \quad z \in \mathcal{D} \subset \mathbb{C}^4. \tag{0.8}$$

Thus, in such cases, the equations for scattering data are linear. This circumstance expressed in terms of a particular language provides the basis for a successful investigation of the Yang–Mills self-dual (and anti-self-dual) equation (see Atiyah [1979]; Atiyah and Hitchin [1988]; Atiyah and Ward [1977]; Beals and Coifman [1986]; Forgacs and Horvath [1981]; Khenkin [1980]; Hitchin [1982]; Manin [1984]; Newman [1986]; Ward [1977; 1981]; Ward and Wells [1990]; Woodhouse [1985]; Yang [1977]; and others).

It is possible to develop these results and also the results of Isenberg and Yasskin [1979], Khenkin [1980], and Witten [1978] to reduce the Yang–Mills full equation to the following equations for the generalized scattering data $\theta(y,z,\lambda_1,\lambda_2)$ (Theorem 7.1),

$$\frac{\partial}{\partial\bar{\lambda}_1}\theta_2(y,z,\lambda_1,\lambda_2) - \frac{\partial}{\partial\bar{\lambda}_2}\theta_2(y,z,\lambda_1,\lambda_2) + [\theta_1,\theta_2] = 0,$$

$$\left(\alpha_\nu \cdot \frac{\partial}{\partial y}\right)\theta_j(y,z,\lambda_1,\lambda_2) = O((y-z)^3),$$

$$\left(\beta_\nu \cdot \frac{\partial}{\partial z}\right)\theta_j(y,z,\lambda_1,\lambda_2) = O((y-z)^3), \tag{0.9}$$

$$\theta_j(y,z,\lambda_1,\lambda_2) = O\left(\frac{1}{1+|\lambda_j|^2}\right),$$

$$\nu,j = 1,2, \quad y,z \in \mathcal{D} \subset \mathbb{C}^4, \quad \lambda_1,\lambda_2 \in \mathbb{C}P^1.$$

The form $\theta = \theta_1(x,x,\lambda_1,\lambda_2)\,d\bar{\lambda}_1 + \theta_2(x,x,\lambda_1,\lambda_2)\,d\bar{\lambda}_2$ is the Radon–Penrose transform of the Yang–Mills field.

In order to obtain a solution of the Yang–Mills equation, it is sufficient to find a solution of the system (0.9) and to do an inverse scattering transform (or the Radon–Penrose transform) of the form $\theta(x,x,\lambda_1,\lambda_2)$ into the gauge field $a(x), x \in \mathcal{D}$. Unfortunately, the first equation in (0.9) is nonlinear.

Nevertheless, Theorem 7.2 makes it possible to find all the forms θ with sufficiently small norm satisfying (0.9) through the forms θ^0 satisfying a linearized variant of (0.9). For this, we shall need the following nonlinear integral equation,

$$\theta = \theta^0 + R(\theta \wedge \theta), \tag{0.10}$$

where R is a suitable integral operator inverting the $\bar{\partial}$-operator on the $(0,1)$-forms defined in the third infinitesimal neighborhood $L^{(3)}(\mathcal{D})$ of the submanifold $L(\mathcal{D})$ in $L_+(\mathcal{D}) \times L_-(\mathcal{D})$. The mapping $\theta \mapsto \theta^0$ (after the Radon–Penrose inverse transform) gives a linearizing mapping of the analytical fields in $\mathcal{D}$

with a small norm that transforms the Yang–Mills equation in its linear part, that is, in the Maxwell equation. By the same token, the Flato–Simon result (Flato and Simon [1980]) for a formal linearization of the Yang–Mills equations is being developed.

Theorem 7.3 determines additional conditions under which the Radon–Penrose transform leads to the complete linearization of the Yang–Mills equations. If, for example, the Yang–Mills equation for a gauge field with values in $\mathrm{gl}(2, \mathbb{C})$ are supplemented with the commutation condition of self-dual and anti-self-dual parts of the curvature of this field, then in terms of the generalized scattering data $\theta(y, z, \lambda_1, \lambda_2)$, such a system may be represented in the form

$$\frac{\partial \theta_2}{\partial \bar{\lambda}_1} = \frac{\partial \theta_1}{\partial \bar{\lambda}_2} = 0, \quad [\theta_1, \theta_2] = 0,$$

$$\left(\alpha_\nu \cdot \frac{\partial}{\partial y} \right)\theta_j = 0, \quad \left(\beta_\nu \cdot \frac{\partial}{\partial z} \right)\theta_j = 0,$$

$$\nu, j = 1, 2, \quad y, z \in \mathcal{D}, \quad \lambda_1, \lambda_2 \in \mathbb{C}P^1.$$

This representation makes it possible, for example, to give a linear procedure of the solution to the Cauchy problem for the system of equations mentioned above.

Theorem 7.4 makes it possible to write the Yang–Mills equation in terms of the partial (non-overdetermined) scattering data (or, otherwise, the Radon–Penrose partial transform) of the form

$$\left((2\bar{\lambda} + 1)\theta_1(x, \lambda^2 + \lambda, \lambda^2 - \lambda) + (2\bar{\lambda} - 1)\theta_2(x, \lambda^2 + \lambda, \lambda^2 - \lambda) \right) d\bar{\lambda}$$

with a one-dimensional spectral parameter. Earlier, an unsuccessful attempt in this direction was made in Forgacs, Horvath and Palla [1982]. Further, in Chap. 8, the interpretation in terms of the Penrose-type transform of the Yang–Mills, Weyl–Dirac, and Klein–Gordon nonhomogeneous equations is given. For this, first of all, we state the result of Khenkin [1980] and Khenkin and Manin [1982] (Theorem 8.1). The result is that the Radon–Penrose transform θ on $L(\mathcal{D})$ of the gauge field $a(x)$, $x \in \mathcal{D} \subset \mathbb{C}M$ (respectively, $\mathcal{M}$) may be extended to the field $L_+(\mathcal{D}) \times L_-(\mathcal{D})$ so that

$$\bar{\partial}\theta + \theta \wedge \theta = \mathcal{J}\langle \zeta \cdot \eta \rangle^3. \tag{0.11}$$

In addition, the elements of the cohomology spaces $L(\mathcal{D})$, with coefficients in $\mathrm{End}\, E_\theta(-1, 0)$, $\mathrm{End}\, E_\theta(0, -1)$ and $\mathrm{End}\, E_\theta(-2, 0)$, $\mathrm{End}\, E_\theta(0, -2)$, may be represented by the $(0, 1)$-forms $\psi_\pm$ and $\phi_\pm$ in $L_+(\mathcal{D}) \times L_-(\mathcal{D})$ with the properties

$$\bar{\partial}\psi_\pm + [\theta, \psi_\pm] = G_\pm \langle \zeta \cdot \eta \rangle^2, \tag{0.12}$$

$$\bar{\partial}\phi_\pm + [\theta, \phi_\pm] = F_\pm \langle \zeta \cdot \eta \rangle. \tag{0.13}$$

The main result of Khenkin [1980], Khenkin and Manin [1982], and Khenkin and Manin [1980] is the following (Theorem 8.2). In order that the forms θ

118 G.M. Khenkin, R.G. Novikov

and J, $\psi_\pm$ and $G_\pm$, $\psi_\pm$ and $F_\pm$, respectively, satisfy the Cauchy–Riemann equations (0.11), (0.12), and (0.13), it is necessary and sufficient that the corresponding inverse Penrose transforms of the $(0,1)$-forms θ, $\psi_\pm$, $\phi_\pm$ and of the $(0,2)$-forms J, $G_\pm$, $F_\pm$, respectively, satisfy the Yang–Mills, Weyl–Dirac and Klein–Gordon nonhomogeneous equations.

The fifth topic is touched upon in Chaps. 7, 8, and 9. Originally, the self-dual and the anti-self-dual solutions $a_\pm$ in $\mathcal{D}$ of the Yang–Mills equations were interpreted in the form of holomorphic vector bundles $E_\pm$ over $L_\pm(\mathcal{D})$ (Ward [1977; 1981]; Atiyah and Ward [1977]). In addition, the solutions of the massless equations (Weyl–Dirac for $\ell = 1$, Klein–Gordon for $\ell = 2$, and so on) in the corresponding gauge field $a_\pm$ (Penrose [1969]; Eastwood, Penrose and Wells [1981]; Hitchin [1980]) correspond to the cohomologies $L_\pm(\mathcal{D})$ with coefficients in $E_\pm \otimes \mathcal{O}(-\ell)$.

The interpretation of solutions in $\mathcal{D}$ of the full Yang–Mills equation in the form of holomorphic vector bundles on $L^{(3)}(\mathcal{D})$ was first obtained in the works Isenberg, Yasskin and Green [1978] and Witten [1978].

In Theorem 8.3, the following results from Khenkin and Manin [1982], Khenkin [1984; 1980] are given. They develop and improve the fundamental works of Penrose, Ward, Witten, Isenberg, Yasskin and Green.

The Penrose-type transforms determine the isomorphism between

(a) the space of the holomorphic (respectively, CR-) bundles E_θ over $L^{(3)}(\mathcal{D}) \subset L_+(\mathcal{D}) \times L_-(\mathcal{D})$, which are trivial on all quadrics $\mathcal{L}(x)$, $x \in \mathcal{D}$, and the space of the holomorphic (respectively, smooth) connections ∇_a in the fixed bundle E over the domain $\mathcal{D} \subset \mathbb{C}M$ (respectively, $\mathcal{M}$) satisfying the homogeneous Yang–Mills equation;

(b) the space of the $\bar\partial$-cohomologies of the manifold $L^{(2)}(\mathcal{D})$ with coefficients in $E_\theta(-1,0)$ or $E_\theta(0,-1)$ and the space of holomorphic (respectively, smooth) solutions in $\psi_A, \psi^{A'}$ in $\mathcal{D}$ with values in $E \otimes S_\pm$ of the homogeneous Weyl–Dirac equations in the gauge field a;

(c) the space of the $\bar\partial$-cohomologies $L^{(1)}(\mathcal{D})$ with coefficients in $E_\theta(-2,0)$ or $E_\theta(0,-2)$ and the space of holomorphic (respectively, smooth) solutions $\varphi_\pm$ in $\mathcal{D}$ with values in E of the the homogeneous Klein–Gordon equation in the field a.

Theorem 7.4 gives a new and, we hope, more effective interpretation of the solutions of the Yang–Mills equations in terms of holomorphic (respectively, CR-) bundles over the manifold of the form

$$\{(\zeta,\eta) \in L^{(3)}(\mathcal{D}) : 2i\zeta^0\eta^{0'}(\zeta^1\eta^{0'} + \zeta^0\eta^{1'}) = (\zeta^1\eta^{0'} - \zeta^0\eta^{1'})^2\}.$$

Theorems 7.1 and 8.4 make it possible to characterize those of the Yang–Mills fields that correspond to the holomorphic (respectively, CR-) bundles E_θ on the kth infinitesimal neighborhood $L^{(k)}(\mathcal{D})$ of the manifold $L(\mathcal{D})$ in $L_+(\mathcal{D}) \times L_-(\mathcal{D})$. In particular, the following important statement is valid. In order for the $(0,1)$-form θ with values in End E and the $(0,2)$-form Ω with values in $\operatorname{End} E(-4,-4)$ to be connected on $L_+(\mathcal{D}) \times L_-(\mathcal{D})$ by the Cauchy–Riemann equations of the form

$$\bar{\partial}\theta + \theta \wedge \theta = \Omega\langle \zeta \cdot \eta \rangle^4,$$

it is necessary and sufficient that the corresponding inverse Penrose transforms $\theta \mapsto a_A^{B'}(x)$, $\Omega \mapsto \omega_{AB}^{A'B'}(x)$ simultaneously satisfy the homogeneous Yang–Mills equations and the commutation relations of the form $[f_{AB}, f^{A'B'}] = \omega_{AB}^{A'B'}$, where

$$f_{AC} = \left[\nabla_{CC'}, \nabla_A^{C'}\right], \qquad f^{A'C'} = \left[\nabla^{CC'}, \nabla_C^{A'}\right], \qquad \nabla_{AB'} = \frac{\partial}{\partial x^{AB'}} + a_{AB'}(x).$$

In terms of bundles, a similar statement was first formulated (without proof) in Buchdahl [1985] and then as a hypothesis in Yasskin [1987].

The meaningful physical models are usually described not by the Yang–Mills, Weyl–Dirac or Klein–Gordon free massless fields but by the interacting (Maxwell–Dirac, Yang–Mills–Higgs, Salam–Weinberg) systems (Faddeev and Slavnov [1988]).

In Chap. 9, the results from Khenkin [1981; 1982] are given. They show that the Yang–Mills fields that interact, for example, with the Higgs scalar fields admit an interpretation in the form of suitable holomorphic bundles over $L^{(2)}(\mathcal{D})$. Namely, we consider holomorphic (respectively, CR-) bundles over $L^{(2,5)}(\mathcal{D}) \subset L_+(\mathcal{D}) \times L_-(\mathcal{D})$ that are topologically equivalent to the bundle $E^0 = n_+\mathcal{O}(-1,0) \oplus n_-\mathcal{O}(0,-1)$ and analytically equivalent to this bundle on each quadric $\mathcal{L}^{(1)}(x) \cap L(\mathcal{D})$. Such bundles E_θ are coded by the forms θ on $L_+(\mathcal{D}) \times L_-(\mathcal{D})$ with values in $\operatorname{End} E^0$ and such that

$$\bar{\partial}\theta + \theta \wedge \theta = O\begin{pmatrix} \langle \zeta \cdot \eta \rangle^4 & \langle \zeta \cdot \eta \rangle^3 \\ \langle \zeta \cdot \eta \rangle^3 & \langle \zeta \cdot \eta \rangle^4 \end{pmatrix}, \qquad \theta = \begin{pmatrix} \theta_+ & \phi_+\langle \zeta \cdot \eta \rangle \\ \phi_-\langle \zeta \cdot \eta \rangle & \theta_- \end{pmatrix}. \tag{0.14}$$

Theorems 9.1 and 9.2 state the following:

(a) the nonabelian Radon–Penrose-type transform $\theta_\pm \mapsto a_\pm$, $\phi_\pm \mapsto \varphi_\pm$ realizes a one-to-one correspondence between the holomorphic (respectively, CR-) bundles E_θ on $L^{(2,5)}(\mathcal{D})$ and the holomorphic (respectively, smooth) solutions $(a_\pm, \varphi_\pm)$ of the Yang–Mills–Higgs system on $\mathcal{D} \subset \mathbb{C}M$ (or $\mathcal{M}$) (see Chaps. 1 and 9);

(b) for a fixed bundle E_θ on $L^{(2,5)}(\mathcal{D})$, the (abelian) Penrose-type transform realizes the isomorphism between the cohomology space of the manifold $L^{(2)}(\mathcal{D})$ with coefficients in E_θ and the space of the holomorphic (respectively, smooth) solutions of the Dirac system in the Yang–Mills–Higgs field $(a_\pm, \varphi_\pm)$ (see Chap. 1.9).

The description (in terms of holomorphic vector bundles of $L^{(2,5)}(\mathcal{D})$) of solutions of the general (more physical but more awkward) system of the Yang–Mills–Higgs–Dirac equations follows from the results of Khenkin [1982],

Khenkin and Manin [1982; 1980]. Unfortunately, in the corresponding statements in Khenkin [1982], due to an inaccuracy, there appears a bundle on $L^{(3)}(\mathcal{D})$ instead of bundles on $L^{(2,5)}(\mathcal{D})$.

With the help of these results (as well as those in Theorem 7.2), it is possible to linearize a sophisticated nonlinear system of the Yang–Mills–Higgs–Dirac equation, i.e., for example, to find all holomorphic solutions in the domain $\mathcal{D}$ with sufficiently small norms starting with solutions in $\mathcal{D}$ of the corresponding linearized homogeneous equations.

Theorem 9.3 describes those solutions of the Yang–Mills–Higgs system that correspond to the holomorphic bundles E_θ extended from $L^{(2,5)}(\mathcal{D})$ to $L^{(3)}(\mathcal{D})$.

Ward's (Ward [1977]) interpretation of self-dual solutions of the Yang–Mills equations in terms of holomorphic vector bundles not only clarified the complex-geometric sense of these solutions but turned out to be especially useful for the construction and for the description of global solutions of these equations on the compactified Euclidean space (for the theory of instantons and monopoles, see Atiyah [1979]; Atiyah and Hitchin [1988]; Atiyah and Ward [1977]; Hitchin [1982]; Manin [1984]; Ward [1981]). The representation in the form of holomorphic vector bundles on the space of the light rays of the Yang–Mills–Higgs–Dirac full equations detects the explicit complex-geometric sense of these fundamental equations of relativistic physics. Such an interpretation may also give interesting new and explicit solutions of these equations. We note here explicit solutions of the full Yang–Mills equations obtained in the complex-geometric way in Kapranov and Manin [1986] and Manin [1984].

In this part, we examine only a small part of the works published recently on the complex interpretation with help of the Radon–Penrose-type transforms of physical fields and equations. One of the most brilliant topics, which we did not examine here, is an interpretation of the Einstein equation in terms of complex structures on the space of the null-geodesic. Such an interpretation was discovered in the self-dual case in Penrose [1976]. The complete results for the full Einstein equation were obtained more recently (see Baston and Mason [1987]; LeBrun [1982; 1983; 1991]; Yasskin [1987]).

Among modern mathematical models describing especially effective interacting physical fields are the so-called Yang–Mills supersymmetric equations and supergravitation. It turned out that supersymmetric equations are also successfully described in complex-geometric terms (see Khenkin [1984]; Khudaverdyan, Roslyj and Schwarz [1986]; Manin [1984]; Witten [1978; 1986]; Harnad and Shnider [1986]).

A number of problems of modern string theory are also well interpreted in the framework of the Penrose twistor theory (see Hughston and Shaw [1990]; Isenberg and Yasskin [1986]).

Let us at least note that a deep connection of the theory of the Yang–Mills fields with holomorphic geometry turns out to be fruitful not only for modern physical models but also for modern mathematical theories (see, for example, Manin [1984]; Okonek and Van de Ven [1990]).

Chapter 1
The Gauge Fields and the
Yang–Mills–Higgs Equations

In this chapter, we state the main facts of the classical theory of Yang–Mills fields on the Minkowski or on the Euclidean space (see also Atiyah [1979]; Gu [1981]; Jaffe and Taubes [1980]; Penrose and Rindler [1984,1986]; Faddeev and Slavnov [1988]).

We denote by G either the group $GL(n, \mathbb{C})$ of all invertible matrices of the nth-order or any subgroup of this group. The most important group to consider as such a subgroup is the group $SU(n)$.

Let $\mathcal{G}$ be the Lie algebra of the group G.

Let $x = (x_0, x_1, x_2, x_3)$ be a point of the space R^4 provided with the Minkowski or the Euclidean metric.

A set of smooth functions $a_\mu(x)$, $\mu = 0, 1, 2, 3$ with values in $\mathcal{G}$ is called a *gauge potential*. To such a set correspond the operators of the covariant differentiation

$$\nabla_\mu = \frac{\partial}{\partial x^\mu} + a_\mu.$$

These operators act on vector functions with values in $\mathbb{C}^n$, i.e., on sections of the (trivial) n-dimensional bundle E over R^4. The one-form

$$a = \sum_{\mu=0}^{3} a_\mu(x)dx^\mu$$

is called the *connection form* (in the bundle E) with the gauge group G. Suppose

$$f = da + a \wedge a.$$

The two-form f is called the *curvature form* of the connection a. It may be written in the form of

$$f = \sum_{\alpha=1}^{3} f_{0\alpha}dx^0 \wedge dx^\alpha + f_{23}dx^2 \wedge dx^3 + f_{31}dx^3 \wedge dx^1 + f_{12}dx^1 \wedge dx^2, \quad (1.2)$$

where

$$f_{\mu\nu} = [\nabla_\mu, \nabla_\nu] = \frac{\partial a_\nu}{\partial x^\mu} - \frac{\partial a_\mu}{\partial x^\nu} + [a_\mu, a_\nu].$$

The field a is called *gauge-equivalent* to the field a' if there exists the function $g(x)$ with values in G such that

$$a' = g^{-1}ag + g^{-1}dg. \tag{1.3}$$

Besides, the forms of the curvature of the fields a and a' are connected by the equality

$$f' = g^{-1}fg.$$

The so-called Bianchi identity

$$df + a \wedge f - f \wedge a = 0 \tag{1.4}$$

is valid. In terms of covariant derivatives, this identity has the form

$$\left[\nabla_\lambda, [\nabla_\mu, \nabla_\nu]\right] + \left[\nabla_\nu, [\nabla_\lambda, \nabla_\mu]\right] + \left[\nabla_\mu, [\nabla_\nu, \nabla_\lambda]\right] = 0.$$

We introduce the following designations

$$E_\alpha = f_{0\alpha}, \quad \alpha = 1, 2, 3; \qquad H_1 = f_{23}, \qquad H_2 = f_{31}, \qquad H_3 = f_{12}.$$

In the particular case when $G = U(1)$, the vector $E = (E_1, E_2, E_3)$ is called the *electric field*, and $H = (H_1, H_2, H_3)$ is called the *magnetic field*.

We consider the dual two-form $*f = \tilde{f}$ whose coefficients are defined by the equalities

$$\tilde{f}_{\mu\nu} = \frac{1}{2}\sum_{\alpha,\beta}\sqrt{|\det g_{\alpha\beta}|}\varepsilon_{\mu\nu\alpha\beta}f^{\alpha\beta}, \qquad f^{\alpha\beta} = \sum_{i,j}g^{\alpha i}g^{\beta j}f_{ij},$$

where the metric $g_{\alpha\beta}$ is considered Euclidean or pseudo-Euclidean, and $\varepsilon_{\mu\nu\alpha\beta}$ is a sign of the transposition $(\mu, \nu, \alpha, \beta)$.

For the case of the Euclidean metric $g_{\alpha\beta} = \delta_{\alpha\beta}$, we have

$$\tilde{f} = \sum_\alpha H_\alpha dx^0 \wedge dx^\alpha + E_1 dx^2 \wedge dx^3 + E_2 dx^3 \wedge dx^1 + E_3 dx^1 \wedge dx^2. \tag{1.5}$$

For the case of the pseudo-Euclidean metric $g_{\alpha\beta} = \varepsilon_\alpha\delta_{\alpha\beta}$, where $\varepsilon_1 = \varepsilon_2 = \varepsilon_3 = -\varepsilon_0 = -1$, we have

$$\tilde{f} = \sum_\alpha H_\alpha dx^0 \wedge dx^\alpha - E_1 dx^2 \wedge dx^3 - E_2 dx^3 \wedge dx^1 - E_3 dx^1 \wedge dx^2.$$

The following differential equation is called the *Yang–Mills equation with a given current*:

$$d\tilde{f} + a \wedge \tilde{f} - \tilde{f} \wedge a = j, \tag{1.6}$$

where $j = \sum_{\alpha<\beta<\gamma}j_{\alpha\beta\gamma}dx^\alpha \wedge dx^\beta \wedge dx^\gamma$. The three-form j is called the *form of the (axial) current*.

We can write equation (1.6) in the following form:

$$\sum_\mu\left[\nabla_\mu, f^{\mu\nu}\right] = j^\nu(x),$$

where $j_0 = -j_{123}$, $j^1 = j_{023}$, $j^2 = -j_{013}$, $j^3 = j_{012}$.

The homogeneous Yang–Mills equation is of special interest:

$$\sum_\mu [\nabla_\mu, f^{\mu\nu}] = 0. \tag{1.7}$$

On the Euclidean space and for the gauge group $\mathrm{SU}(n)$, equation (1.7) is the Euler equation for the Lagrangian of the form

$$||f||^2 = -\int_{R^4} \mathrm{Trace}(f \wedge \tilde{f}).$$

The density of this Lagrangian (and, consequently, of the Yang–Mills equation) depends only on the conformal class of the metric on R^4.

The form f as any two-form on R^4 is expanded in the sum of the form $f = f^+ + f^-$, where

$$2f^+ = f + \tilde{f}, \qquad 2f^- = f - \tilde{f}, \qquad \tilde{f}^\pm = \pm f^\pm \tag{1.8}$$

for the case of the Euclidean metric, and

$$2f^+ = f - i\tilde{f}, \qquad 2f^- = f + i\tilde{f}, \qquad \tilde{f}^\pm = \pm i f^\pm, \quad i = \sqrt{-1}, \tag{1.9}$$

for the case of the Minkowski metric.

The fields a with the property $f^+ \equiv 0$ or $f^- \equiv 0$ are automatically solutions of the equation (1.7). Besides, the fields with the property $f^- \equiv 0$ are called *self-dual*, but with the property $f^+ \equiv 0$, *anti-self-dual*.

On the Euclidean space, the fields a with the gauge group $\mathrm{SU}(n)$ with finite action $||f||^2$ and with the fixed topological invariant (charge)

$$8\pi^2 k = -\int_{R^4} \mathrm{Trace}(f \wedge f), \quad k = \pm 1, \pm 2, \ldots,$$

are of the most physical and mathematical interest.

It turned out (see Belavin, Polyakov, Schwartz and Tyupkin [1975]) that for a fixed k, the minimum of the action $||f||^2$ is reached on the fields with the property $\tilde{f} = (\mathrm{sign}\, k)f$, i.e., on the Yang–Mills self-dual ($k > 0$) or anti-self-dual ($k < 0$) fields (with topological charge k). These fields are called *instantons* ($k > 0$) or *anti-instantons* ($k < 0$), respectively. A more detailed presentation of the mathematical instanton theory can be found in Atiyah [1979] and Manin [1984].

One of the most interesting problems in connection with the instanton theory is to find all (including non-self-dual) solutions of the Yang–Mills equation (1.7) on R^4 with the gauge group $\mathrm{SU}(n)$, with the finite action $||f||^2$ and with the fixed charge $k > 0$. The first examples of such non-self-dual solutions are given in Sibner, Sibner and Uhlenbeck [1989] and are based on Taubes [1985]. In connection with this problem, it is shown in Bourguignon and Lawson [1981] that any local minimum for the Lagrangian $||f||^2$ is at the same time a global minimum, i.e., it is realized again on the instantons.

On the Euclidean space, equation (1.6) may be written in the form

$$\sum_{\mu}[\nabla_{\mu},[\nabla_{\mu},\nabla_{\nu}]] = j_{\nu}(x), \tag{1.10}$$

and equation (1.7) in the form

$$\sum_{\mu}[\nabla_{\mu},[\nabla_{\mu},\nabla_{\nu}]] = 0, \tag{1.11}$$

respectively. It is very useful to consider these equations on $\mathbb{C}^4$ as well. Besides, the functions $a_{\mu}(x)$, $j_{\nu}(x)$ are considered to be analytical with respect to the variables x^{μ}, $\mu = 0,1,2,3$. If in the equations (1.10) and (1.11), $x^0 = \chi^0$, $x^1 = i\chi^1$, $x^2 = i\chi^2$, and $x^3 = i\chi^3$, where $\{\chi^{\mu}\}$ are real variables, then the equations (1.10) and (1.11) with respect to the variables $\{\chi^{\mu}\}$ are the Yang–Mills equations on the Minkowski space.

In $\mathbb{C}^4$, we now go over to the so-called spinor coordinates $x^{\alpha\alpha'}$ where $\alpha = 0,1$, $\alpha' = 0',1'$,

$$x^{\alpha\alpha'} = \begin{pmatrix} x^0 + ix^1 & -i(x^2 + ix^3) \\ -i(x^2 - ix^3) & x^0 - ix^1 \end{pmatrix}.$$

We provide the space $\mathbb{C}^4$ with such coordinates with the metric $\det|dx^{\alpha\alpha'}|$. In accordance with this metric, the indices α and α' are raised and lowered by the formula $x_{\alpha\alpha'} = x^{\beta\beta'}\varepsilon_{\beta\alpha}\varepsilon_{\beta'\alpha'}$, where $\varepsilon^{01} = -\varepsilon^{0'1'} = -1$; $\varepsilon^{\alpha\beta} = -\varepsilon^{\beta\alpha}$; $\varepsilon^{\alpha'\beta'} = -\varepsilon^{\beta'\alpha'}$.

In spinor coordinates, the form a from (1.1) may be written as

$$a = \sum_{\alpha,\alpha'} a_{\alpha\alpha'} dx^{\alpha\alpha'}. \tag{1.13}$$

The following formula is valid

$$a_{\alpha\alpha'} = \frac{1}{2}\begin{pmatrix} a_0 - ia_1 & i(a_2 - ia_3) \\ i(a_2 + ia_3) & a_0 + ia_1 \end{pmatrix}.$$

The form f from (1.2) may be written as

$$f = f_{00'01'}dx^{00'} \wedge dx^{01'} + f_{00'10'}dx^{00'} \wedge dx^{10'} + f_{00'11'}dx^{00'} \wedge dx^{11'}$$
$$+ f_{10'11'}dx^{10'} \wedge dx^{11'} + f_{11'01'}dx^{11'} \wedge dx^{01'} + f_{01'10'}dx^{01'} \wedge dx^{10'}.$$

The following formulae are also true:

$$4f_{00'01'} = if_{02} + if_{31} + f_{03} + f_{12}; \quad 4f_{00'10'} = if_{02} - if_{31} - f_{03} + f_{12};$$
$$4f_{00'11'} = 2if_{01}; \quad 4f_{10'11'} = -if_{02} - if_{31} + f_{03} + f_{12};$$
$$4f_{11'01'} = if_{02} - if_{31} + f_{03} - f_{12}; \quad 4f_{01'10'} = -2if_{23}.$$

In spinor coordinates, the form f turns into the sum of the two addends:

$$f_{\alpha\alpha'\beta\beta'} = \tfrac{1}{2}(f_{\alpha\alpha'\beta\beta'} - f_{\beta\alpha'\alpha\beta'} + f_{\beta\alpha'\alpha\beta'} - f_{\beta\beta'\alpha\alpha'})$$
$$= \tfrac{1}{2}(\varepsilon_{\alpha\beta}f^{\mu}_{\mu\alpha'\beta'} + \varepsilon_{\alpha'\beta'}f^{\mu'}_{\beta\mu'\alpha}),$$

of which one is self-dual, the other anti-self-dual. Suppose

$$f_{\alpha\beta} = \tfrac{1}{2}f^{\mu'}_{\beta\mu'\alpha}; \qquad f_{\alpha'\beta'} = \tfrac{1}{2}f^{\mu}_{\mu\alpha'\beta'}.$$

Then

$$f_{\alpha\alpha'\beta\beta'} = \varepsilon_{\alpha\beta}f_{\alpha'\beta'} + \varepsilon_{\alpha'\beta'}f_{\alpha\beta}.$$

We have

$$\tilde{f}_{\alpha\alpha'\beta\beta'} = -\varepsilon_{\alpha\beta}f_{\alpha'\beta'} + \varepsilon_{\alpha'\beta'}f_{\alpha\beta},$$
$$f = -f_{\alpha\beta}dx^{\alpha0'} \wedge dx^{\beta1'} + f_{\alpha'\beta'}dx^{0\alpha'} \wedge dx^{1\beta'}. \tag{1.14}$$

In spinor coordinates, the Yang–Mills equation (1.6), (1.10) may be written as

$$j^{\beta'}_{\alpha} = \left[\nabla_{\alpha\alpha'}, f^{\alpha'\beta'}\right] = \left[\nabla^{\beta\beta'}, -f_{\alpha\beta}\right], \tag{1.15}$$

where

$$\nabla^{\alpha\alpha'} = \frac{\partial}{\partial x_{\alpha\alpha'}} + a^{\alpha\alpha'}, \qquad \nabla_{\alpha\alpha'} = \frac{\partial}{\partial x^{\alpha\alpha'}} + a_{\alpha\alpha'},$$
$$j = 2j^{\beta'}_{\alpha}dx^{\alpha0'} \wedge dx^{\alpha1'} \wedge dx_{\alpha\beta'}.$$

For details, see Gindikin and Khenkin [1981] and Penrose and Rindler [1984, 1986].

The Yang–Mills equations are a rather sophisticated nonlinear system of equations. In most physical applications, these equations are solved and investigated only by perturbation theory by fixing a calibration and introducing boundary conditions at infinity (see, for example, Faddeev and Slavnov [1988]).

It was first discovered in Belavin, Polyakov, Schwartz and Tyupkin [1975] that the Yang–Mills equations on the Euclidean space have very interesting explicit solutions ("instantons") from the physical point of view, which can be found on the basis of algebraic-topological constructions outside of the framework of perturbation theory. After this, a very large and interesting class of solutions of the Yang–Mills vacuum (currentless) equation with the group SL(2, $\mathbb{C}$) was obtained with the help of the following substitution (see Corrigan and Fairlie [1977]):

$$a_{\beta0'} = \frac{1}{2\varphi}\begin{pmatrix} \nabla_{\beta0'}\varphi & 0 \\ -2\nabla_{\beta1'}\varphi & -\nabla_{\beta0'}\varphi \end{pmatrix}; \qquad a_{\beta1'} = \frac{1}{2\varphi}\begin{pmatrix} -\nabla_{\beta1'}\varphi & -2\nabla_{\beta0'}\varphi \\ 0 & \nabla_{\beta1'}\varphi \end{pmatrix},$$

$$\tag{1.16}$$

where $\nabla_{\alpha\beta'} = \frac{\partial}{\partial x^{\alpha\beta'}}$.

Lack of a current in the Yang–Mills equations for the field a of the form (1.16) turned out to be equivalent to the following scalar φ^4 equation:

$$\Box\varphi + \lambda\varphi^3 = 0, \tag{1.17}$$

where $\Box = \nabla^{\alpha\beta'}\nabla_{\alpha,\beta'}$, $\lambda \in \mathbb{C}$.

In order that the Yang–Mills field a corresponding to (1.17) would be self-dual, it turned out that the condition $\lambda = 0$ is necessary and sufficient. In addition to the scalar functions φ of the form

$$\varphi(x) = \frac{c}{\det|x^{\alpha\alpha'} - c^{\alpha\alpha'}|},$$

where $c \in R$, $c^{\alpha\alpha'}$ is a unitary matrix, the corresponding Yang–Mills fields exactly coincide with SU(2) instantons with the topological charge 1. To obtain all the SU(2) anti-instantons (with the charge -1) in this way, it turned out to be sufficient to take the special solutions of (1.17) (with constant $\lambda = 1$) of the form

$$\varphi(x) = \frac{c}{c^2 + \det|x^{\alpha\alpha'} - c^{\alpha\alpha'}|}.$$

Not only are the Yang–Mills fields of great interest but also the gauge fields interacting with matter fields (see, for example, the Salam–Weinberg model in Faddeev and Slavnov [1988]). We give here a system of Yang–Mills–Higgs equations describing the interaction of the Yang–Mills fields with the so-called Higgs scalar fields. Let

$$\nabla^{\pm}_{\alpha\beta'} = \frac{\partial}{\partial x^{\alpha\beta'}} + a^{\pm}_{\alpha\beta'}(x)$$

be holomorphic $GL(n_{\pm}, \mathbb{C})$ connections in the fixed (for simplicity, trivial) $n_{\pm}$-dimensional bundles $E_{\pm}$ over some domain $\mathcal{D}$ in $\mathbb{C}^4$. The $(n_{\pm} \times n_{\mp})$-matrix functions $\varphi_{\pm}$ are considered to be gauge-equivalent to the matrix functions $\varphi'_{\pm}$ if $\varphi'_{\pm} = g_{\pm}\varphi_{\pm}g_{\mp}^{-1}$, where $g_{\pm}$ are the gauge transforms for the fields $a_{\pm}$. Suppose

$$\nabla^{\pm\mp}_{\alpha\alpha'}\varphi_{\pm} = \frac{\partial\varphi_{\pm}}{\partial x^{\alpha\alpha'}} + a^{\pm}_{\alpha\alpha'}\varphi_{\pm} - \varphi_{\pm}a^{\mp}_{\alpha\alpha'},$$

$$\Box_{\pm\mp}\varphi_{\pm} = \left[\nabla^{\alpha\alpha'}_{\mp\pm}, [\nabla^{\pm\mp}_{\alpha\alpha'}, \varphi_{\pm}]\right].$$

The gauge fields $a_{\pm}$ and $\varphi_{\pm}$ are called the *Yang–Mills–Higgs interacting fields* if they satisfy a system of equations of the following form on $\mathcal{D}$,

$$\nabla^{++}_{\alpha\beta'}f^{\beta'}_{+\alpha'} = \tfrac{1}{2}\left([\nabla^{+-}_{\alpha\alpha'}, \varphi_+]\varphi_- - \varphi_+[\nabla^{-+}_{\alpha\alpha'}, \varphi_-]\right),$$

$$\nabla^{--}_{\alpha\beta'}f^{\beta'}_{-\alpha'} = \tfrac{1}{2}\left([\nabla^{-+}_{\alpha\alpha'}, \varphi_-]\varphi_+ - \varphi_-[\nabla^{+-}_{\alpha\alpha'}, \varphi_+]\right), \tag{1.18}$$

$$\Box_{+-}\varphi_+ + \lambda\varphi_+\varphi_-\varphi_+ = 0,$$

$$\Box_{-+}\varphi_- + \lambda\varphi_-\varphi_+\varphi_- = 0, \quad \lambda \in \mathbb{C},$$

where $f_{\pm}$ are curvature forms of the connections $\nabla_{\pm}$.

The classical Yang–Mills–Higgs equations are obtained from the equations (1.18) if we restrict ourselves to the case when $n_+ = n_-$, $\mathcal{D}$ is a domain of the real Minkowski space $\varphi_+ = \varphi_- = \varphi_+^*$, $a_+ = a_- = -a_+^*$ and the gauge group $G = \mathrm{SU}(n)$ (see Atiyah and Hitchin [1988]; Faddeev and Slavnov [1988]).

The meaningful (from the physical and mathematical point of view) solutions of the Yang–Mills–Higgs system outside the framework of perturbation theory were first found in 1974 and were called *Polyakov–'t Hooft monopoles* (see Jaffe and Taubes [1980]).

The interaction of the Yang–Mills fields simultaneously with both scalar and spinor fields is described by a more awkward system of the Yang–Mills–Higgs–Dirac equations (see, for example, Faddeev and Slavnov [1988]).

The introduction in recent years of serious algebraic-geometric techniques led to the description of fundamental results concerning the instantons and monopoles (in Bogomolov's limit $\lambda \to 0$) with any topological charge (see Atiyah [1979]; Atiyah and Hitchin [1988]; Forgacs, Horvath and Palla [1981]; Hitchin [1982]; Manin [1984]; Taubes [1985]; Ward [1981]). The most important role in this description belongs to the interpretation of the Yang–Mills self-dual fields in the form of holomorphic vector bundles over the space of the Penrose twistors, which appeared in Ward [1977].

Among the most important and interesting analytical problems concerning the Yang–Mills–Higgs–Dirac equations, there is an investigation of the solvability of the Cauchy problem. Sufficient conditions of existence of global solutions to the Cauchy problem for these equations on the Minkowski space were obtained in the important papers Segal [1979], Choquet-Bruhat and Christodoulou [1981], and Eardley and Moncrief [1982]. There are group-theoretical and analytical arguments (see Flato and Simon [1980] and here Chap. 7) that the Cauchy problem for the Yang–Mills equation must admit a linearization. The linearization (by the Radon–Penrose transform) of the Cauchy problem for the Yang–Mills self-dual equation follows from constructions given in Gindikin and Khenkin [1981], Khenkin [1984; 1980], and Manakov and Zakharov [1981]. For further results, see Chaps. 7, 8, and 9.

Chapter 2
The Yang–Mills Equations as the Compatibility Equations of a Linear System

Developing the works Belavin and Zakharov [1978], Witten [1978], and Forgacs, Horvath and Palla [1982], we give here an expanded interpretation of the Yang–Mills equations as the compatibility equations of a system with complex spectral parameters. The results of this chapter are mainly stated for the Yang–Mills equations (1.10) and (1.11) in the domain $\mathcal{D} \in \mathbb{C}^4$. For the case of the equations on the real Euclidean (or pseudo-Euclidean) space, the corresponding results are specialized almost automatically.

We consider the diagonal $y = z$ in the space $\mathbb{C}^8$ (respectively, R^8) with the coordinates $y \in \mathbb{C}^4$ and $z \in \mathbb{C}^4$. On this four-dimensional diagonal, we take $x = \frac{1}{2}(y + z)$ as coordinates. We also consider the antidiagonal $y = -z$ on which we take $w = \frac{1}{2}(y - z)$ as coordinates.

In the trivial n-dimensional bundle over $\mathbb{C}^8$, we introduce the operators of covariant differentiation of the form

$$\nabla_{x^j} = \frac{\partial}{\partial x^j} + a_{x^j}(x, w); \qquad \nabla_{w^j} = \frac{\partial}{\partial w^j} + a_{w^j}(x, w),$$

$$\nabla_{y^j} = \frac{\partial}{\partial y^j} + a_{y^j}(y, z); \qquad \nabla_{z^j} = \frac{\partial}{\partial z^j} + a_{z^j}(y, z),$$

where the functions a_{x^j}, a_{w^j}, a_{y^j}, a_{z^j} with values in $\mathrm{gl}(n, \mathbb{C})$ depend on their arguments holomorphically (respectively, smoothly).

Furthermore, $a_{x^j} = a_{y^j} + a_{z^j}$; $a_{w^j} = a_{y^j} - a_{z^j}$. Then the following equalities are valid:

$$\nabla_{x^j} = \nabla_{y^j} + \nabla_{z^j}; \qquad \nabla_{w^j} = \nabla_{y^j} - \nabla_{z^j}.$$

All equations in this chapter are considered in convex domains of the form $x + w \in \mathcal{D}$, $x - w \in \mathcal{D}$, where $\mathcal{D}$ is a domain in $\mathbb{C}^4$ (respectively, R^4).

Based on Belavin and Zakharov [1978], Forgacs, Horvath and Palla [1982], and Witten [1978], we consider on $\mathbb{C}^8$ (respectively, R^8) the following linear system of equations depending on two spectral parameters $\lambda_1, \lambda_2 \in \mathbb{C}^1$,

$$(\alpha_\nu \cdot \nabla_y)\eta(y, z, \lambda_1, \lambda_2) = O\big((y - z)^k\big),$$
$$(\beta_\nu \cdot \nabla_z)\eta(y, z, \lambda_1, \lambda_2) = O\big((y - z)^k\big),$$

$$(2.1)$$

where

$$\nu = 1, 2;$$
$$\alpha_1 = (1, -i, -\lambda_1, -i\lambda_1); \qquad \alpha_2 = (\lambda_1, i\lambda_1, 1, -i),$$
$$\beta_1 = (1, -i, -\lambda_2, i\lambda_2); \qquad \beta_2 = (\lambda_2, i\lambda_2, 1, i),$$

with k being a natural number. Here we denoted by $O(w^k)$, $w \in \mathbb{C}^4$ the expression of the form

$$\sum_{\nu_0 + \nu_1 + \nu_2 + \nu_3 = k} C_{\nu_0 \nu_1 \nu_2 \nu_3}(x, w) w_0^{\nu_0} w_1^{\nu_1} w_2^{\nu_2} w_3^{\nu_3}.$$

It is convenient to write the system (2.1) in the variables x and w. Thus we have

$$(m \cdot \nabla_x)\mu(x, w, \lambda) = 0(w^k); \quad (m \cdot \nabla_w)\mu(x, w, \lambda) = O(w^k);$$

$$(m_1 \nabla_{x^0} - m_0 \nabla_{x^1} - m_3 \nabla_{w^2} + m_2 \nabla_{w^3})\mu(x, w, \lambda) = O(w^k); \qquad (2.1^1)$$

$$(m_1 \nabla_{w_0} - m_0 \nabla_{w^1} - m_3 \nabla_{x^2} + m_2 \nabla_{x^3})\mu(x, w, \lambda) = O(w^k),$$

where

$$m_0(\lambda) = (1 - \lambda_1\lambda_2), \qquad m_1(\lambda) = -i(1 + \lambda_1\lambda_2),$$

$$m_2(\lambda) = -(\lambda_1 + \lambda_2), \qquad m_3(\lambda) = -i(\lambda_1 - \lambda_2).$$

It is useful to write the system (2.1) in spinor coordinates:

$$(\nabla_{y^{00'}} + i\lambda_1 \nabla_{y^{10'}})\eta(y, z, \lambda) = O(w^k),$$

$$(\lambda_1 \nabla_{y^{11'}} - i\nabla_{y^{01'}})\eta(y, z, \lambda) = O(w^k),$$

$$(\nabla_{z^{00'}} + i\lambda_2 \nabla_{z^{01'}})\eta(y, z, \lambda) = O(w^k), \qquad (2.1^2)$$

$$(\lambda_2 \nabla_{z^{11'}} - i\nabla_{z^{10'}})\eta(y, z, \lambda) = O(w^k).$$

The compatibility equations for the system (2.1) have the form

$$[\nabla_{y^\mu}, \nabla_{y^\nu}] = \tfrac{1}{2}\varepsilon_{\mu\nu\alpha\beta}[\nabla_{y\alpha}, \nabla_{y\beta}] + O(w^{k-1}),$$

$$[\nabla_{z^\mu}, \nabla_{z^\nu}] = \tfrac{1}{2}\varepsilon_{\mu\nu\alpha\beta}[\nabla_{z^\alpha}, \nabla_{z^\beta}] + O(w^{k-1}), \qquad (2.2)$$

$$[\nabla_{y^\mu}, \nabla_{z^\nu}] = O(w^{k-1}).$$

In the equations (2.1) and (2.2), the fields $a_{x^j}(x, w)$ and $a_{w^j}(x, w)$ may be considered to be the polynomials of (k-1)th degree in the variable w. The function $\mu(x, w, \lambda)$ in the equation (2.1^1) may be considered to be the polynomial of kth degree in w.

If the compatibility equations (2.2) hold, then on the basis of a suitable linear integral equation, it is possible to find a function $\mu(x, w, \lambda)$, holomorphic (respectively, smooth) in x, w and smooth in $\lambda \in \mathbb{C}P^1 \times \mathbb{C}P^1$ with values in $GL(n, \mathbb{C})$, which satisfies the equations (2.2^1). We explain in detail how to obtain the corresponding integral equations, since the operators that occur here are useful for the further presentation.

We consider a system of linear differential equations of the form

$$Df = h + O(w^k), \qquad (2.3)$$

or, in greater detail,

$$D_j f(x, w, \lambda) = h_j(x, w, \lambda) + O_j(w^k), \quad j = 0, 1, 2, 3,$$

where

$$D_0 = \left(m(\lambda) \cdot \frac{\partial}{\partial x}\right), \qquad D_1 = \left(m(\lambda) \cdot \frac{\partial}{\partial w}\right),$$

$$D_2 = m_1(\lambda)\frac{\partial}{\partial x^0} - m_0(\lambda)\frac{\partial}{\partial x^1} - m_3(\lambda)\frac{\partial}{\partial w^2} + m_2(\lambda)\frac{\partial}{\partial w^3},$$

$$D_3 = m_1(\lambda)\frac{\partial}{\partial w^0} - m_0(\lambda)\frac{\partial}{\partial w^1} - m_3(\lambda)\frac{\partial}{\partial x^2} + m_2(\lambda)\frac{\partial}{\partial x^3}.$$

The function $f(x, w, \lambda)$ is a polynomial of kth degree in w; the functions $h_j(x, w, \lambda)$ are polynomials of $(k-1)$th degree in w and linearly increasing in λ_1 and in λ_2.

For the solvability of the system (2.3) in the functions f holomorphic (respectively, smooth) in x and smooth in $(\lambda_1, \lambda_2) \in \mathbb{C}P^1 \times \mathbb{C}P^1$, the following conditions are necessary and sufficient: The functions h_j should also be holomorphic (respectively, smooth) in x, and the following compatibility conditions are valid

$$\hat{D}h = O(w^{k-1}), \tag{2.4}$$

or, in greater detail,

$$D_j h_\ell - D_\ell h_j = O(w^{k-1}), \quad j, \ell = 0, 1, 2, 3.$$

Moreover, it is possible to construct the integral operator G with values in the space of the functions $f(x, w, \lambda)$ so that

$$DGh = h \tag{2.3^1}$$

in the space of the functions $h(x, w, \lambda)$ satisfying (2.4). The existence of such an operator follows, for example, from Ehrenpreis's "fundamental principle" or from explicit formulae for the solution of the $\bar{\partial}$-equation (see Part 1 of this volume).

Formula (2.3^1) makes it possible to find the necessary solution of the system (2.1^1) (or (2.1)) on the basis of the integral equation of the form

$$\mu = I + GA\mu$$

where

$$-A = \begin{cases} m(\lambda)a_x(x, w) \\ m(\lambda)a_w(x, w) \\ m_1(\lambda)a_{x^0} - m_0(\lambda)a_{x^1} - m_3(\lambda)a_{w^2} + m_2(\lambda)a_{w^3} \\ m_1(\lambda)a_{w^0} - m_0(\lambda)a_{w^1} - m_3(\lambda)a_{x^2} + m_2(\lambda)a_{x^3}. \end{cases}$$

It is useful to do a change of variables in order to find an explicit form of the operator G:

$$\begin{pmatrix} x^0 & w^0 \\ x^1 & w^1 \end{pmatrix} = \begin{pmatrix} m_0 & m_1 \\ m_1 & -m_0 \end{pmatrix} \begin{pmatrix} \tilde{x}^0 & \tilde{w}^0 \\ \tilde{x}^1 & \tilde{w}^1 \end{pmatrix},$$

$$\begin{pmatrix} x^2 & w^2 \\ x^3 & w^3 \end{pmatrix} = \begin{pmatrix} m_2 & m_3 \\ m_3 & -m_2 \end{pmatrix} \begin{pmatrix} \tilde{x}^2 \tilde{w}^2 \\ \tilde{x}^3 \tilde{w}^3 \end{pmatrix},$$

and furthermore,

$$\chi^0 = \tilde{x}^0 + \tilde{x}^2, \qquad \chi^1 = \tilde{x}^0 - \tilde{x}^2, \qquad \chi^2 = \tilde{x}^1 + \tilde{w}^3, \qquad \chi^3 = \tilde{w}^1 + \tilde{x}^3,$$
$$\omega^0 = \tilde{w}^0 + \tilde{w}^2, \qquad \omega^1 = \tilde{w}^0 - \tilde{w}^2, \qquad \omega^2 = -\tilde{w}^3, \qquad \omega^3 = \tilde{w}^1.$$

After this change, the system (2.3) takes the form

$$D'_j f'(\chi, \omega, \lambda) = h'_j(\chi, \omega, \lambda) + O_j(w^k), \quad j = 0, 1, 2, 3,$$

where

$$D'_0 = \frac{\partial}{\partial \chi^0}, \qquad D'_1 = \frac{\partial}{\partial \omega^0}, \qquad D'_2 = \frac{\partial}{\partial \omega^2}, \qquad D'_3 = \frac{\partial}{\partial \omega^3},$$

and the functions f' and h'_j are holomorphic (smooth) in the variables χ and ω. The variables χ^1, χ^2, χ^3, ω^1 are parameters with respect to the operators D'_j. To find the solving operator G for the obtained system (for the case of the convex domain $\mathcal{D}$) is a standard problem of function theory (see Part 1 or Khenkin [1985]).

Further, we consider the system (2.2) of the special form

$$[\nabla_{y^\mu}, \nabla_{y^\nu}] = \tfrac{1}{2}\varepsilon_{\mu\nu\alpha\beta}[\nabla_{y^\alpha}, \nabla_{y^\beta}]$$
$$- \tfrac{1}{3}\big(\mathcal{J}_{\mu\nu j}(x) - \tfrac{1}{2}\varepsilon_{\mu\nu\alpha\beta}\mathcal{J}_{\alpha\beta j}(x)\big)w^j + O(w^2),$$
$$[\nabla_{z^\mu}, \nabla_{z^\nu}] = - \tfrac{1}{2}\varepsilon_{\mu\nu\alpha\beta}[\nabla_{z^\alpha}, \nabla_{z^\beta}] \qquad\qquad (2.5)$$
$$- \tfrac{1}{3}\big(\mathcal{J}_{\mu\nu j}(x) + \tfrac{1}{2}\varepsilon_{\mu\nu\alpha\beta}\mathcal{J}_{\alpha\beta j}(x)\big)w^j + O(w^2),$$
$$[\nabla_{y^\mu}, \nabla_{z^\nu}] = \tfrac{1}{6}\mathcal{J}_{j\nu\mu}w^j + O(w^2).$$

The following equality follows from these equations,

$$\sum_{\mu=0}^{3} [\nabla_{\chi^\mu}, [\nabla_{\chi^\mu}, \nabla_{\chi^\nu}]] = \mathcal{J}_\nu(x) + O(w), \qquad (2.5^1)$$

which means that the field $a(x) = a_\mu(x)dx^\mu$ so that

$$a_j(x) = a_{x^j}(x, 0) \qquad\qquad (2.6)$$

satisfies the Yang–Mills equation (1.10).

Conversely, let the field $a(x)$ satisfy the Yang–Mills equations (1.10). Then it is possible to define $a_{x^j}(x, w)$ and $a_{w^j}(x, w)$ so that the equations (2.5) and (2.6) will hold. For example, it is possible to define these fields by the formulae

$$
\begin{aligned}
a_{x^\mu}(x, w) &= a_\mu(x) - \tilde{f}_{\mu j}(x)w^j - \tfrac{1}{2}w^j w^k\left[\nabla_j, f_{\mu k}(x)\right], \\
a_{w^\mu}(x, w) &= -\tfrac{1}{2}f_{\mu j}(x)w^j - \tfrac{1}{3}w^j w^k\left[\nabla_j, \tilde{f}_{\mu k}(x)\right],
\end{aligned}
\tag{2.7}
$$

where $f_{\mu j}, \tilde{f}_{\mu j}, \nabla_j$ are defined by the formulae (1.2) and (1.5). The following theorem implies from the mentioned relations.

Theorem 2.1 (Witten [1978]). *If the compatibility equations (2.2) for the system (2.1) are valid for $k = 3$, then the field $a(x)$, $x \in \mathcal{D}$ defined by the formula (2.6) satisfies the Yang–Mills vacuum equation (1.11). The inverse statement is valid: If the field $a(x)$ satisfies (1.11), then the fields $a_{x^\mu}(x, w)$ and $a_{w^\mu}(x, w)$ defined by formulae (2.7) satisfy the equations (2.2) for $k = 3$.*

This important result further leads (see Chaps. 7 and 8) to the meaningful interpretation of the Yang–Mills equations for the gauge fields by $\bar{\partial}$-equations for scattering data of these fields.

The following result describes all those solutions of the Yang–Mills equation that are generated by the compatibility equations (2.2) for $k \geq 4$.

Theorem 2.2. *If the compatibility equations (2.2) are valid for $k = 4$, then the field $a(x)$, $x \in \mathcal{D}$ defined by formula (2.6) satisfies the Yang–Mills vacuum equation (1.11); and, moreover, the self-dual $f_{\alpha\beta}$ and the anti-self-dual $f_{\alpha'\beta'}$ components of the curvature of the field a of the form (1.14) commutate*

$$
\left[f_{\alpha\beta}(x), f_{\alpha'\beta'}(x)\right] = 0, \quad x \in \mathcal{D}.
\tag{2.8}
$$

The inverse statement is also valid: For any solution $a(x), x \in \mathcal{D}$ of equation (1.11) with the additional property (2.8), there exist the fields $a_{x^j}(x, w)$, $a_{w^j}(x, w)$ satisfying the condition (2.6) and such that the equalities (2.2) will hold for $k = 4$.

Realization of the compatibility conditions (2.2) for $k \geq 5$ leads to the solutions of the Yang–Mills equations with the commutation conditions (2.8) and with the following additional conditions: For any $(\alpha_1, \dots, \alpha_\ell)$ and $(\beta'_1, \dots, \beta'_\ell)$, $\alpha_i = 0, 1$, $\beta'_j = 0', 1'$,

$$
\sum \left[\nabla^{\alpha_{i_\ell}\beta'_{j_\ell}}\left[\cdots\left[\nabla^{\alpha_{i_3}\beta'_{j_3}}f_{\alpha_{i_1}\alpha_{i_2}}(x), f_{\beta'_{j_1}\beta'_{j_2}}(x)\right]\cdots\right]\right] = 0,
\tag{2.8^1}
$$

where $\ell = 3, \dots, k - 3$; the sum is taken for all transpositions $(i_1, \dots, i_\ell)$, $(j_1, \dots, j_\ell)$ of the sequence $(1, 2, \dots, \ell)$.

Now we give a proof of the first part of the theorem. In order to prove that (2.8) follows from (2.2) for $k = 4$, it is sufficient to prove the equality

$$
[f_{x^i x^j} + \tilde{f}_{x^i x^j}, f_{x^\alpha x^\beta} - \tilde{f}_{x^\alpha x^\beta}] = O(w),
\tag{2.9}
$$

where
$$f_{x^i x^j} = [\nabla_{x^i}, \nabla_{x^j}], \qquad \tilde{f}_{x^\mu x^\nu} = \tfrac{1}{2}\varepsilon_{\mu\nu\alpha\beta}[\nabla_{x^\alpha}, \nabla_{x^\beta}].$$

From (2.1) and (2.2) for $k = 4$, one finds

$$[\nabla_{x^\mu}, \nabla_{x^\nu}] = [\nabla_{y^\mu} + \nabla_{z^\mu}, \nabla_{y^\nu} + \nabla_{z^\nu}] = [\nabla_{y^\mu}, \nabla_{y^\nu}] + [\nabla_{z^\mu}, \nabla_{z^\nu}] + O(w^3).$$

Hence and from (2.2), it follows that

$$\begin{aligned}
f_{x^\mu x^\nu} + \tilde{f}_{x^\mu x^\nu} &= \big([\nabla_{y^\mu}, \nabla_{y^\nu}] + [\nabla_{z^\mu}, \nabla_{z^\nu}]\big) \\
&\quad + \tfrac{1}{2}\varepsilon_{\mu\nu\alpha\beta}\big([\nabla_{y^\alpha}, \nabla_{y^\beta}] + [\nabla_{z^\alpha}, \nabla_{z^\beta}]\big) + O(w^3) \\
&= 2[\nabla_{y^\mu}, \nabla_{y^\nu}] + O(w^3), \\
f_{x^i x^j} - \tilde{f}_{x^i x^j} &= 2[\nabla_{z^i}, \nabla_{z^j}] + O(w^3).
\end{aligned}$$

From these equalities, one finds

$$[f_{x^\mu x^\nu} + \tilde{f}_{x^\mu x^\nu}, f_{x^i x^j} - \tilde{f}_{x^i x^j}] = \big[[\nabla_{y^\mu}, \nabla_{y^\nu}], [\nabla_{z^i}, \nabla_{z^j}]\big] + O(w^3).$$

In view of the Jacobi identity, it follows that

$$\begin{aligned}
-\big[[\nabla_{y^\mu}, \nabla_{y^\nu}], [\nabla_{z^i}, \nabla_{z^j}]\big] &= \big[[[\nabla_{z^i}, \nabla_{z^j}], \nabla_{y^\mu}], \nabla_{y^\nu}\big] \\
&\quad + \big[[\nabla_{y^\nu}, [\nabla_{z^i}, \nabla_{z^j}]], \nabla_{y^\mu}\big].
\end{aligned}$$

Due to the Jacobi identity and (2.2), it follows that

$$-\big[[\nabla_{z^i}, \nabla_{z^j}], \nabla_{y^\mu}\big] = \big[[\nabla_{y^\mu}, \nabla_{z^i}], \nabla_{z^j}\big] + \big[[\nabla_{z^j}, \nabla_{y^\mu}], \nabla_{z^i}\big] = O(w^2).$$

From the last four equalities, one obtains the equality (2.9).

The following result shows that for the fields $a_\mu(x)$ with values in gl(2, $\mathbb{C}$) and satisfying the Yang–Mills equations, the equalities (2.8) (or (2.2) for $k = 4$) under the corresponding definition of the fields $a_{x^j}(x, w)$ and $a_{w^j}(x, w)$ are a criterion of fulfilment of the equalities (2.2) for any $k \geq 4$.

Theorem 2.3. *Let $a_\mu(x)dx^\mu$ be a solution of equation (1.11), with the property (2.8) and with the fields $a_\mu(x)$ taking values in gl(2, $\mathbb{C}$). Then, for some calibration, the field a has one of the following forms*

$$a = \begin{pmatrix} a^{11} & 0 \\ 0 & a^{22} \end{pmatrix}; \qquad a = \begin{pmatrix} a^{11} & a^{12} \\ 0 & a^{11} \end{pmatrix};$$

$$a = \begin{pmatrix} a^{11} & 0 \\ 0 & a^{11} \end{pmatrix} + a^{\pm},$$

where a^{ij} are solutions of the Maxwell equations, and $a^{\pm}$ are self-dual or anti-self-dual solutions of the Yang–Mills equation.

Remark. An analogous result, undoubtedly, takes place and for the more general case. For example, it has already been proved for the case of the fields $a(x)$ with values in su(3).

Consequence. The compatibility equations (2.2) for any fixed $k \geq 4$ for the fields $a_x(x, w)$ and $a_w(x, w)$ with values in su(2) imply self-duality or anti-self-duality or commutativity of the Yang–Mills fields $a(x) = a_x(x, 0)$. S.V. Manakov told us that the result of this consequence is one of his unpublished results. Note, also, that the statement of this consequence for $k \geq 6$ follows from the published results Buchdahl [1985], Isenberg and Yasskin [1979], and Ponomarev [1984].

The proof of Theorem 2.3 uses several additional statements.

Lemma 2.1. *Let* $\chi = \chi_i dx^i$; $f = f_{ij} dx^i \wedge dx^j$, $i < j$, *where* $\chi_i(x)$, $f_{ij}(x)$ *are complex-valued functions. Let* $f = \tilde{f}$ *or* $f = -\tilde{f}$. *Let* $\sum_{ij} |f_{ij}(x)| \neq 0$ *for* $x \in \mathcal{D}$. *If* $\chi \wedge f = 0$, *then* $\chi = 0$ *in* $\mathcal{D}$.

Lemma 2.2. *Let* Q *and* N *be* (2×2)-*matrices, with* Q *not being proportional to the identity, i.e.,* $Q \neq \lambda \cdot I, \forall \lambda \in \mathbb{C}$. *Let* $[Q, N] = 0$. *Then there are the numbers* $\alpha, \beta \in \mathbb{C}$ *such that* $N = \alpha \cdot I + \beta \cdot Q$.

Lemma 2.3. *Any solution* a *of the Yang-Mills equation can be represented in the form*

$$a_j = a'_j + (\operatorname{tr} a_j)I; \qquad f_{ij} = f'_{ij} + (\operatorname{tr} f_{ij})I.$$

In addition, $\operatorname{tr} a$ *is a solution of the Maxwell equation,* a' *is a solution of the Yang–Mills equations, and the matrices* a'_j *and* f'_{ij} *have a zero trace.*

Lemma 2.4. *Let* a *satisfy equation* (1.11). *Then the following equalities are valid:*

$$df^+ + a \wedge f^+ - f^+ \wedge a = 0,$$
$$df^- + a \wedge f^- - f^- \wedge a = 0,$$

where $f^\pm = f \pm \tilde{f}$.

Proof of Theorem 2.3. First, suppose that all f^+_{ij} or all f^-_{ij} are proportional to the unit matrix. Then, taking into account the equality from Lemma 2.3, we obtain the third equality from Theorem 2.3. Thus, we can suggest that among the matrices f^+_{ij}, there is the matrix that is not proportional to the unit matrix. Let it be $f^+_{01} = Q(x)$. Then due to Lemma 2.2 and to the equality (2.8), the following equalities are valid:

$$f^-_{ij} = \alpha^-_{ij} I + \beta^-_{ij} Q. \tag{2.10}$$

It is possible to think that not all β^-_{ij} are equal to zero. (Otherwise, the already analyzed case is realized.) Hence and from the equality (2.8) and Lemma 2.2, it follows that the following equalities are valid:

$$f^+_{ij} = \alpha^+_{ij} I + \beta^+_{ij} Q \tag{2.11}$$

(in addition, $a_{01}^+ = 0$, $\beta_{01}^+ = 1$). We carry out the gauge transform after which the matrix $Q(x)$ will be reduced to the Jordan normal form. (For a gauge transform, the matrix $Q(x)$ is transformed as the components of the curvature $Q \mapsto g^{-1}Qg$.) Two variants are possible:

$$Q = \begin{pmatrix} q_1 & 0 \\ 0 & q_2 \end{pmatrix} \quad \text{and} \quad Q = \begin{pmatrix} q_1 & 1 \\ 0 & q_1 \end{pmatrix}.$$

In the first case, in view of (2.10) and (2.11), we have

$$f^+ = \begin{pmatrix} f_1^+ & 0 \\ 0 & f_2^+ \end{pmatrix}; \qquad f^- = \begin{pmatrix} f_1^- & 0 \\ 0 & f_2^- \end{pmatrix}. \tag{2.12}$$

We prove that in this calibration the form a also has a diagonal form (to within the already examined case),

$$a = \begin{pmatrix} a_{11} & a_{12} \\ a_{21} & a_{22} \end{pmatrix} = \begin{pmatrix} a_{11} & 0 \\ 0 & a_{22} \end{pmatrix}. \tag{2.13}$$

For that, we shall need Lemma 2.4 and Lemma 2.1. For the forms a_{12} and a_{21} due to Lemma 2.4 and to (2.12), one obtains

$$a_{12} \wedge f_2^\pm - f_1^\pm \wedge a_{12} = a_{12} \wedge (f_2^\pm - f_1^\pm) = 0, \tag{2.14}$$
$$a_{21} \wedge f_1^\pm - f_2^\pm \wedge a_{21} = a_{21} \wedge (f_1^\pm - f_2^\pm) = 0.$$

If $f_1^+ = f_2^+$ or $f_1^- = f_2^-$, then such a case was already analyzed at the very beginning of the proof. Thus we can consider that $f_1^+ \neq f_2^+$ and $f_1^- \neq f_2^-$. Then, from (2.14) and Lemma 2.1, it follows that $a_{12} = a_{21} = 0$, and the equality (2.13) is proved. Thus, the first case in Theorem 2.3 corresponds to the case

$$Q = \begin{pmatrix} q_1 & 0 \\ 0 & q_2 \end{pmatrix}.$$

Now we have to consider the second variant. Due to Lemma 2.3,

$$a = a' + (\operatorname{tr} a)I; \qquad f = f' + (d\operatorname{tr} a)I, \tag{2.15}$$

where

$$f_{ij}'^\pm = \beta_{ij}^\pm Q'; \qquad Q' = \begin{pmatrix} 0 & 1 \\ 0 & 0 \end{pmatrix}.$$

Hence and from the equality of Lemma 2.4 for a' and f', it follows that

$$a' = \begin{pmatrix} a_{11}' & a_{12} \\ 0 & -a_{11}' \end{pmatrix}.$$

Furthermore, from the equation $f' = da' + a' \wedge a'$, one obtains the equality $da_{11}' = 0$. Hence it follows that there are the functions g_1 and g_2 such that

$g_1^{-1}dg_1 = a'_{11}$; $g_2^{-1}dg_2 = -a'_{11}$. We do a gauge transform of the solution a of the form (2.15) with the help of the matrix function

$$G = \begin{pmatrix} q_1 & 0 \\ 0 & q_2 \end{pmatrix}.$$

We have

$$G^{-1}aG + G^{-1}dG = \begin{pmatrix} 0 & \frac{q_2}{q_1}a_{12} \\ 0 & 0 \end{pmatrix} + (\operatorname{tr} a)I,$$

$$G^{-1}fG = \begin{pmatrix} 0 & \frac{q_2}{q_1}f_{12} \\ 0 & 0 \end{pmatrix} + (d\operatorname{tr} a)I.$$

Hence it follows that the variant

$$Q = \begin{pmatrix} q_1 & 1 \\ 0 & q_1 \end{pmatrix}$$

corresponds to the second case in Theorem 2.3.

We show the simplest way of constructing the solutions of the Yang–Mills equation (1.11) for which the equalities (2.8) are not valid. Let the field a take the form

$$a = \begin{pmatrix} a^{11} & a^{12} \\ 0 & a^{22} \end{pmatrix}. \tag{2.16}$$

The curvature of this field has the form

$$f = \begin{pmatrix} f^{11} & f^{12} \\ 0 & f^{22} \end{pmatrix},$$

where

$$f^{11} = da^{11}, \qquad f^{22} = da^{22}, \qquad f^{12} = da^{12} + (a^{11} - a^{22}) \wedge a^{12}.$$

Proposition 2.1. *In order that the field a of the form (2.16) will satisfy the Yang–Mills equations (1.11), it is necessary and sufficient that the components a^{11} and a^{22} satisfy the Maxwell equations $d\tilde{f}^{11} = 0$ and $d\tilde{f}^{22} = 0$, and that the component a^{12} satisfies the linear equation of the form*

$$d\tilde{f}^{12} + (a^{11} - a^{22}) \wedge \tilde{f}^{12} + (\tilde{f}^{11} - \tilde{f}^{22}) \wedge a^{12} = 0.$$

In addition, the commutator of the self-dual f^+ and of the anti-self-dual f^- components of the curvature takes the form

$$[f^+_{\alpha\beta}, f^-_{\mu\nu}] = \begin{pmatrix} 0 & (f^{12})^+_{\alpha\beta}[(f^{22})^-_{\mu\nu} - (f^{11})^-_{\mu\nu}] + (f^{12})^-_{\mu\nu}[(f^{11})^+_{\alpha\beta} - (f^{22})^+_{\alpha\beta}] \\ 0 & 0 \end{pmatrix}.$$

From this proposition, it follows that, in particular, the fields $a(x)$ of the form (2.16) satisfying the Yang–Mills equations, in general, do not satisfy the additional equalities (2.8), and all the more (2.8^1).

We give here another example illustrating on the one hand Theorems 2.1–2.3, and on the other hand, the difficulties of finding nontrivial solutions of the Yang–Mills equations depending only on two variables from the compatibility conditions.

Proposition 2.2. *Let the form* $a = a_\mu(x)dx^\mu$ *satisfy the Yang–Mills equations* (1.11), *where the fields* $a_\mu(x)$ *have values in* $\mathrm{gl}(2, \mathbb{C})$ *and* $a_\mu(x) = a_\mu(x_0, x_1)$; $\mu = 0, 1, 2, 3$. *Then the fields* $a_{x^j}(x, w)$ *and* $a_{w^j}(x, w)$ *satisfying* (2.2) *and* (2.6) *for* $k = 3$ *can be chosen in the form*

$$a_{x^j}(x, w) = a_{x^j}(x_0, x_1, w_0, w_1)$$

and

$$a_{w^j}(x, w) = a_{w^j}(x_0, x_1, w_0, w_1)$$

iff the equalities (2.8) *and* (2.8^1) *are valid.*

One of the main difficulties in the practical use of the stated interpretations of the Yang–Mills equations in the form of the compatibility equations of a linear system consists of the existence of two spectral parameters λ_1 and λ_2 in the system (2.1). In Forgacs, Horvath and Palla [1982] it is noted that the same compatibility equations (2.2) occur if one uses only one spectral parameter in the form

$$\lambda_1 = \lambda, \quad \lambda_2 = \lambda^{k+1}, \quad k = 1, 2, \ldots \tag{2.17}$$

in the linear system (2.1).

This idea led the authors of Forgacs, Horvath and Palla [1982] to the wrong statement about the possibility of a linearization of the Yang–Mills equations with the help of the Riemann–Hilbert problem with respect to the parameter $\lambda \in \mathbb{C}P^1$.

We will show in Chap. 7 that a reduction of the Yang–Mills equations to the Riemann–Hilbert one-dimensional problem (or to the $\bar{\partial}$-equation) is possible, in principle, not on the sphere $\mathbb{C}P^1$, but on the degenerated tori for the following special selections of one spectral parameter:

$$\lambda_1 = \lambda + \frac{1}{\lambda}, \qquad \lambda_2 = \lambda - \frac{1}{\lambda}, \tag{2.18^1}$$

$$\lambda_2 = \lambda^2 + \lambda, \qquad \lambda_2 = \lambda^2 - \lambda. \tag{2.18^2}$$

We formulate here the following simple but important statement.

Proposition 2.3. *Suppose that for the system* (2.1), *one of the substitutions* (2.17), (2.18^1), (2.18^2), *or* $\lambda_1 = \bar{\lambda}_2$ *is made. Then the compatibility equations for the obtained system with one spectral parameter are as the equations* (2.2) *were earlier.*

For the substitution (2.17), this statement was first made in Forgacs, Horvath and Palla [1982].

Chapter 3
The Holomorphic Bundles and
Cohomologies in Terms of $\bar{\partial}$-Equations

We review here (see also Khenkin [1985]; Leiterer [1986]; Onishchik [1986]) an interpretation of the vector bundles and $\bar{\partial}$-cohomologies on the complex manifold in terms of the Cauchy–Riemann equations.

Let X be a complex manifold; let $\{X_j\}$ be a local finite covering of X by the holomorphy domains X_j, $j \in \mathcal{J}$; let $\mathrm{GL}(n, \mathbb{C})$ be a group of invertible complex $n \times n$-matrices. Suppose that in the domains $X_j \cap X_k$ we have the holomorphic functions $F_{jk} = F_{kj}^{-1}$ with values in $\mathrm{GL}(n, \mathbb{C})$ forming a multiplicative cocycle, i.e.,

$$F_{jk}(x)F_{k\ell}(x)F_{\ell j}(x) = I, \qquad x \in X_j \cap X_k \cap X_\ell,$$

where I is the unit matrix. The cocycle $\{F_{jk}\}$ defines a holomorphic n-dimensional vector bundle F over X. The system of the mappings h_j of the domains X_j to $\mathbb{C}^n$, such that

$$h_j(x) = F_{jk}(x)h_k(x), \quad x \in X_j \cap X_k,$$

is called the section of the bundle F over X.

The bundles F with n smooth sections that are linearly independent at each point $x \in X$ will be called *topologically trivial*. This means that there exist smooth matrix functions A_j on X_j with values in $\mathrm{GL}(n, \mathbb{C})$ such that

$$F_{jk}(x) = A_j(x)(A_k(x))^{-1}, \quad x \in X_j \cap X_k.$$

We consider the $(0, 1)$-form θ with the $(n \times n)$-matrix coefficients with values in $\mathrm{gl}(n, \mathbb{C})$:

$$\theta(z) = \{(A_j(x))^{-1}\bar{\partial}A_j(x), \quad x \in X_j, \quad j \in \mathcal{J}\}.$$

This definition is correct, since in the domain $X_j \cap X_k$, we have

$$A_j^{-1}\bar{\partial}A_j = A_j^{-1}F_{jk}\bar{\partial}A_k = A_k^{-1}\bar{\partial}A_k.$$

First let us note that the form θ is defined by the vector bundle F up to the $\bar{\partial}$-gauge equivalence

$$\theta \sim B^{-1}\bar{\partial}B + B^{-1}\theta B, \tag{3.1}$$

where B is any smooth function with values in $\mathrm{GL}(n, \mathbb{C})$, and secondly, that the form θ satisfies the Cauchy–Riemann nonlinear equation

$$\bar{\partial}\theta + \theta \wedge \theta = 0. \tag{3.2}$$

Conversely, for any $(0, 1)$-form θ with $(n \times n)$-matrix coefficients that satisfies (3.2) and is defined up to $\bar{\partial}$-gauge equivalence (3.1) on X in view of the generalized Grothendieck–Dolbeault lemma (see Onishchik [1986]) in the domains X_j, there will be matrix functions A_j with values in $\mathrm{GL}(n, \mathbb{C})$ such that $A_j^{-1}\bar{\partial}A_j = \theta$ on X_j. Furthermore, suppose $F_{jk} = A_j A_k^{-1}$ on $X_j \cap X_k$; then we have

$$\bar{\partial}F_{jk} = A_j(A_j^{-1}\bar{\partial}A_j - A_k^{-1}\bar{\partial}A_k)A_k^{-1} = 0$$

on $X_j \cap X_k$.

The holomorphic matrix functions F_{jk} with values in $\mathrm{GL}(n, \mathbb{C})$ constructed in such a way give the (topologically trivial) bundle F over X.

The space of $(0,1)$-forms θ on X satisfying the condition (3.2) and being defined up to the $\bar{\partial}$-gauge equivalence (3.1) is denoted by $H^{0,1}(X, \mathrm{GL}(n, \mathbb{C}))$. This space is called the space of $\bar{\partial}$-connections in the fixed (in this case, trivial) n-dimensional bundle $E = X \times \mathbb{C}^4$ with the gauge group $\mathrm{GL}(n, \mathbb{C})$. The n^2-dimensional bundle $\mathrm{End}\,E$ is associated with an arbitrary holomorphic n-dimensional bundle E over X. As before, we can consider the $(0,1)$-forms θ on X with values in $\mathrm{End}\,E$ and with the properties (3.1) and (3.2), where in (3.1) B is a smooth function with values in the nondegenerated endomorphisms E. The space of such forms is denoted by

$$H^{0,1}(X, \mathrm{GL}(E)).$$

We denote by E_θ the holomorphic bundle obtained from the fixed bundle E with the help of the fixed $\bar{\partial}$-connection θ. The bundle E_θ is the so-called deformation of the bundle E. The holomorphic sections of the bundle E_θ are those smooth connections h of the bundle E that satisfy the Cauchy–Riemann equation of the form

$$\bar{\partial}h + \theta h = 0.$$

Furthermore, let X be a hypersurface in the complex manifold Y. The divisor on Y corresponding to the hypersurface X generates the linear bundle S on Y in which there is a section s equal to zero exactly on X (see Hartshorne [1977]; Onishchik [1986]).

In the following considerations, a very important role is played by the vector bundles on the infinitesimal neighborhoods $X^{(\nu)}$ of the manifold X in Y, $\nu = 0, 1, 2, \dots$. Such bundles are usually obtained with help of the locally-finite covering $\{Y_j\}$ of the manifold Y and of the holomorphic $(n \times n)$-matrix functions $F_{jk} = (F_{kj})^{-1}$ in the domains $Y_j \cap Y_k$ forming a holomorphic cocycle on $X^{(\nu)}$, i.e.,

$$F_{jk}F_{k\ell}F_{\ell j} = I + O(s^{\nu+1}) \quad \text{on } Y_j \cap Y_k \cap Y_\ell.$$

The following variant of the Grothendieck–Dolbeault theorem takes place.

Proposition 3.1 (Khenkin [1980]; Khenkin and Manin [1982]). *Let E be a fixed n-dimensional holomorphic bundle over Y (in particular, trivial). There is a canonical isomorphism between the space of the holomorphic vector bundles of rank n over $X^{(\nu)}$, topologically equivalent to E, and the space $H^{0,1}(X^{(\nu)}, \mathrm{GL}(E))$ of the smooth $(0,1)$-forms θ on Y with values in $\mathrm{End}\,E$ defined up to the gauge equivalence*

$$\theta \sim B^{-1}\bar{\partial}B + B^{-1}\theta B + O(s^{\nu+1}) \quad \text{on } Y \tag{3.3}$$

$(B, B^{-1}$ takes values in $\operatorname{End} E)$ and satisfying the relation

$$\bar{\partial}\theta + \theta \wedge \theta = \mathcal{J}_{\nu+1}s^{\nu+1} \quad \text{on } Y. \tag{3.4}$$

In addition, the space of the cohomologies $H^k(X^{(\nu)}, E_\theta)$ (or $H^k(X^{(\nu)}, \operatorname{End} E_\theta)$) of the manifold $X^{(\nu)}$ with coefficients in the sheaf of the holomorphic sections E_θ (or $\operatorname{End} E_\theta$) is isomorphic to the factor space of the space of the smooth $(0, k)$-forms ψ (or Ψ) on Y, with the coefficient sections E (or $\operatorname{End} E$) satisfying the relations

$$\bar{\partial}\psi + \theta \wedge \psi = \omega_{\nu+1}s^{\nu+1} \tag{3.5}$$

or

$$\bar{\partial}\Psi + [\theta, \Psi] = \Omega_{\nu+1}s^{\nu+1} \quad \text{on } Y \tag{3.5^1}$$

by the subspace of forms

$$\psi = \bar{\partial}\varphi + \theta \wedge \varphi + O(s^{\nu+1}), \tag{3.6}$$

or

$$\Psi = \bar{\partial}\phi + [\theta, \phi] + O(s^{\nu+1}), \tag{3.6^1}$$

where

$$[\theta, \phi] = \theta \wedge \phi - (-1)^k \phi \wedge \theta,$$

with k being the order of the form ϕ.

Note, that the forms $\mathcal{J}_\nu$ and Ω_ν on the right-hand sides of (3.4) and (3.5^1) belong to the space $H^2(X, \operatorname{End} E_\theta \otimes S^{-\nu})$, but the form ω_ν on the right-hand side of (3.5) belongs to the space $H^2(X, E_\theta \otimes S^{-\nu})$.

The result of Proposition 3.1 may be taken (if desired) for the definition of the corresponding spaces of the cohomologies of the complex space $X^{(\nu)}$.

Furthermore, we shall need the definition of the $\bar{\partial}$-cohomologies of partially complex manifolds (or, otherwise, CR-manifolds).

Let M be a real submanifold of codimension ℓ in the complex manifold Y and let $T_x^c(M)$ be the largest complex subspace in the tangent space $T_x(M)$ in the point $x \in M$. The manifold M is called a CR-manifold if the number $\dim T_x^c(M)$ does not depend on $x \in M$. Suppose that M is a generating manifold, i.e., in any coordinate neighborhood $\Omega \subset Y$, we have

$$M \cap \Omega = \{x \in \Omega : \rho_1(x) = \cdots = \rho_\ell(x) = 0\},$$

where $\{\rho_\nu(x)\}$ are smooth real functions with the property $\bar{\partial}\rho_1 \wedge \bar{\partial}\rho_2 \wedge \cdots \wedge \bar{\partial}\rho_\ell \neq 0$ on M.

By $C_{p,q}^{(\alpha)}(M, E)$ we denote the space of differential forms of the type (p, q) on M whose coefficients take values in E and have the smoothness of the class $C^{(\alpha)}$. The two forms g and $\tilde{g}$ from $C_{p,q}^{(\alpha)}(M, E)$ are considered to be equal if $\int_M g \wedge \varphi = \int_M \tilde{g} \wedge \varphi$ for any form with compact support $\varphi \in C_{n-p,n-\ell-q}^{(\infty)}(Y, E^*)$, where E^* is a bundle dual to E.

If the forms $f \in C_{0,q-1}(M,E)$ and $g \in C_{0,q}(M,E)$ such that for any form with compact support $\varphi \in C^{(\infty)}_{n,n-\ell-q}(Y,E^*)$, we have $\int_M g \wedge \varphi = (-1)^q \int_M f \wedge \bar{\partial}\varphi$, then by definition, we consider that $\bar{\partial}_\tau f = g$, where $\bar{\partial}_\tau$ is the Cauchy–Riemann tangent operator (see Khenkin [1984; 1985]).

Suppose that the complex hypersurface $X \subset Y$ mentioned above intersects with M in a general position, i.e.,

$$\dim T^c_x(M \cap X) = \dim T^c_x(M) - 1$$

for any $x \in M \cap X$. Suppose $N^{(\nu)} = M \cap X^{(\nu)}$.

We consider the space $H^{0,1}(N^{(\nu)}, \mathrm{GL}(E))$ of the $(0,1)$-forms θ on M with coefficients in $\mathrm{End}\, E$ of the smoothness $C^{(\infty)}$ satisfying the relation

$$\bar{\partial}_\tau \theta + \theta \wedge \theta = \mathcal{J}_{\nu+1} s^{\nu+1} \quad \text{on } M$$

and being considered up the gauge equivalence

$$\theta \sim B^{-1} \bar{\partial}_\tau B + B^{-1} \theta B + O(s^{\nu+1}) \quad \text{on } M.$$

Each element from $H^{0,1}(N^{(\nu)}, \mathrm{GL}(E))$ defines a so-called CR-bundle of the rank n over the ν-th infinitesimal neighborhood of the CR-manifold N in M. In addition, the spaces of the cohomologies $H^k(N^{(\nu)}, E_\theta)$ and $H^k(N^{(\nu)}, \mathrm{End}\, E_\theta)$ are defined by the relations (3.5) and (3.6) in which the forms have the smoothness of the class $C^{(\infty)}$, and the operator $\bar{\partial}$ is replaced by $\bar{\partial}_\tau$.

In contrast to the complex manifolds $X^{(\nu)}$, an analog of the Grothendieck–Dolbeault lemma, in general, is not valid on the CR-manifolds $N^{(\nu)}$ (the Lewy effect). Besides, the spaces of the cohomologies $H^k(N^{(\nu)}, E_\theta)$ are not even always separated (see Khenkin [1984; 1985]).

The following variant of the result of Griffiths [1966] for the extension of bundles and $\bar{\partial}$-cohomologies from X to Y is proved on the basis of Proposition 3.1.

Proposition 3.2 (Khenkin and Manin [1982]). *In order for the form $\theta \in H^{0,1}(X^{(\nu)}, \mathrm{GL}(E))$ to be gauge-equivalent on $X^{(\nu)}$ to the form extendable to the element $\tilde{\theta} \in H^{0,1}(X^{(\nu+1)}, \mathrm{GL}(E))$, it is necessary and sufficient that the obstruction form $\mathcal{J}_{\nu+1}$ on the right-hand side of (3.4) is the zero element $[\mathcal{J}_{\nu+1}]$ in $H^2(X, \mathrm{End}\, E_\theta \otimes s^{-\nu-1})$. If $[\mathcal{J}_{\nu+1}] = 0$, then the group $H^1(X, \mathrm{End}\, E_\theta \otimes s^{-\nu-1})$ acts effectively and transitively on the set of such extensions $\tilde{\theta}$. A similar result relating to an extension is also valid for CR-bundles.*

Now we give the simplest examples of complex manifolds, bundles over them, and some calculations of their $\bar{\partial}$-cohomologies, which are in turn useful for later examples.

The bundle $\mathcal{O}(k,\ell)$ over $\mathbb{C}P^n \times \mathbb{C}P^n$. Let $\zeta^0, \ldots, \zeta^n$ be coordinates in $\mathbb{C}^{n+1}$, or, otherwise, homogeneous coordinates in $\mathbb{C}P^n$. Suppose $X_j = \{\zeta \in$

$\mathbb{C}P^n : \zeta^j \neq 0\}$. We denote by $\mathcal{O}(\ell)$ the linear bundle over $\mathbb{C}P^n$ being given by the transition functions $(\zeta^k/\zeta^j)^\ell$ on $X_j \cap X_k$. On $\mathbb{C}^{n+1}$, the functions of the variables $\zeta^0, \ldots, \zeta^n$ of the homogeneity ℓ are sections of the bundle $\mathcal{O}(\ell)$. We denote by $\mathcal{O}(k, \ell)$ the one-dimensional bundle over $\mathbb{C}P^n \times \mathbb{C}P^n$, sections of which are functions of the homogeneity (k, ℓ) on $\mathbb{C}^{n+1} \times \mathbb{C}^{n+1}$.

Suppose $E(k, \ell) = E \otimes (k, \ell)$ for the case when $X \subset \mathbb{C}P^n \times \mathbb{C}P^n$ and E is a holomorphic vector bundle over X.

The $\bar{\partial}$-cohomologies of $\mathbb{C}P^n$ and quadrics in $\mathbb{C}P^n$. Let $\mathcal{L}$ be a quadric in $\mathbb{C}P^n$ of the following form in the homogeneous coordinates

$$\mathcal{L} = \{z \in \mathbb{C}P^n : (z^0)^2 + (z^1)^2 + \cdots + (z^n)^2 = 0\}. \tag{3.7}$$

For the important particular case $n = 3$, we have the isomorphism $\mathcal{L} \simeq \mathbb{C}P^1 \times \mathbb{C}P^1$, which is defined by the formulae

$$\begin{aligned}
z^0 &= \zeta^0\eta^0 + \zeta^1\eta^1, & z^1 &= -i(\zeta^0\eta^0 - \zeta^1\eta^1), \\
z^2 &= i(\zeta^1\eta^0 + \zeta^0\eta^1), & z^3 &= -(\zeta^1\eta^0 - \zeta^0\eta^1),
\end{aligned} \tag{3.8}$$

where $(\zeta^0, \zeta^1, \eta^0, \eta^1)$ are homogeneous coordinates in $\mathbb{C}P^1 \times \mathbb{C}P^1$.

We shall need the classical facts about the $\bar{\partial}$-cohomologies of $\mathbb{C}P^n$ and of the quadric $\mathcal{L}$ in $\mathbb{C}P^n$, which will be formulated in the form of the following statement (for details, see for example, Hartshorne [1977]; Khenkin and Polyakov [1990]).

Proposition 3.3 (J.P. Serre). *There are the following equalities for the $\bar{\partial}$-cohomologies of $\mathbb{C}P^n$ and the quadric $\mathcal{L}$ in $\mathbb{C}P^n$ with coefficients in $\mathcal{O}(\ell)$:*

$$\begin{aligned}
&H^k(\mathbb{C}P^n, \mathcal{O}(\ell)) = 0, \quad k = 1, 2, \ldots, n-1; \quad \ell = 0, \pm 1, \pm 2, \ldots, \\
&H^n(\mathbb{C}P^n, \mathcal{O}(\ell)) = \left[H^0(\mathbb{C}P^n, \mathcal{O}(-\ell - n - 1)) \right]^*,
\end{aligned} \tag{3.9}$$

$$\left.\begin{aligned}
&H^k(\mathcal{L}, \mathcal{O}(\ell)) = 0, \quad k = 1, 2, \ldots, n-2; \quad \ell = 0, \pm 1, \pm 2, \ldots, \\
&H^{n-1}(\mathcal{L}, \mathcal{O}(\ell)) = [H^0(\mathcal{L}, \mathcal{O}(-\ell - n + 1))]^*,
\end{aligned}\right\} \tag{3.10}$$

where $H^0(\mathbb{C}P^n, \mathcal{O}(\ell))$ is the space of the polynomials of $z^0, \ldots, z^n$ of the homogeneity ℓ, and $H^0(\mathcal{L}, \mathcal{O}(\ell))$ is the space of the same polynomials restricted to $\mathcal{L}$.

The $\bar{\partial}$-cohomologies with coefficients in the bundle $\mathcal{O}(k, \ell)$ are also possible for the quadric $\mathcal{L}$ in $\mathbb{C}P^3$ due to the isomorphism $\mathcal{L} \simeq \mathbb{C}P^1 \times \mathbb{C}P^1$. If $k = \ell$, then the bundle $\mathcal{O}(\ell, \ell)$ on $\mathcal{L}$ is a restriction of the bundle $\mathcal{O}(\ell)$ from $\mathbb{C}P^3$ in view of (3.8), and we can immediately use (3.10). If $k \neq \ell$, then the bundle $\mathcal{O}(k, \ell)$ does not extend to a bundle on $\mathbb{C}P^3$. For this case, for the computation of cohomologies $\mathcal{L}$, we can use Künneth's formula for cohomologies of the direct product (see Hartshorne [1977]) and (3.9) for $n = 1$.

In particular, for $\mathcal{L} \subset \mathbb{C}P^3$, we shall need the following equalities

$$\begin{aligned}
&H^1(\mathcal{L}, \mathcal{O}(\ell, \ell)) = 0, & &H^1(\mathcal{L}, \mathcal{O}(-1, 0)) = 0, \\
&H^1(\mathcal{L}, \mathcal{O}(-2, -1)) = 0, & &H^1(\mathcal{L}, \mathcal{O}(-2, 0)) \simeq \mathbb{C},
\end{aligned} \tag{3.11}$$

$$H^2(\mathcal{L}, \mathcal{O}(\ell, \ell)) = 0, \quad \ell \geq -1, \qquad H^2(,\mathcal{O}(-2, -2)) \simeq \mathbb{C},$$
$$H^2(\mathcal{L}, \mathcal{O}(-1, 0)) = 0, \qquad H^2(\mathcal{L}, \mathcal{O}(-2, 0)) = 0, \qquad H^2(\mathcal{L}, \mathcal{O}(-2, -1)) = 0.$$
$$(3.12)$$

The equalities $H^1(\mathcal{L}, \mathcal{O}) = 0$ and $H^2(\mathcal{L}, \mathcal{O}) = 0$ mean, for example, that any $\bar{\partial}$-equation of the form $\bar{\partial}g = f$ is solvable on $\mathcal{L}$ for f being a $\bar{\partial}$-closed $(0, j+1)$-form with continuous coefficients ($j = 0, 1$).

We give here the useful explicit formulae from Khenkin [1985] and Khenkin and Polyakov [1990] for operators solving the equation $\bar{\partial}g = f$ on $\mathcal{L} \subset \mathbb{C}P^3$:

$$g = (R^j f)(z)$$
$$= \frac{(-1)^j}{(2\pi i)^3} \times \int f(\zeta) \frac{\det[\zeta + z, \bar{\zeta}, \bar{z}, (1-j)d\bar{\zeta} + jd\bar{z}] \wedge (\zeta d\zeta \rfloor \omega(\zeta))}{(1 - \bar{\zeta}z)(1 - \zeta\bar{z})^2},$$
$$\{\zeta \in \mathbb{C}^4 : |\zeta| = 1, \zeta^2 = 0\}, \tag{3.13}$$

where

$$z \in \mathbb{C}^4 : |z| = 1, \ z^2 = 0, \qquad \omega(\zeta) = d\zeta_0 \wedge d\zeta_1 \wedge d\zeta_2 \wedge d\zeta_3,$$

and $(\zeta d\zeta)\rfloor\omega(\zeta)$ is a holomorphic three-form such that

$$(\zeta d\zeta) \wedge (\zeta d\zeta \rfloor \omega(\zeta)) = \omega(\zeta).$$

Chapter 4
The Nonabelian Radon Transform

By developing the results of original works Penrose [1969], Ward [1977], Isenberg and Yasskin [1979], Witten [1978], and also the results of later papers, such as Khenkin [1980; 1984] and Manakov and Zakharov [1981], we shall give here a new more general and at the same time more constructive version of the nonabelian Radon transform along complex straight lines in $\mathbb{C}^4$. Under the restriction to complex null lines in $\mathbb{C}^4$ (with the complex metric $(dx)^2$), this transform turns into the so-called Radon–Penrose transform transferring gauge fields on the Minkowski space $\mathcal{M}$ or the Euclidean $\mathcal{E}$ into $\bar{\partial}$-gauge fields (CR-bundle) on the space of the complex null lines crossing $\mathcal{M}$ or $\mathcal{E}$. The characterizations of the image and inversion formulae for the suitable Radon–Penrose partial transforms are given.

We will define the complex line in $\mathbb{C}^4$ crossing through a point $X^0 \in \mathbb{C}^4$ by the equation

$$x = ms + x^0, \tag{4.1}$$

where the vector $m \in \mathbb{C}^4$, $m \neq 0$ is fixed and $s \in \mathbb{C}$ is a complex parameter.

The lines of the form (4.1), in which the vectors m differ from each other, by multiplying by a constant, coincide, so a set of complex lines in $\mathbb{C}^4$ crossing through a fixed point x^0 form a three-dimensional complex manifold $\mathbb{C}P^3_{x_0}$. The set of all complex lines in $\mathbb{C}^4$ forms a complex manifold P of dimension 6.

Let $\mathcal{D}$ be a convex domain in $\mathbb{C}^4$ (respectively, in $R^4 \subset \mathbb{C}^4$). Let us consider the equation

$$\sum_{j=0}^{3} m_j \left(\frac{\partial}{\partial x^j} + a_j(x) \right) \mu(x, m) = 0, \tag{4.2}$$

where $m \in \mathbb{C} \backslash \{0\}$, $a_j(x)$ are functions with values in $\mathrm{gl}(n, \mathbb{C})$ holomorphic (respectively, smooth) in $x \in \mathcal{D} \subset \mathbb{C}^4$ (respectively, R^4).

In view of the general theory (see Chap. 2), there exists a function $\mu(x, m)$ holomorphic (respectively, smooth) in $x \in \mathcal{D}$ and smooth in $m \in \mathbb{C}^4 \backslash \{0\}$, with values in $\mathrm{GL}(n, \mathbb{C})$ satisfying (4.2). In addition, $\mu(x, m) = \mu(x, cm)$ for any complex $c \neq 0$. Suppose

$$\phi_j(x, m) = \mu^{-1}(x, m) \frac{\partial}{\partial \bar{m}_j} \mu(x, m). \tag{4.3}$$

These functions are holomorphic (respectively, smooth) in $x \in \mathcal{D}$. From (4.2) and (4.3), it follows that

$$\frac{\partial \phi_k(x, m)}{\partial \bar{m}_j} - \frac{\partial \phi_j(x, m)}{\partial \bar{m}_k} + [\phi_k, \phi_j] = 0, \tag{4.4a}$$

$$\sum_{j=0}^{3} m_j \frac{\partial}{\partial x^j} \phi_k(x, m) = 0, \tag{4.4b}$$

$$\bar{c} \phi_j(x, cm) = \phi_j(x, m), \quad c \neq 0. \tag{4.4c}$$

We shall call the form $\phi = \sum_{j=0}^{3} \phi_j(x, m) d\bar{m}_j$ the nonabelian Radon transform of the field $a(x) = \sum a_j(x) dx^j$.

If the field $a(x)$ is given in the domain $\mathcal{D}$, then the form ϕ is the $(0, 1)$-form on the manifold $P(\mathcal{D})$ of all complex lines in $\mathbb{C}^4$ crossing $\mathcal{D}$. The field a should be considered up to the gauge transform (1.3), and the form ϕ up to the $\bar{\partial}$-gauge transform of the form

$$\phi \sim h^{-1} \bar{\partial} h + h^{-1} \phi h, \tag{4.4d}$$

where h is a function with values in $\mathrm{GL}(n, \mathbb{C})$ satisfying the equation

$$\sum_{j=0}^{3} m_j \frac{\partial}{\partial x^j} h(x, m) = 0.$$

Theorem 4.1. *The nonabelian Radon transform of the form* (4.2), (4.3) *realizes the one-to-one correspondence between the holomorphic (respectively, smooth) gauge fields $a(x)$ in the convex domain $\mathcal{D} \subset \mathbb{C}^4$ (respectively, R^4) and* *the holomorphic (respectively, smooth) ones in the x $(0,1)$-forms $\phi(x,m)$ with the properties (4.3) and (4.4).*

The forms ϕ with such properties give (see Chap. 3) topologically trivial holomorphic vector bundles E_ϕ over $P(\mathcal{D})$, analytically trivial on each $\mathbb{C}P^3_{x^0} \subset P(\mathcal{D})$, $x^0 \in \mathcal{D}$.

The complex line of the form (4.1) is called a *complex light ray* if $m^2 = \sum m_j^2 = 0$.

The equation $m^2 = 0$ gives the surface $\mathcal{L}(x^0) \simeq \mathbb{C}P^1 \times \mathbb{C}P^1$ in $\mathbb{C}^3_{x^0}$ (see Chap. 3). It is possible to introduce the coordinates λ_1 and λ_2 in the affine part of this surface so that the following equalities are valid (cf. (2.1^1)):

$$\begin{aligned}
m_0(\lambda) &= 1 - \lambda_1\lambda_2; &\quad m_1(\lambda) &= -i(1 + \lambda_1\lambda_2); \\
m_2(\lambda) &= -(\lambda_1 + \lambda_2); &\quad m_3(\lambda) &= -i(\lambda_1 - \lambda_2).
\end{aligned} \tag{4.5}$$

The set of complex light rays forms a complex manifold L of dimension 5.

Let us further consider the equation

$$\sum_{j=0}^{3} m_j(\lambda)\left(\frac{\partial}{\partial x_j} + a_j(x)\right)\mu(x,\lambda) = 0, \quad x \in \mathcal{D}, \tag{4.6}$$

and let $\mu(x,\lambda)$ be a solution of (4.6) with values in $\mathrm{GL}(n,\mathbb{C})$ smooth in $\lambda \in \mathcal{L}$ and holomorphic in $x \in \mathcal{D}$. Suppose

$$\theta_j(x,\lambda) = \mu^{-1}(x,\lambda)\frac{\partial}{\partial \bar\lambda_j}\mu(x,\lambda), \quad j = 1, 2. \tag{4.7}$$

From (4.3), (4.5), and (4.7), it follows that

$$\begin{aligned}
\theta_1 &= -\bar\lambda_2\phi_0 + i\bar\lambda_2\phi_1 - \phi_2 + i\phi_3, \\
\theta_2 &= -\bar\lambda_1\phi_0 + i\bar\lambda_1\phi_1 - \phi_2 + i\phi_3.
\end{aligned} \tag{4.7^1}$$

From (4.6) and (4.7), it follows that

$$\frac{\partial}{\partial\bar\lambda_1}\theta_2(x,\lambda) - \frac{\partial}{\partial\bar\lambda_2}\theta_1(x,\lambda) + [\theta_1,\theta_2] = 0, \tag{4.8a}$$

$$\sum_{j=0}^{3} m_j(\lambda)\frac{\partial}{\partial x^j}\theta_k(x,\lambda) = 0, \quad k = 1, 2, \tag{4.8b}$$

$$\theta_j(x,\lambda) = O\left(\frac{1}{|\lambda_j|^2}\right) \quad \text{for } |\lambda_j| \to \infty, \quad x \in \mathcal{D} \subset \mathbb{C}^4. \tag{4.8c}$$

We shall call the form $\theta = \theta_1(x,\lambda)d\bar\lambda_1 + \theta_2(x,\lambda)d\bar\lambda_2$ the Radon–Penrose transform of the gauge field $a(x)$. The form θ is the $(0,1)$-form on the manifold

$L(\mathcal{D})$ of all complex light rays crossing a domain $\mathcal{D} \subset \mathbb{C}^4$. This form should be considered up to the $\bar{\partial}$-gauge transform

$$\theta \sim h^{-1}\bar{\partial}h + h^{-1}\theta h, \qquad (4.8d)$$

where $(m(\lambda)\frac{\partial}{\partial x})h = 0$.

The following (refined) variant of the Ward–Witten–Isenberg–Yasskin–Green result is valid (cf. Theorem 4.4).

Theorem 4.2. *Let $\mathcal{D}$ be a convex domain in $\mathbb{C}^4$. The Radon–Penrose transform realizes the one-to-one correspondence between the holomorphic gauge fields $a(x), x \in \mathcal{D}$, and the $(0,1)$-forms $\theta(x,\lambda)$ holomorphic in x with the properties (4.7) and (4.8).*

The further refinement of Theorem 4.2 is possible for gauge fields over the Euclidean real space $\mathcal{E} = \{z \in \mathbb{C}^4 : \operatorname{Im} z = 0\}$ or the Minkowski space $\mathcal{M} = \{z \in \mathbb{C}^4 : \operatorname{Im} z_0 = 0, \ \operatorname{Re} z_j = 0, \ j = 1,2,3\}$.

Theorem 4.2^1. *Let $\mathcal{D}$ be a convex domain in $\mathcal{E}$ or $\mathcal{M}$. Then the Radon–Penrose transform of the form (4.6), (4.7) determines a one-to-one correspondence between $C^{(\infty)}$-smooth gauge fields in the domain $\mathcal{D}$ and $(0,1)$-forms $\theta(x,\lambda)$, $C^{(\infty)}$-smooth in $x \in \mathcal{D}$, with the properties (4.7) and (4.8).*

The forms θ with the properties (4.7) and (4.8) give (see Chap. 3) topologically trivial n-dimensional holomorphic vector bundles (or CR-bundles) over $L(\mathcal{D})$ that are analytically trivial on each quadric $\mathcal{L}(x) \subset L(\mathcal{D})$.

Below, we shall show (Theorem 4.4) that such forms θ give all holomorphic vector bundles with the described properties.

Unexpectedly, it turned out that similar (to Theorems 4.1 and 4.2) results take place also for some special four-dimensional submanifolds in the manifold of all complex light rays.

Firstly, we can restrict ourselves to complex light rays satisfying in each two-dimensional quadric $\mathcal{L}(x)$ the equation

$$\{m \in \mathcal{L}(x) : m_2 m_3 = i(m_0 + im_1)^2\} = \mathcal{L}'(x). \qquad (4.9^1)$$

The equation (4.9^1) defines a degenerate torus obtained from the Riemann sphere by the identification of two different points. It is possible to introduce the parameter λ on this torus by formulae (4.5), where $\lambda_1 = \lambda + \frac{1}{\lambda}$, $\lambda_2 = \lambda - \frac{1}{\lambda}$.

The values of the parameter λ that are equal to zero and to infinity correspond to the same point on this torus.

Secondly, it is also possible to consider *complex light rays* defined by the equality

$$\{m \in \mathcal{L}(x) : (m_0 + im_1)m_2 = m_3^2\} = \mathcal{L}''(x). \qquad (4.9^2)$$

The surface (4.9^2) is also a degenerate torus. This torus is obtained from the Riemann sphere by introducing a double point at the point ∞. It is possible to introduce the parameter λ on the surface (4.9^2) by formulae (4.5), where

$$\lambda_1 = \lambda^2 + \lambda, \qquad \lambda_2 = \lambda^2 - \lambda.$$

This value of the parameter $\lambda = \infty$ corresponds to a special (double) point on this torus.

By $L'(\mathcal{D})$ and $L''(\mathcal{D})$, respectively, we denote the complex manifolds of light rays of the form (4.9^1) or (4.9^2) crossing the domain $\mathcal{D} \subset \mathbb{C}^4$. These manifolds have the complex dimension 4.

Let us consider the Radon–Penrose partial transform along light rays of the form (4.9^1) or (4.9^2). Using a solution of the equation (4.6) in which the substitutions (4.9^1) or (4.9^2) are made, suppose

$$\mu^{-1} \frac{\partial}{\partial \bar{\lambda}} \mu(x, \lambda) = \theta'(x, \lambda), \tag{4.10^1}$$

where $\mu(x, 0) = \mu(x, \infty)$ for the case (4.9^1),

$$\mu^{-1} \frac{\partial}{\partial \bar{\lambda}} \mu(x, \lambda) = \theta''(x, \lambda), \tag{4.10^2}$$

where

$$\lambda^2 \frac{\partial}{\partial \bar{\lambda}} \mu(x, \lambda)|_{\lambda=\infty} = \lambda^2 \frac{\partial}{\partial \lambda} \mu(x, \lambda)\bigg|_{\lambda=\infty} = 0$$

for the case (4.9^2).

The forms $\theta' d\bar{\lambda}$ and $\theta'' d\bar{\lambda}$ are defined up to the suitable $\bar{\partial}$-gauge equivalence. The following formula is valid for the case (4.9^1):

$$\theta'(x, \lambda) = \left(1 - \frac{1}{\bar{\lambda}^2}\right) \theta_1\big(x, \lambda_1(\lambda), \lambda_2(\lambda)\big) + \left(1 + \frac{1}{\bar{\lambda}^2}\right) \theta_2\big(x, \lambda_1(\lambda), \lambda_2(\lambda)\big). \tag{4.11^1}$$

The following formula is valid for the case (4.9^2):

$$\theta''(x, \lambda) = (1 + 2\bar{\lambda})\theta_1\big(x, \lambda_1(\lambda), \lambda_2(\lambda)\big) - (1 - 2\bar{\lambda})\theta_2\big(x, \lambda_1(\lambda), \lambda_2(\lambda)\big). \tag{4.11^2}$$

From (4.6) and (4.10), it follows that

$$\left(m(\lambda)\frac{\partial}{\partial x}\right) \theta'(x, \lambda) = 0, \qquad \theta'(x, \lambda) = O\left(\frac{1}{|\lambda|^2}\right) \tag{4.12^1}$$

for the case (4.9^1), and

$$\left(m(\lambda)\frac{\partial}{\partial x}\right) \theta''(x, \lambda) = 0, \qquad \theta''(x, \lambda) = O\left(\frac{1}{|\lambda|^3}\right) \tag{4.12^2}$$

for the case (4.9^2).

Theorem 4.3. *The Radon–Penrose partial transform of the form (4.6) and (4.10^1) (respectively, (4.10^2)) realizes the one-to-one correspondence between the holomorphic gauge fields $a(x)$ in the convex domain $\mathcal{D} \subset \mathbb{C}^4$ and the $(0,1)$-forms $\theta'(x, \lambda)d\bar{\lambda}$ holomorphic in x (respectively, $\theta''(x, \lambda)d\bar{\lambda}$) with the properties (4.10^1) and (4.12^1) (respectively, (4.10^2) and (4.12^2)). Such forms*

148   G.M. Khenkin, R.G. Novikov

$\theta' d\bar{\lambda}$ (respectively, $\theta'' d\bar{\lambda}$) defined up to the $\bar{\partial}$-gauge equivalence are in one-to-one correspondence with the topologically trivial n-dimensional holomorphic vector bundles $E_{\theta'}$ (respectively, $E_{\theta''}$) over $\Lambda'(\mathcal{D})$ (respectively, $L''(\mathcal{D})$) and analytically trivial in each degenerated torus $\mathcal{L}'(x)$ (respectively, $\mathcal{L}''(x)$).

We represent a function $\mu(x, \lambda)$ in the neighborhood of the point $\lambda = \infty$ in the form

$$\mu(x, \lambda) = g(x)\left(1 + \sum_{j=2}^{\infty} \frac{\mu_{-j}(x)}{\lambda^j} + \frac{\mu'_{-j}(x)}{\bar{\lambda}^j}\right) \tag{4.13}$$

for the inversion of the Radon–Penrose partial transform, for example, in the case (4.9^2). Let us represent a gauge field $a(x)$ in the form

$$a(x) = (dg(x))g^{-1}(x) + g(x)b(x)g^{-1}(x). \tag{4.14}$$

Then for the field $b(x) = \sum b_\nu(x) dx^\nu$ in view of (4.6), (4.13), and (4.14), we have the relation

$$b_0(x) + ib_1(x) = 0; \qquad b_2(x) = \frac{1}{2}\left(\frac{\partial}{\partial x^0} + i\frac{\partial}{\partial x^1}\right)\mu_{-2}(x),$$

$$b_3(x) = \frac{i}{2}\left(\frac{\partial}{\partial x^0} + i\frac{\partial}{\partial x^1}\right)\mu_{-3}(x), \tag{4.15}$$

$$b_0(x) - ib_1(x) = \left(\frac{\partial}{\partial x^0} + i\frac{\partial}{\partial x^1}\right)\mu_{-4}(x) - 2b_2(x)(1 + \mu_{-2}(x)).$$

Note, that the property (4.3) in Theorem 4.1 (the triviality of the bundle E_ϕ on each $\mathbb{C}P_x^3$) and the property (4.7) in Theorem 4.2 (the triviality of the bundle E_θ on each quadric $\mathcal{L}(x)$) are automatic consequences of the Cauchy–Riemann (nonlinear) equations (4.4a) and (4.8a) and of the classical facts about the triviality of the $\bar{\partial}$-cohomologies $H^1(\mathbb{C}P^3, \mathcal{O})$ and $H^1(\mathcal{L}, \mathcal{O})$ (see Proposition 3.3) for $n = 1$ and for the forms ϕ and θ with small norms for any n.

In Theorem 4.3, the Cauchy–Riemann (nonlinear) differential equations for the forms $\theta' d\bar{\lambda}$ and $\theta'' d\bar{\lambda}$ are absent. The properties (4.10^1) and (4.10^2) about the triviality of the bundles $E_{\theta'}$ and $E_{\theta''}$ on the degenerated tori $\mathcal{L}'(x)$ and $\mathcal{L}''(x)$ are not automatically valid (and thus obtain more significance), since the equalities occur (see Hartshorne [1977]):

$$H^1(\mathcal{L}', \mathcal{O}) \cong H^1(\mathcal{L}'', \mathcal{O}) \cong C \neq 0.$$

Furthermore, it is important to state the Radon–Penrose transform introduced earlier in terms of the Penrose twistor theory (Penrose and Rindler [1984,1986]) and in spinor coordinates.

Let us introduce the main notions of the twistor theory. Let $\mathbb{C}_+^4$ and $\mathbb{C}_-^4$ be two mutually dual four-dimensional complex spaces (twistors and dual twistors) with the coordinates $\zeta = (\zeta^A, \zeta_{B'})$ and $\eta = (\eta_A, \eta^{B'})$ and with the bilinear form

$$\langle \zeta \cdot \eta \rangle = \zeta^A \eta_A - \zeta_{A'} \eta^{A'},$$

where

$$A = 0, 1; \qquad B' = 0', 1'; \qquad \zeta^A \eta_A = \zeta^0 \eta_0 + \zeta^1 \eta_1.$$

The coordinates in $\mathbb{C}_+^4$ and $\mathbb{C}_-^4$ will be considered as homogeneous coordinates of the corresponding points in the three-dimensional projective spaces $\mathbb{C}P_+^3$ and $\mathbb{C}P_-^3$. Let us further consider the five-dimensional complex hypersurface

$$L = \{(\zeta, \eta) \in \mathbb{C}P_+^3 \times \mathbb{C}P_-^3 : \langle \zeta \cdot \eta \rangle = 0\}. \tag{4.16}$$

If ζ is interpreted as a point, and η as a hypersurface in $\mathbb{C}P^3$, then the relation $\langle \zeta \cdot \eta \rangle = 0$ means that $\zeta \in \eta$. Suppose $L(\mathbb{C}M_0) = \{(\zeta, \eta) \in L : \zeta^A \neq 0, \eta^{B'} \neq 0\}$. The points $L(\mathbb{C}M_0)$ parametrize complex null lines in the Minkowski complex space $\mathbb{C}M_0 \simeq \mathbb{C}^4$ with the spinor coordinates $x = \{x_{AB'}\}$ and the metric $\det |dx_{AB'}|$. Namely, the complex null line in $\mathbb{C}M_0$ defined by the equations

$$\zeta_{B'} = x_{AB'} \zeta^A, \tag{4.17^1}$$

$$\eta_{A'} = x_{AB'} \eta^{B'}, \tag{4.17^2}$$

corresponds to a fixed point $(\zeta, \eta) \in L(\mathbb{C}M_0)$.

Separately, the equations (4.17^1) and (4.17^2) define null planes in the space $\mathbb{C}M_0$, called α and β-planes, respectively. Conversely, for a fixed point $x \in \mathbb{C}M_0$, the equations (4.17^1) and (4.17^2) considered as the equations for ζ and η form two mutually orthogonal two-dimensional subspaces in $\mathbb{C}_+^4$ and $\mathbb{C}_-^4$ that define on L the compact submanifold

$$\mathcal{L}(x) = \mathcal{L}_+(x) \times \mathcal{L}_-(x) \simeq \mathbb{C}P^1 \times \mathbb{C}P^1,$$

where $\mathcal{L}_\pm \subset \mathbb{C}P_\pm^3$.

The construction given by Penrose in 1967 (see Penrose and Rindler [1986]) makes it possible to define the Minkowski compactified (and complexified) space $\mathbb{C}M$ as the Grassmannian manifold of all two-dimensional subspaces in $\mathbb{C}_+^4$ (or $\mathbb{C}_-^4$). The corresponding tautological two-dimensional bundles over $\mathbb{C}M$ denoted by $S_\pm$ are called *spinor bundles*. The bundle $S_+ \otimes S_-$ is canonically isomorphic to the cotangent bundle in $\mathbb{C}M$.

In view of (4.17), there is a natural trivilization of these bundles over $\mathbb{C}M_0$. Besides, ζ^A are coordinates in the fibre $S_+(x)$, and $\eta^{B'}$ are coordinates in the layer $S_-(x), x \in \mathbb{C}M_0$. The metrics $\det |dx_{AB'}|$ are fixed sections of the bundle $\bigwedge^2 S_+ \otimes \bigwedge^2 S_-$, where $\bigwedge^2$ denotes the exterior product of degree 2.

For the domain $\mathcal{D} \subset \mathbb{C}M$, suppose that

$$L(\mathcal{D}) = \bigcup_{x \in \mathcal{D}} \mathcal{L}(x); \qquad L_\pm(\mathcal{D}) = \bigcup_{x \in \mathcal{D}} \mathcal{L}_\pm(x).$$

The manifold $L(\mathcal{D})$ consists of all complex null lines crossing the domain $\mathcal{D}$. The manifolds $L_\pm(\mathcal{D})$ consist, respectively, of all two-dimensional α (or β) null planes crossing the domain $\mathcal{D}$.

Let us consider the real Minkowski spaces $\mathcal{M}_0$ consisting of those points $x \in \mathbb{C}M_0$ for which $x_{AB'}$ is the Hermite matrix and the real Euclidean spaces

150　　　　　　　　　　G.M. Khenkin, R.G. Novikov

$\mathcal{E}_0$ consisting of those points $x \in \mathbb{C}M_0$ for which $x_{AB'}$ is an orthogonal matrix of the form (1.12).

Let $\mathcal{M}$ be a compactification of $\mathcal{M}_0$, and let $\mathcal{E}$ be a compactification of $\mathcal{E}_0$ in $\mathbb{C}M$. Then the corresponding manifolds $L(\mathcal{E})$, $L_{\pm}(\mathcal{E})$ and $L(\mathcal{M})$, $L_{\pm}(\mathcal{M})$ are CR-submanifolds in the complex manifolds L or $\mathbb{C}P^3_{\pm}$ of the form

$$
\begin{aligned}
L(\mathcal{E}) &= \{(\zeta, \eta) \in L : \bar{\zeta}_0\eta_0 + \bar{\zeta}_1\eta_1 - \bar{\zeta}_{0'}\eta_{0'} - \bar{\zeta}_{1'}\eta_{1'} = 0\}; \\
L_+(\mathcal{E}) &= \mathbb{C}P^3_+; \qquad L_-(\mathcal{E}) = \mathbb{C}P^3_-; \\
L_+(\mathcal{M}) &= \{\zeta \in \mathbb{C}P^3_+ : \mathrm{Im}(\zeta^0\bar{\zeta}_{0'} + \zeta^1\bar{\zeta}_{1'}) = 0\}, \\
L_-(\mathcal{M}) &= \{\zeta \in \mathbb{C}P^3_- : \mathrm{Im}(\eta_0\bar{\eta}^{0'} + \eta_1\bar{\eta}_{1'}) = 0\}, \\
L(\mathcal{M}) &= \{(\zeta, \eta) \in L : \zeta \in L_+(\mathcal{M}), \eta \in L_-(\mathcal{M})\}.
\end{aligned}
$$

By $F(\mathcal{D})$, we denote the flag submanifold in $L(\mathcal{D}) \times \mathcal{D}$, defined by the equations (4.17^1) and (4.17^2) or, geometrically, by the incidence relations $\zeta \in x \subset \eta$, where ζ is a point in $\mathbb{C}P^3_+$, x is a projective line in $\mathbb{C}P^3_+$, and η is a hyperplane in $\mathbb{C}P^3_+$, $x \in \mathcal{D}, (\zeta, \eta) \in L(\mathcal{D})$. We have the natural double bundle

$$
\begin{array}{ccc}
 & F(\mathcal{D}) & \\
{}^{\pi_1}\swarrow & & \searrow^{\pi_2} \\
L(\mathcal{D}) & & \mathcal{D}
\end{array}
$$

Using this bundle and the spinor coordinates, we define the Radon–Penrose transform of the connections ∇ in a fixed bundle over $\mathcal{D}$ into holomorphic vector bundles over $L(\mathcal{D})$.

Suppose, that $\mathcal{D}$ is a domain in $\mathbb{C}M$ such that any nonempty intersection of the domain $\mathcal{D}$ with a null line is contractible.

For the holomorphic one-form $a = a_{AA'}dx^{AA'}$ defining a connection in the trivial n-dimensional bundle E over $\mathcal{D}$, we consider lifting π_2^*a of this form on $F(\mathcal{D})$. The fibres of the bundle $\pi_1 : F(\mathcal{D}) \to L(\mathcal{D})$ are given in spinor coordinates by the equations

$$
\zeta^A dx_{AA'} = 0 \quad \text{and} \quad \eta^{A'} dx_{AA'} = 0.
$$

So, the operator

$$
\partial_1 = \sum_{AA'} \zeta^A \eta^{A'} \frac{\partial}{\partial x^{AA'}} \tag{4.18}
$$

is exactly an operator of differentiation along the fibres π_1. In the usual coordinates, this operator appeared earlier as $m(\lambda)\frac{\partial}{\partial x}$. In view of the contractibility of the fibres π_1, there exists a smooth section of the fibering π_1 over $L(\mathcal{D})$. So, we can solve the equation with respect to μ on $F(\mathcal{D})$:

$$
(\partial_1\mu)\mu^{-1} = \zeta^B \eta^{B'} a_{BB'}, \tag{4.19}
$$

where μ is a smooth function with values in $\mathrm{GL}(n, \mathbb{C})$, which is holomorphic along the fibres π_1. Let us now consider the $(0,1)$-form

$$
\theta^* = \mu^{-1}\bar{\partial}\mu \tag{4.20}
$$

(cf. (4.17)) on $F(\mathcal{D})$. In view of (4.17^1), the differentials in the form θ^* depend only on (ζ, η). Thus, in order for the form θ^* to be the lift $\pi_1^*\theta$ of some form θ on $L(\mathcal{D})$, it is sufficient to check the equality $\partial_1\theta^* = 0$ (cf. (4.8b)). We have

$$\mu(\partial_1\theta^*)\mu^{-1} = \mu(\partial_1(\mu^{-1}\bar{\partial}\mu))\mu^{-1}$$
$$= \bar{\partial}((\partial_1\mu)\mu^{-1}) = \bar{\partial}(\zeta^B\eta^{B'}a_{BB'}) = 0.$$

The form θ constructed on $L(\mathcal{D})$ in view of (4.20) satisfies the Cauchy–Riemann nonlinear equation. In addition, the form θ is not uniquely defined by the connection a (see (4.19) and (4.20)), but up to the $\bar{\partial}$-gauge equivalence (3.1). One of the calibrations is fixed in (4.8).

We constructed the $\bar{\partial}$-connection in the topologically trivial n-dimensional bundle over $L(\mathcal{D})$ so that (see Chap. 3) the holomorphic vector bundle $E_\theta = E_\nabla$ is defined over $L(\mathcal{D})$. The relation (4.20) means that the bundle E_θ is analytically trivial on all quadrics $\mathcal{L}(x)$, $x \in \mathcal{D}$. Now we define an inverse transform (for simplicity) under the additional suggestion that $\mathcal{D}$ is a domain of holomorphy (the Stein manifold; see Onishchik [1986]). Let θ be a form of the $\bar{\partial}$-connection in the topologically trivial bundle over $L(\mathcal{D})$ satisfying the condition

$$\pi_1^*\theta = \mu^{-1}\bar{\partial}\mu, \qquad (4.20^1)$$

where μ is a smooth function with values in $\mathrm{GL}(n, \mathbb{C})$. Due to the contractibility of the fibres π_1, condition (4.20^1) follows from the analytical triviality of the bundle E_θ on the quadrics $\mathcal{L}(x)$ and from the Stein property of $\mathcal{D}$. The obstruction to transforming the function μ, satisfying (4.20^1), from $F(\mathcal{D})$ to $L(\mathcal{D})$ is a possible nonvanishing of the function $\partial_1\mu$. Let us show that the function $(\partial_1\mu)\mu^{-1}$ is holomorphic on $F(\mathcal{D})$. In fact,

$$\bar{\partial}((\partial_1\mu)\mu^{-1}) = \mu(\partial_1\pi_1^*\theta)\mu^{-1}.$$

Since the fibres of the projection π_1 are compact, the equality (4.19) follows from the holomorphy and from the homogeneity of the first order with respect to ζ and η of the function $(\partial_1\mu)\mu^{-1}$, and $a_{AA'}$ are holomorphic functions of $x \in \mathcal{D}$. Given the $\bar{\partial}$-connection θ, the holomorphic form $a = a_{AA'}dx^{AA'}$ is defined on $\mathcal{D}$ by this construction only up to the gauge equivalence (1.3). Thus, the form a defines the $\mathrm{GL}(n, \mathbb{C})$-connection in the trivial n-dimensional bundle E over $\mathcal{D}$. Thus, we obtained the following geometric variant of Theorem 4.2.

Theorem 4.4 (Isenberg and Yasskin [1979]; Ward [1977]; Witten [1978]). *Let the intersections of the domain $\mathcal{D}$ with all null lines in $\mathbb{C}M$ be contractible. Then the Radon–Penrose transform (4.19), (4.20) determines the one-to-one correspondence between a set of holomorphic connections ∇ in the n-dimensional trivial bundle E over $\mathcal{D}$ and a set of all n-dimensional topologically trivial holomorphic vector bundles E_∇ over $L(\mathcal{D})$ that, in addition, are analytically trivial on all quadrics $\mathcal{L}(x)$, $x \in \mathcal{D}$.*

Now we state a refinement from Khenkin [1980] of Theorem 4.4 for fields over the real Minkowski or Euclidean space.

Theorem 4.4[1]. *Let the intersections of the domain $\mathcal{D} \subset \mathcal{M}$ with all light rays be connected, or let $\mathcal{D} \subset \mathcal{E}$. Then the Radon–Penrose transform (4.19), (4.20) determines a one-to-one correspondence between a set of smooth connections ∇ in the n-dimensional trivial bundle E over $\mathcal{D}$ and a set of all n-dimensional topologically trivial CR-bundles E_∇ over the CR-manifold $L(\mathcal{D}) \subset \mathbb{C}P_+^3 \times \mathbb{C}P_-^3$ that, in addition, are analytically trivial on all quadrics $\mathcal{L}(x)$, $x \in \mathcal{D}$.*

Similar results are also valid for the partial Radon–Penrose transforms when the manifold $\mathbb{C}P_+^3 \times \mathbb{C}P_-^3$ is replaced by the manifold of the form (see (3.8), (4.9[1]), and (4.9[2])):

$$\{(\zeta,\eta) : (\zeta^1\eta^{0'} + \zeta^0\eta^{1'})(-\zeta^1\eta^{0'} + \zeta^0\eta^{1'}) = (2\zeta^0\eta^{0'})^2\} = (\mathbb{C}P_+^3 \times \mathbb{C}P_-^3)' \quad (4.21^1)$$

or

$$\{(\zeta,\eta) : 2i\zeta^0\eta^{0'}(\zeta^1\eta^{0'} + \zeta^0\eta^{1'}) = (\zeta^1\eta^{0'} - \zeta^0\eta^{1'})^2\} = (\mathbb{C}P_+^3 \times \mathbb{C}P_-^3)'', \quad (4.21^2)$$

where the quadrics $\mathcal{L}(x)$ are replaced with the degenerate tori

$$\mathcal{L}'(x) = \mathcal{L}(x) \cap (\mathbb{C}P_+^3 \times \mathbb{C}P_-^3)' \quad \text{or} \quad \mathcal{L}''(x) = \mathcal{L}(x) \cap (\mathbb{C}P_+^3 \times \mathbb{C}P_-^3)'',$$

and where the manifold $L(\mathcal{D})$ is replaced with the manifold

$$L'(\mathcal{D}) = L(\mathcal{D}) \cap (\mathbb{C}P_+^3 \times \mathbb{C}P_-^3)' \quad \text{or} \quad L''(\mathcal{D}) = L(\mathcal{D}) \cap (\mathbb{C}P_+^3 \times \mathbb{C}P_-^3)''.$$

Due to Proposition 3.1 (or to the definition of the space $H^{0,1}(N^{(\nu)}, \mathrm{GL}(E))$ from Chap. 3), the space of the n-dimensional topologically trivial holomorphic (or CR-) bundles over the manifold $L(\mathcal{D})$ analytically trivial on the quadrics $\mathcal{L}(x)$ is described by such $(0,1)$-forms θ on $L_+(\mathcal{D}) \times L_-(\mathcal{D})$ with values in $\mathrm{gl}(n, \mathbb{C})$ that satisfy the relation

$$\bar{\partial}\theta + \theta \wedge \theta = \mathcal{I}_1\langle \zeta \cdot \eta \rangle, \tag{4.22}$$

that are considered up to the gauge equivalence

$$\theta \sim B^{-1}\bar{\partial}B + B^{-1}\theta B + O(\langle \zeta \cdot \eta \rangle), \tag{4.23}$$

and are $\bar{\partial}$-trivial in the sense of (4.20) on all quadrics $\mathcal{L}(x)$, $x \in \mathcal{D}$.

By $\tilde{H}^{0,1}(L^{(0)}(\mathcal{D}), \mathrm{GL}(n, \mathbb{C}))$ we denote the space of such forms.

In the following, we shall need the additional property of bundles corresponding to elements of this space.

Proposition 4.1 (Khenkin and Manin [1982]). *For any form $\theta \in \tilde{H}^{0,1}(\Lambda(\mathcal{D}), \mathrm{GL}(n, \mathbb{C}))$, the bundle E_θ is analytically trivial not only on the quadrics $\mathcal{L}(x)$, $x \in \mathcal{D}$, but also on their infinitesimal neighbourhoods $\mathcal{L}^{(1)}(x) \cap L(\mathcal{D})$.*

To prove this proposition, is sufficient to know that a set of extensions of the trivial bundle $E_\theta|_{\mathcal{L}(x)}$ from $\mathcal{L}^{(0)}(x)$ to $\mathcal{L}^{(1)}(x)$ consists of the only (trivial) extension. Due to Proposition 3.2, it is equivalent to the fact that the group $H^1(\mathcal{L}(x), N^* \otimes \mathrm{End}\, E)$, which acts on the set of extensions of the bundle

$E_\theta|_{\mathcal{L}(x)}$ from $\mathcal{L}^{(0)}(x)$ to $\mathcal{L}^{(1)}(x) \cap L(\mathcal{D})$, is equal to zero, and N^* is a conormal bundle for $\mathcal{L}(x)$ in $L(\mathcal{D})$. The last sentence immediately follows from the elementary equality

$$N^* \simeq (2\mathcal{O}(-1,0) \oplus 2\mathcal{O}(0,-1))/\mathcal{O}(-1,-1)$$

(see (3.7) and (3.8)) and from equalities (3.11) for the bundles $\mathcal{O}(-1,0)$, $\mathcal{O}(0,-1)$, $\mathcal{O}(-1,-1)$.

Chapter 5
The Radon–Penrose Transform as the Limit Case of the Faddeev-type Scattering Data

In this chapter, the characterization of the Faddeev-type scattering data (see Ablowitz and Nachman [1986]; Beals and Coifman [1986]; Faddeev [1974]; Khenkin and Novikov [1988; 1987]; Newton [1989]) for the Laplacian in the gauge field a on the Euclidean space $\mathcal{E}_0$ is given. On this basis, it is proved that these scattering data in a suitable limit become the Radon–Penrose transform of the gauge field a given on $\mathcal{E}_0$. Similar results are also valid for the d'Alembertian in the gauge field on the Minkowski space $\mathcal{M}_0$.

Let us consider the equation on R^4,

$$\sum_{j=0}^{3}\left(\frac{\partial}{\partial x^j} + a_j\right)^2 \varphi + v\varphi = 0, \tag{5.1}$$

where $a_j(x), v(x)$, and $\varphi(x)$ are smooth functions with values in $\mathrm{gl}(n, \mathbb{C})$; moreover, $a_j(x)$ and $v(x)$ decrease sufficiently rapidly for $x \to \infty$.

For any $k \in \mathbb{C}^4 : k^2 = \sum_{j=0}^{3} k_j^2 = 0, \ k \neq 0$, the solution $\varphi(x, k)$ of equation (5.1) is defined from the following integral equation

$$\phi(x, k) = e^{ikx} I - \int_{y \in R^4} G(x - y, k)(2a(y)\nabla\psi(y, k)$$
$$+ \left(a^2(y) + \nabla \cdot a(y) + v(y)\right)\psi(y, k)) \, dy,$$

where

$$G(x, k) = -\left(\frac{1}{2\pi}\right)^4 \int_{\zeta \in R^4} \frac{e^{i(\zeta + k)x}}{\zeta^2 + 2k\zeta} d\zeta.$$

The Green function $G(x, k)$ for the operator $\triangle + k^2$ was introduced by Faddeev in 1965.

The function $\psi(x, k)$, in general, is increasing exponentially if $|x| \to \infty$ or $|k| \to \infty$. So, it is convenient to introduce the function $\varphi(x, k) = \psi(x, k)e^{-ikx}$, which in view of (5.1) satisfies the following equation:

$$\triangle\varphi + 2i(k \cdot \nabla\varphi) = A(\varphi), \tag{5.1^1}$$

where $A(\varphi) = -[2a\nabla\varphi + (2iak + \nabla \cdot a + a^2 + v)\varphi]$.

The following function $H(k, p)$ with values in $\mathrm{gl}(n, \mathbb{C})$ can naturally be called the generalized scattering data for equation (5.1),

$$H(k, p) = \left(\frac{1}{2\pi}\right)^4 \int_{x \in R^4} e^{ipx} A(\varphi)(x)\, dx, \tag{5.3}$$

where $k \in \mathbb{C}^4 : k^2 = 0$, $p \in R^4$.

The scattering data of such a type for the classical Schrödinger potential equation when $a = 0$ were introduced by L.D. Faddeev.

The formulae (5.2) and (5.3) need to be refined for $k = 0$. It is possible to show, for example, that the limit

$$\lim_{k \to 0 : k^2 = 0} H(k, p) = H(0, p)$$

exists.

We denote by S the set of exceptional points $k \in \mathbb{C}^4$ (the zero measure set on the manifold $k^2 = 0$) for which the equation (5.2) loses unique solvability. It is possible to show (see Khenkin and Novikov [1988; 1987]) that for fixed $p \in R^4$, $x \in R^4$ on the three-dimensional complex manifold $\{z \in \mathbb{C}^4 : k^2 = 0, \ k \notin S\}$, the following $\bar{\partial}$-equations in k for the scattering data $H(k, p)$ and for the function $\varphi(x, k)$ are valid:

$$\bar{\partial}\varphi(x, k) = -2\pi \int_{\zeta \in R^4} (\zeta \cdot d\bar{k})e^{i\zeta x}\varphi(x, k + \zeta)H(k, -\zeta)\delta(\zeta^2 + 2k\zeta)\, d\zeta, \tag{5.4}$$

$$\bar{\partial}H(k, p) = -2\pi \int_{\zeta \in R^4} (\zeta \cdot d\bar{k})H(k, -\zeta)H(k + \zeta, p + \zeta) \cdot \delta(\zeta^2 + 2k\zeta)\, d\zeta, \tag{5.5}$$

where

$$2\pi(\zeta \cdot d\bar{k})\delta(\zeta^2 + 2k\zeta) = \bar{\partial}\frac{1}{\zeta^2 + 2k\zeta},$$

that is, the integrals on the right-hand side have the following meaning,

$$\int_{\zeta \in R^4} f(\zeta)\delta(\zeta^2 + 2k\zeta)d\zeta = \int_{\zeta^2 + 2k\zeta = 0} |\mathcal{J}(k, \zeta)|^{-1}f(\zeta)|d\zeta_3 \wedge d\zeta_4|,$$

where $\mathcal{J}(k, \zeta)$ is the Jacobian of the mappings

$$(\zeta_0, \zeta_1, \zeta_2, \zeta_3) \mapsto (\zeta^2 + 2\zeta \operatorname{Re} k, 2\zeta \operatorname{Im} k, \zeta_2, \zeta_3).$$

This type of $\bar{\partial}$-equations for generalized scattering data for the classical Schrödinger potential operator was first introduced in Beals and Coifman [1986] and Ablowitz and Nachman [1986].

The equations (5.4) and (5.5) make it possible (see Khenkin and Novikov [1988; 1987]) to express the generalized scattering data $H(k,p)$ restricted to the manifolds

$$\Omega = \{(k,p) \in \mathbb{C}^4 \times R^4 : k^2 = 0, 2kp = p^2\} \tag{5.6}$$

by "physical" scattering data (scattering amplitude, and so on). Due to this, it is possible to show that by replacing the fields $a(x)$ and $v(x)$ in (5.1) with the gauge equivalent fields $a' = g^{-1}ag + g^{-1}dg$ and $v' = g^{-1}vg$, where g is a smooth rapidly decreasing function with values in $\mathrm{GL}(n, \mathbb{C})$, the corresponding scattering data $H(k,p)$ do not change on the manifold Ω.

We denote by $C_B^{(\alpha)}(R^4)$ the space of smooth functions $f(x)$ on R^4 with the following property: Any derivative of the order $|\alpha'| \leq \alpha$ admits an estimate of the form $|D^{\alpha'} f(x)| = O((1 + |x|)^{-B})$. We denote by $C_B^{(\alpha)}(\Omega)$ the space of smooth functions $h(k,p)$ on Ω with the following property: Any derivative of the order $|\alpha'| \leq \alpha$ with respect to $(k,p) \in \Omega$ admits an estimate of the form

$$|D^{\alpha'} h(k,p)| = O_\varepsilon\big((1 + |p|)^{-B}\big), \quad |k| > \varepsilon > 0.$$

The following statement proved by the methods of Khenkin and Novikov [1988; 1987] gives a characterization of the scattering data for equation (5.1) with smooth rapidly decreasing fields $a(x)$ and $v(x)$.

Theorem 5.1. *Let the fields $a(x)$ and $v(x)$ with values in $\mathrm{gl}(n, \mathbb{C})$ belong to $C_\infty^{(\infty)}(R^4)$ and let them have a sufficiently small norm in some space $C_B^{(\alpha)}$, $\alpha, B > 5$. Then the exceptional set S for equation (5.2) is empty, and the corresponding scattering data $H(k,p)$ satisfy equation (5.5) on Ω, and in addition, the function $(1 + |k|)^{-1}H(k,p)$ belongs to $C_\infty^{(\infty)}(\Omega)$. Conversely, if the function $(1 + |k|)^{-1}H(k,p) \in C_{\infty)}^{(\infty)}(\Omega)$ with values in $\mathrm{gl}(n, \mathbb{C})$ has a sufficiently small norm in some space $C_\beta^{(\alpha)}$, $\alpha, \beta > 5$, and satisfies (5.5), then there exist the fields $a(x)$ and $v(x)$ with values in $\mathrm{gl}(n, \mathbb{C})$ and from the class $C_\infty^{(\infty)}(R^4)$ (unique up to the gauge equivalence) for which $H(k,p)$ are scattering data of the corresponding equation (5.1). In addition, the field $a(x)$ is gauge-trivial iff the function $H(k,p)$ is bounded on Ω.*

Theorem 5.1 develops results relating to a characterization of scattering data for the classical Schrödinger potential operator at fixed energy (see Beals and Coifman [1986]; Khenkin and Novikov [1987]; Novikov [1989]).

From Theorem 5.1, one obtains that for smooth rapidly decreasing fields the scattering data for a fixed p have no more than linear growth with respect to $k \to \infty$, $k^2 = 0$.

The following asymptotic formulae (Khenkin and Novikov [1988; 1987]) are valid:

$$\varphi(x,k) = \mu(x,m)\left(I + \frac{\mu_1(x,m)}{|k|} + o\left(\frac{1}{|k|}\right)\right),$$

$$|k|^{-1}H(k,p) = H_0(m,p) + \frac{H_1(m,p)}{|k|} + o\left(\frac{1}{|k|}\right),$$

$$(5.7)$$

for fixed p and m such that $m^2 = 0$, $pm = 0$, $|m| = 1$, and for $k \to \infty$ such that $k^2 = 0$, $\rho^2 = 2k\rho$, $\frac{k}{|k|} \to m$.

The functions H_0, H_1, μ, and μ_1 in (5.7) are homogeneous of the orders, $1, 0, -1$, respectively, with respect to the variable m:

$$H_0(e^{i\alpha}m,p) = e^{i\alpha}H_0(m,p); \qquad \mu(x,e^{i\alpha}m) = \mu(x,m), \qquad (5.8a)$$

$$H_1(e^{i\alpha}m,p) = H_1(m,p); \qquad \mu_1(x,e^{i\alpha}m) = e^{-i\alpha}\mu_1(x,m). \quad (5.8b)$$

Thus, these functions may be considered to be sections of the corresponding one-dimensional bundles over manifolds of the variables

$$m \in \mathbb{C}P^3, \quad p \in R^4, \quad x \in R^4 : m^2 = 0, \quad m \cdot p = 0.$$

Suppose that

$$\Phi(x,m) = -2\pi \int_{\zeta \in R^4} (\zeta \cdot d\bar{m})H_0(m,-\zeta)e^{i\zeta x}\delta(2\zeta \cdot m)d\zeta, \qquad (5.9.a)$$

$$\Psi(x,m) = -2\pi \int_{R^n} (\zeta \cdot d\bar{m})e^{i\zeta x}\Big[H_0(m,-\zeta)\Big(\zeta \cdot \frac{\partial}{\partial m}\Big)\delta(2\zeta \cdot m)$$

$$+ H_1(m,-\zeta)\delta(2\zeta \cdot m)\Big]d\zeta. \qquad (5.9b)$$

The $(0,1)$-forms $\Phi(x,m)$ and $\Psi(x,m)$ are homogeneous of the orders $0, -1$, respectively, with respect to the variable m. From the relations (5.9), it immediately follows that

$$\left(m\frac{\partial}{\partial x}\right)\Phi = 0 \quad \text{and} \quad \left(m\frac{\partial}{\partial x}\right)\Psi = 0, \quad m^2 = 0.$$

These equalities mean that the forms Φ and Ψ are defined on the manifold of the complex light rays $L(\mathcal{E}_0)$.

Theorem 5.2. *The equations (5.4), (5.5), and (5.1) for the scattering data $\varphi(x,k)$ and $H(k,p)$ of the form (5.7) for $k \to \infty$ ($k^2 = 0$), turn into the following $\bar{\partial}$-equations for the functions $\mu(x,m)$, $\mu_1(x,m)$, and the forms $\Phi(x,m)$, $\Psi(x,m)$, $m^2 = 0$, $x \in R^4$:*

$$\bar{\partial}\mu = \mu\Phi, \tag{5.10a}$$

$$\bar{\partial}\mu_1 + \Phi\mu_1 - \mu_1\Phi = \Psi, \tag{5.10b}$$

$$\bar{\partial}\Phi + \Phi \wedge \Phi = 0, \qquad m \cdot \frac{\partial\Phi}{\partial x} = 0, \tag{5.11a}$$

$$\bar{\partial}\Psi + \Phi \wedge \Psi + \Psi \wedge \Phi = 0, \qquad m \cdot \frac{\partial\Psi}{\partial x} = 0, \tag{5.11b}$$

$$\left(m\frac{\partial}{\partial x}\right)\mu(x,m) + (m \cdot a)\mu(x,m) = 0, \tag{5.12a}$$

$$- 2i\mu\left[\left(m\frac{\partial}{\partial x}\right)\mu_1\right]\mu^{-1} = v. \tag{5.12b}$$

The equations (5.11a) and (5.11b) mean that the form Φ determines the CR-bundle E_Φ over the manifold $L(\mathcal{E}_0)$, and the form Ψ determines an element of the cohomology space $H^1(L(\mathcal{E}_0), \operatorname{End} E_\Phi(-1,-1))$ (see Chap. 6). Equation (5.10a) means that the bundle E_Φ is trivial on each quadric $\mathcal{L}(x) \subset L(\mathcal{E}_0)$. Equation (5.10b) means that the form Ψ is $\bar{\partial} + \Phi$-exact on each quadric $\mathcal{L}(x)$.

The equations (5.10a) and (5.12a) are obtained from (5.4), (5.1), and (5.7) by immediate passage to the limit. In order to obtain (5.11a) from (5.5), it is sufficient to note first that the following equality is obtained from (5.5) and (5.7),

$$\bar{\partial}H_0(m,p) = -2\pi \int_{\zeta \in R^4} (\zeta \cdot d\bar{m})H_0(m,-\zeta)H_0(m,p+\zeta)\delta(2\zeta \cdot m)\, d\zeta.$$

This equality together with (5.9a) gives (5.11a). Due to the use of (5.10a)–(5.12a), it is also possible to obtain, although in a more awkward manner, the equations (5.10b)–(5.12b).

The equalities (5.8a)–(5.11a) may be written in the form of (4.3) and (4.4), and equation (5.12a) in the form of (4.2). Since $m^2 = 0$, these equations may also be written in the form of (4.6), (4.7), and (4.8).

Thus, formula (5.9a) shows how the generalized scattering data are related to the Radon–Penrose transform $\Phi(x,m)$ of the field $a(x)$.

Now it is useful to note that the equations (5.2) and (5.3) have the following (relatively simple) limiting form for $k \to \infty$,

$$\mu(x,m) = I + \int_{y \in R^4} iG_0(x-y,m)(m \cdot a)\mu(y,m)\, dy, \tag{5.13}$$

where

$$G_0(x,m) = \left(\frac{1}{2\pi}\right)^4 \int_{\zeta \in R^4} \frac{e^{i\zeta x}d\zeta}{m \cdot \zeta},$$

$$H_0(m,p) = -\left(\frac{1}{2\pi}\right)^4 \int_{x \in R^4} 2i(m \cdot a)\mu(x,m)dx. \tag{5.14}$$

The integral equation (5.13) and formulae (5.9a) and (5.14) provide, in particular, the method of construction of the concrete Radon–Penrose transform (among gauge-equivalent possibilities) of the form (4.7) and (4.8) for the field a given on the Euclidean space.

In Theorem 5.2, the formulae were obtained in Khenkin and Novikov [1988; 1987] (for the scalar case $n = 1$) for the purpose of solving the following inverse scattering problem: Given the scattering data $H(k, p)$ on the manifold Ω, construct a "magnetic" field $a(x)$ and a potential $v(x)$.

It turned out that for finding the field $a(x)$, it is necessary to carry out the inverse nonabelian Radon transform of the scattering data $H_0(m, p)$, which for this case is as follows. First, the form Φ is being introduced by formula (5.9a). Then the $\bar{\partial}$-equation (5.10a) is solved on the basis of the linear integral equation of the form

$$\mu = I + R^1(\mu\Phi), \tag{5.15}$$

where R^1 is an operator inverting the $\bar{\partial}$-operator on the quadric $\{m \in \mathbb{C}P^3 : m^2 = 0\}$ (for example, of the form (3.13)). In view of (5.11a), the solution of the integral equation (5.15) also satisfies the differential $\bar{\partial}$-equation (5.10a). At last, an unknown field $a(x)$ is obtained from formula (5.12).

It is necessary to carry out the following transform of the scattering data $H_0(m, p)$ and $H_1(m, p)$ to find the potential $v(x)$. First, the forms Φ and Ψ are being introduced by formulae (5.9a) and (5.9b). Then the $\bar{\partial} + \Phi$-equation (5.10b) is being solved on the basis of the integral equation of the form

$$\mu_1 = R^1(\Psi + \mu_1\Phi - \Phi\mu_1). \tag{5.16}$$

Furthermore, an unknown field $v(x)$ is being obtained by formula (5.12b). In what follows, the transform defined here, $\Phi(x, m) \mapsto v(x)$, plays an important role. This transform will be introduced anew when twistor and spinor coordinates are discussed in Chap. 6, where it will be called a Penrose transform of the cohomologies space $H^1(L(\mathcal{E}_0), \operatorname{End} E_\Phi(-1, -1))$.

Chapter 6
The $\bar{\partial}$-Cohomologies of a Manifold of Complex Light Rays

The detailed information about the spaces of $\bar{\partial}$-cohomologies of the manifold of complex null lines $L(\mathcal{D})$, where $\mathcal{D}$ is a domain in $\mathbb{C}M$, turned out to be important for the interpretation of the Yang–Mills equations in terms of Cauchy–Riemann equations for scattering data. The computation of these cohomology spaces, which goes back to Penrose (Penrose [1969]), uses a computation of their restrictions to the quadrics $\mathcal{L}(x)$, $x \in \mathcal{D}$. Our presentation sums up—and to some extent generalizes—the results and constructions in

Atiyah [1979], Eastwood, Penrose and Wells [1981], Gindikin and Khenkin [1981], Khenkin [1984; 1980], Khenkin and Manin [1982; 1981], Hitchin [1980], and Woodhouse [1985].

For the domain $\mathcal{D}$ in $\mathbb{C}M$, we have

$$L(\mathcal{D}) = \left\{ (\zeta, \eta) \in L_+(\mathcal{D}) \times L_-(\mathcal{D}) : \langle \zeta \cdot \eta \rangle = 0 \right\}.$$

The divisor corresponding to the hypersurface $\langle \zeta \cdot \eta \rangle = 0$ in $\mathbb{C}P_+^3 \times \mathbb{C}P_-^3$ gives the linear bundle $S = \mathcal{O}(1,1)$, whose section $s = \langle \zeta \cdot \eta \rangle$ is equal to zero exactly on $L(\mathbb{C}M)$.

For the νth infinitesimal neighbourhood $L^{(\nu)}(\mathcal{D}) \subset L_+(\mathcal{D}) \times L_-(\mathcal{D})$, write

$$\tilde{H}^{0,1}(L^{(\nu)}(\mathcal{D}), \mathrm{GL}(n, \mathbb{C}))$$
$$= \tilde{H}^{0,1}(L^{(0)}(\mathcal{D}), \mathrm{GL}(n, \mathbb{C})) \cap H^{0,1}(L^{(\nu)}(\mathcal{D}), \mathrm{GL}(n, \mathbb{C})),$$

where the definitions from Chaps. 3 and 4 are used.

Let E_θ denote the topologically trivial holomorphic bundle over $L^{(\nu)}(\mathcal{D})$ corresponding to the fixed form $\theta \in \tilde{H}^{0,1}(L^{(\nu)}(\mathcal{D}), \mathrm{GL}(n, \mathbb{C}))$. Suppose $E_\theta(k,\ell) = E_\theta \otimes \mathcal{O}(k,\ell)$. Let us consider spaces of the $\bar{\partial}$-cohomologies of the form $H^j(L(\mathcal{D}), E_\theta(k,\ell))$, where $k \leq 0, \ell \leq 0$.

Such cohomologies are computed for many compact manifolds and for many holomorphically convex manifolds (see Hartshorne [1977]; Onishchik [1986]).

However, the manifold $L(\mathcal{D})$ does not belong to these well-investigated classes. Due to the use of the more general Andreotti–Grauert theory of cohomologies of p-convex manifolds (see, for example, Khenkin [1985]), it is possible to show that for the pseudoconvex $\mathcal{D}$, the spaces $H^j(L(\mathcal{D}),\ E_\theta(k,\ell)) = 0$ for $j = 3, 4, 5$ (Khenkin and Manin [1981]).

If $j = 1, 2$, then the corresponding (more important for us) cohomology spaces may not only be different from zero but may even be infinite-dimensional.

The application of the Leray method of spectral sequences (see Onishchik [1986]; Hartshorne [1977]) to the double bundle

$$
\begin{array}{ccc}
 & F(\mathcal{D}) & \\
{\scriptstyle \pi_1}\swarrow & & \searrow{\scriptstyle \pi_2} \\
L(\mathcal{D}) & & \mathcal{D}
\end{array}
$$

introduced in Chap. 4 makes it nevertheless possible to easily obtain the following general statement about the vanishing of the $\bar{\partial}$-cohomologies.

Proposition 6.1 (Khenkin and Manin [1981]). *Let $\mathcal{D}$ be a pseudoconvex domain in $\mathbb{C}M$. Then $H^1(L(\mathcal{D}), E_\theta(k,\ell)) = 0$ if $k + \ell \leq -3$, $k < 0, \ell < 0$; $H^2(L(\mathcal{D}), E_\theta(k,\ell)) = 0$ if $0 \geq k \geq -2$, $0 \geq \ell \geq -2$, $k + \ell \geq -3$.*

If the k, ℓ do not satisfy the conditions of Proposition 6.1, then the corresponding cohomology spaces, in general, are infinite-dimensional. It is possible to obtain an exact description of them in terms of the Radon–Penrose-type transforms.

Further, let $\mu_x(\zeta, \eta)$ be a smooth function on $L(\mathcal{D})$ with values in GL$(n, \mathbb{C})$, which depends on a parameter $x \in \mathcal{D}$ and such that

$$\theta|_{\mathcal{L}^{(1)}(x) \cap L(\mathcal{D})} = \mu_x^{-1} \bar{\partial} \mu_x, \quad x \in \mathcal{D}. \tag{6.1}$$

The existence of such a function μ_x follows from Proposition 4.1.

Let the $(0,1)$-form Ω be an element of the space $H^1(L(\mathcal{D}), \text{End} \, E_\theta(-1,-1))$. Let us introduce the Penrose transform of the form

$$\omega_+(x, \eta^{c'}) = \int_{\zeta \in \mathcal{L}_+(x)} \sum_{A'} \eta^{A'} \frac{\partial}{\partial \zeta^{A'}} (\mu_x \Omega \mu_x^{-1}) \wedge \zeta^c d\zeta_c,$$
$$\omega_-(x, \zeta^c) = \int_{\eta \in \mathcal{L}_-(x)} \sum_{A} \zeta^{A} \frac{\partial}{\partial \eta^{A}} (\mu_x \Omega \mu_x^{-1}) \wedge \eta^{c'} d\eta_{c'}. \tag{6.2}$$

These definitions are correct, since the vector fields $\sum \eta^{A'} \frac{\partial}{\partial \zeta^{A'}}$ and $\sum \zeta^{A} \frac{\partial}{\partial \eta^{A}}$ are tangent to $L(\mathcal{D})$. Since the functions $\omega_\pm$ are holomorphic with respect to their variables, then due to Liouville's theorem, they depend, in fact, only on $x \in \mathcal{D}$.

It is possible to give another definition of the Penrose transform for Ω (cf. (5.10b) and (5.12b)).

Let us lift Ω on $F(\mathcal{D})$, that is, let us consider the form $\pi_1^*(\Omega)$. Due to Proposition 3.3 on $F(\mathcal{D})$, there is a function $h(x, \zeta^c, \eta^{c'})$ of homogeneity $(-1,-1)$ with respect to $\zeta^c, \eta^{c'}$ such that

$$\pi_1^* \Omega = \bar{\partial} h + [\pi_1^* \theta, h].$$

Applying the differentiation operator ∂_1 along the fibres π_1 of the form (4.18) to both parts of this equality, we obtain

$$\bar{\partial}(\partial_1 h + [\pi_1^* \theta, \partial_1 h]) = 0.$$

Hence, and from (6.1), it follows that the function of the form

$$\mu(\partial_1 h)\mu^{-1} = \omega \tag{6.2^1}$$

is holomorphic on $F(\mathcal{D})$, that is, it depends only on $x \in \mathcal{D}$.

Proposition 6.2 (Khenkin and Manin [1982]). *The Penrose transforms of the form $\Omega \in H^1(L(\mathcal{D}), \text{End} \, E_\theta(-1,-1))$ of the type (6.2) and (6.2^1) are connected by the relations $\omega_+ = \omega_- = -2\pi i \omega$.*

We define the Penrose transform in the following way for the forms Ψ_+ and Ψ_-, which are elements of the spaces $H^1(L(\mathcal{D}), \text{End} \, E_\theta(-1,0))$ and $H^1(L(\mathcal{D}), \text{End} \, E_\theta(0,-1))$. Let us consider the forms $\pi_1^* \Psi_\pm$. Due to Proposition 3.3, on $F(\mathcal{D})$ there will be smooth matrix functions $h_\pm(x, \zeta^c, \eta^{c'})$ of $(-1,0)$ and $(0,-1)$ homogeneity, respectively, with respect to $\zeta^c, \eta^{c'}$ such that $\pi_1^* \Psi_\pm = \bar{\partial} h_\pm + [\pi_1^* \theta, h_\pm]$. Hence and from (6.1), it follows that the functions

$\mu(\partial_1 h_\pm)\mu^{-1}$ are holomorphic and have $(0,1)$ and $(1,0)$ homogeneity, respectively, with respect to $\zeta^c, \eta^{c'}$. From Liouville's theorem, it follows that

$$\mu(\partial_1 h_+)\mu^{-1} = \eta_{A'}\psi^{A'}$$
$$\mu(\partial_1 h_-)\mu^{-1} = \zeta_A\psi^A, \tag{6.3}$$

where $\psi^{A'}$ and ψ^A are holomorphic functions of $x \in \mathcal{D}$ with values in End $E \otimes S_\mp \otimes \bigwedge^2 S_\pm$, where the $S_\pm$ are spinor bundles (see Chap. 3).

If the bundle E_θ corresponds to the form $\theta \in \tilde{H}^{0,1}(L^{(1)}(\mathcal{D}), \mathrm{GL}(n, \mathbb{C}))$, then there is another useful definition of the Penrose transform of the forms $\Psi_\pm$. In order to introduce it, let us note that in view of Propositions 3.2 and 6.1, the following proposition is valid.

Proposition 6.3 (Khenkin and Manin [1982]). *Under the conditions of* Proposition 6.1, *the mapping*

$$H^1(L^{(1)}(\mathcal{D}), \mathrm{End}\, E_\theta(-1,0)) \to H^1(L(\mathcal{D}), \mathrm{End}\, E_\theta(-1,0)),$$

generated by the restriction of cohomologies from $L^{(1)}(\mathcal{D})$ to $L(\mathcal{D})$, is an isomorphism.

Let us further define the Penrose transforms of the forms $\Psi_\pm$

$$\mathcal{P}^{A'}\Psi_+(x, \eta^{c'}) = \int_{\zeta \in \mathcal{L}_+(x)} \frac{\partial}{\partial \zeta_{A'}} \mu_x \Psi_+ \mu_x^{-1} \wedge \zeta^c d\zeta_c,$$
$$\mathcal{P}^{A}\Psi_-(x, \zeta^c) = \int_{\eta \in \mathcal{L}_-(x)} \frac{\partial}{\partial \eta_A} \mu_x \Psi_- \mu_x^{-1} \wedge \eta^{c'} d\eta_{c'}. \tag{6.3^1}$$

Due to Liouville's theorem, these transforms depend only on $x \in \mathcal{D}$.

Proposition 6.4 (Khenkin and Manin [1982]). *The Penrose transforms of the forms* (6.3) *and* (6.3^1) *are connected by the relations*

$$\mathcal{P}^{A'}\Psi_+(x) = -2\pi i \psi^{A'}; \qquad \mathcal{P}^{A}\Psi_-(x) = -2\pi i \psi^A.$$

For the forms Φ_+ and Φ_-, which are elements of the spaces $H^1(L(\mathcal{D}),$ End $E_\theta(-2,0))$ and $H^1(L(\mathcal{D}), \mathrm{End}\, E_\theta(0,-2))$, suppose

$$\varphi_+(x, \eta^{c'}) = \int_{\zeta \in \mathcal{L}_+(x)} \mu \Phi_+ \mu^{-1} \wedge \zeta^c d\zeta_c,$$
$$\varphi_-(x, \zeta^c) = \int_{\eta \in \mathcal{L}_-(x)} \mu \Phi_- \mu^{-1} \wedge \eta^{c'} d\eta_{c'}. \tag{6.4}$$

Again due to Liouville's theorem, these Penrose transforms, in fact, depend only on $x \in \mathcal{D}$.

Let us now define the Penrose transforms for some spaces of two-dimensional cohomologies on $L(\mathcal{D})$.

For the form Γ, which is an element of the space $H^2(L(\mathcal{D}), \operatorname{End} E_\theta(-2, -2))$, suppose

$$\gamma(x) = \int_{(\zeta, \eta) \in \mathcal{L}(x)} \mu_x \Gamma \mu_x^{-1} \wedge \zeta^c d\zeta_c \wedge \eta^{c'} d\eta_{c'}. \tag{6.5}$$

For the forms F_+ and F_-, which are elements of the spaces $H^2(L(\mathcal{D}), \operatorname{End} E_\theta(-3, -1))$ and $H^2(L(\mathcal{D}), \operatorname{End} E_\theta(-1, -3))$, respectively, suppose

$$\begin{aligned}
\mathcal{P}F_+(x) &= \int_{\mathcal{L}(x)} \sum_A \zeta^A \frac{\partial}{\partial \eta^A}(\mu_x F_+ \mu_x^{-1}) \wedge \zeta^c d\zeta_c \wedge \eta^{c'} d\eta_{c'}, \\
\mathcal{P}F_-(x) &= \int_{\mathcal{L}(x)} \sum_{A'} \eta^{A'} \frac{\partial}{\partial \zeta^{A'}}(\mu_x F_- \mu_x^{-1}) \wedge \zeta^c d\zeta_c \wedge \eta^{c'} d\eta_{c'}.
\end{aligned} \tag{6.6}$$

We shall need another definition of the Penrose transforms of the forms F_+ and F_-. Let us consider the forms $\pi_1^* F_\pm$. Due to Proposition 3.3, there will be smooth $(0, 1)$-forms $u_\pm$ on $F(\mathcal{D})$ whose coefficients have values in $\operatorname{End} E_\theta(-3, -1)$ or $\operatorname{End} E_\theta(-1, -3)$ such that

$$\pi_1^* F_\pm = \bar{\partial} u_\pm + [\pi_1^* \theta, u_\pm].$$

Hence and from (6.1), it follows that the $(0, 1)$-form $v_\pm = \mu(\partial_1 u_\pm)\mu^{-1}$ with homogeneity $(-2, 0)$ is $\bar{\partial}$-closed on $F(\mathcal{D})$. Suppose

$$f_+(x, \eta) = \int_{\zeta \in \mathcal{L}_+(x)} v_+(x, \zeta, \eta)\zeta^c d\zeta_c; \quad f_-(x, \zeta) = \int_{\eta \in \mathcal{L}_-(x)} v_-(x, \zeta, \eta)\eta^{c'} d\eta_{c'}. \tag{6.6^1}$$

Proposition 6.5 (Khenkin and Manin [1982]). *The Penrose transforms of the forms (6.6) and (6.6^1) are connected by the relations*

$$\mathcal{P}F_+(x) = -2\pi i f_+(x); \qquad \mathcal{P}F_-(x) = -2\pi i f_-(x).$$

For the forms G_+ and G_-, which are elements of the spaces $H^2(L(\mathcal{D}), \operatorname{End} E_\theta(-3, -2))$ and $H^2(L(\mathcal{D}), \operatorname{End} E_\theta(-2, -3))$, suppose

$$\begin{aligned}
g_A(x) &= \int_{(\zeta, \eta) \in \mathcal{L}(x)} \zeta_A \mu_x G_+ \mu_x^{-1} \wedge \zeta^c d\zeta_c \wedge \eta^{c'} d\eta_{c'}, \\
g_{A'}(x) &= \int_{(\zeta, \eta) \in \mathcal{L}(x)} \eta_{A'} \mu_x G_- \mu_x^{-1} \wedge \zeta^c d\zeta_c \wedge \eta^{c'} d\eta_{c'}.
\end{aligned} \tag{6.7}$$

For the forms

$$\mathcal{J} \in H^2(L(\mathcal{D}), \operatorname{End} E_\theta(-3, 3))$$

and

$$\Omega \in H^2(L(\mathcal{D}), \operatorname{End} E_\theta(-4, -4)),$$

put

$$j_A^{B'}(x) = \iint\limits_{(\zeta,\eta)\in\mathcal{L}(x)} \mu_x \mathcal{J} \mu_x^{-1} \zeta_A \mu^{B'} \wedge \zeta^c d\zeta_c \wedge \eta^{c'} d\eta_{c'}, \tag{6.8}$$

$$\omega_{AB}^{A'B'}(x) = \iint\limits_{(\zeta,\eta)\in\mathcal{L}(x)} \mu_x \Omega \mu_x^{-1} \zeta_A \zeta_B \eta^{A'} \eta^{B'} \wedge \zeta^c d\zeta_c \wedge \eta^{c'} d\eta_{c'}. \tag{6.9}$$

From the definitions (6.2)–(6.9), it immediately follows that the functions $\psi^{A'}, \psi^A, \omega, \varphi_{\pm}, \gamma, f_{\pm}, g_A, g_{A'}, j_A^{B'}, \omega_{AB}^{A'B'}$ are holomorphic, depend only on the variable $x \in \mathcal{D}$, and (for a fixed μ_x) are defined by classes of cohomologies of the corresponding forms $\Psi_{\pm}, \Omega, \Phi_{\pm}, \Gamma, F_{\pm}, G_{\pm}, \mathcal{J}$.

Theorem 6.1. *Let $\mathcal{D}$ be a pseudoconvex domain in $\mathbb{C}M$ such that its intersections with complex null lines are contractible. Then for the fixed form $\theta \in \tilde{H}^{0,1}(L(\mathcal{D}), \mathrm{GL}(n,\mathbb{C}))$, the Penrose transforms of the form (6.2)–(6.9) determine the isomorphisms of:*
(a) spaces of one-dimensional $\bar{\partial}$-cohomologies $L(\mathcal{D})$ with coefficients in

$$\mathrm{End}\, E_\theta(-1,0), \mathrm{End}\, E_\theta(0,-1), \mathrm{End}\, E_\theta(-1,-1),$$
$$\mathrm{End}\, E_\theta(-2,0), \mathrm{End}\, E_\theta(0,-2),$$

respectively, with spaces of holomorphic sections over $\mathcal{D}$ of the bundles

$$\mathrm{End}\, E \otimes S_- \otimes \bigwedge\nolimits^2 S_+, \mathrm{End}\, E \otimes S_+ \otimes \bigwedge\nolimits^2 S_-,$$
$$\mathrm{End}\, E \otimes \bigwedge\nolimits^2 S_+ \otimes \bigwedge\nolimits^2 S_-, \mathrm{End}\, E \otimes \bigwedge\nolimits^2 S_+, \mathrm{End}\, E \otimes \bigwedge\nolimits^2 S_-;$$

(b) spaces of two-dimensional $\bar{\partial}$-cohomologies $L(\mathcal{D})$ with coefficients in

$$\mathrm{End}\, E_\theta(-2,-2), \mathrm{End}\, E_\theta(-3,-1), \mathrm{End}\, E_\theta(-1,-3),$$
$$\mathrm{End}\, E_\theta(-3,-2), \mathrm{End}\, E_\theta(-2,-3),$$

respectively, with spaces of holomorphic sections over $\mathcal{D}$ of the bundles

$$\mathrm{End}\, E \otimes \bigwedge\nolimits^2 S_+ \otimes \bigwedge\nolimits^2 S_-, \mathrm{End}\, E \otimes (\bigwedge\nolimits^2 S_+)^2 \otimes (\bigwedge\nolimits^2 S_-),$$
$$\mathrm{End}\, E \otimes (\bigwedge\nolimits^2 S_-)^2 \otimes (\bigwedge\nolimits^2 S_+), \mathrm{End}\, E \otimes S_+ \otimes \bigwedge\nolimits^2 S_+ \otimes \bigwedge\nolimits^2 S_-,$$
$$\mathrm{End}\, E \otimes S_- \otimes \bigwedge\nolimits^2 S_+ \otimes \bigwedge\nolimits^2 S_-;$$

(c) spaces of two-dimensional $\bar{\partial}$-cohomologies $L(\mathcal{D})$ with coefficients in $\mathrm{End}\, E_\theta(-3,-3), \mathrm{End}\, E_\theta(-4,-4)$, *respectively, with spaces of those holomorphic sections $j_A^{B'}$ and $\omega_{AB}^{A'B'}$ over $\mathcal{D}$ of the bundles*

$$\mathrm{End}\, E \otimes S_+ \otimes S_- \otimes \bigwedge\nolimits^2 S_+ \otimes \bigwedge\nolimits^2 S_-;$$
$$\mathrm{End}\, E \otimes S^2(S_+) \otimes S^2(S_-) \otimes \bigwedge\nolimits^2 S_+ \otimes \bigwedge\nolimits^2 S_-$$

that satisfy the equations

$$\left[\nabla^A_{B'}, j^{B'}_A\right] = 0, \tag{6.10}$$

$$\left[\nabla^A_{A'}, \omega^{A'B'}_{AB}\right] = 0. \tag{6.11}$$

If $\mathcal{D}$ is a domain in the real Minkowski space $\mathcal{M}$ or in the real Euclidean space $\mathcal{E}$, then the following additional result is valid.

Theorem 6.1[1] (Khenkin [1984]). *Let $\theta \in \tilde{H}^{0,1}(L(\mathcal{D}), \mathrm{GL}(n, \mathbb{C}))$, where $\mathcal{D}$ is an arbitrary domain in $\mathcal{E}$ or any domain in $\mathcal{M}$ whose intersections with light rays are connected. Then Theorem 6.1 is valid in replacing the $\bar{\partial} + \theta$-cohomologies with $\bar{\partial}_\tau + \theta$-cohomologies of the CR-manifold $L(\mathcal{D})$ and in replacing holomorphic sections of the corresponding bundles over $\mathcal{D}$ with smooth sections of the same bundles. Besides, the sections of these bundles,*

$$\psi^{A'}, \psi^A, \omega, \varphi_\pm, \gamma, f_\pm, g_A, g_{A'}, j^{B'}_A, \omega^{A'B'}_{AB},$$

are really analytical on $\mathcal{D}$ iff the forms $\Psi_\pm, \Omega, \Phi_\pm, \Gamma, F_\pm, G_\pm, \mathcal{J}$ corresponding to them are locally (but not globally) $(\bar{\partial}_\theta + \theta)$-exact on $L(\mathcal{D})$.

Theorem 6.1[1] shows that Lewy's effect on $L(\mathcal{D})$, i.e., the existence of $\bar{\partial}_\tau$-closed but locally not $\bar{\partial}_\tau$-exact forms on $L(\mathcal{D})$ has a simple physical interpretation: nonanalyticity of the corresponding fields on $\mathcal{D}$.

One of the consequences of Theorem 6.1 most useful for us is the following criterion of the $\bar{\partial}$-exactness of forms that are elements of the spaces $H^2(L^{(\nu)}(\mathcal{D}), \mathrm{End}\, E_\theta)$.

Proposition 6.6. *Let $\theta \in \tilde{H}^{0,1}(L^{(\nu)}(\mathcal{D}), \mathrm{GL}(n, \mathbb{C}))$. In order that the $(0,2)$-form G_ν, which is an element of the space $H^2(L^{(\nu)}(\mathcal{D}), \mathrm{End}\, E_\theta)$, $\nu = 2, 3, \ldots$, be $(\bar{\partial} + \theta)$-exact on $L^{(\nu)}(\mathcal{D})$, i.e., a zero element, it is sufficient that for any $x \in \mathcal{D}$, the restriction $G_\nu|_{\mathcal{L}^{(\nu)}(x)}$ is $(\bar{\partial} + \theta)$-exact on $\mathcal{L}^{(\nu)}(x)$.*

In fact, due to Proposition 6.1, we have $H^2(L^{(1)}(\mathcal{D}), \mathrm{End}\, E_\theta) = 0$. Thus, due to Proposition 3.1, the form G_ν is $(\bar{\partial} + \theta)$-gauge-equivalent to the form $\Gamma\langle \zeta \cdot \eta \rangle^2$, where the form $\Gamma|_{L(\mathcal{D})}$ is an element of the space $H^2(L(\mathcal{D}), \mathrm{End}\, E_\theta(-2, -2))$. Due to Theorem 6.1, the form Γ is $(\bar{\partial} + \theta)$-exact on $L(\mathcal{D})$ iff its Radon–Penrose transform $\gamma(x)$ of the form (6.5) is equal to zero on $\mathcal{D}$. Hence it follows that for $(\bar{\partial} + \theta)$-exactness of the form G_ν on $L^{(2)}(\mathcal{D})$, it is sufficient that for all $x \in \mathcal{D}$, the restrictions G_ν are $(\bar{\partial} + \theta)$-exact on $\mathcal{L}^{(2)}(x)$. Furthermore, let $\nu = 3$. If the form G_3 is $(\bar{\partial} + \theta)$-exact on $L^{(2)}(\mathcal{D})$, then it is $(\bar{\partial} + \theta)$-gauge equivalent to the form $\mathcal{J}\langle \zeta \cdot \eta \rangle^3$ on $L^{(3)}(\mathcal{D})$, where the form $\mathcal{J}|_{L(\mathcal{D})}$ is an element of the space $H^2(L(\mathcal{D}), \mathrm{End}\, E_\theta(-3, -3))$. Due to Theorem 6.1, the form $\mathcal{J}$ is $(\bar{\partial} + \theta)$-exact on $L(\mathcal{D})$ iff its Penrose transform of the form (6.8) is equal to zero. Consequently, for $(\bar{\partial} + \theta)$-exactness of the form G_3 on $L^{(3)}(\mathcal{D})$, it is sufficient that all restrictions G_3 are $(\bar{\partial} + \theta)$-exact on $\mathcal{L}^{(3)}(x)$, $x \in \mathcal{D}$.

Chapter 7
The Yang–Mills Equations in Terms of Scattering Data

Here we discuss on the characterization of scattering data corresponding to the Yang–Mills fields by $\bar{\partial}$-equations based on Chap. 4 and the papers Buchdahl [1985], Khenkin [1980], Isenberg and Yasskin [1979], Ward [1977], and Witten [1978]. With this characterization, the conditions of linearization of the Yang–Mills equations and the integral equation for finding all local solutions of this equation are obtained.

Suppose, as in Chap. 2, the holomorphic (respectively, smooth) fields $a_{x^j}(x, w)$ and $a_{w^j}(x, w)$ in the domain $(x + w) \in \mathcal{D}$, $(x - w) \in \mathcal{D}$, with values in $\mathrm{gl}(n, \mathbb{C})$ satisfying the compatibility equations (2.2), are defined. Let $\mu(x, w, \lambda)$ be a holomorphic (respectively, smooth) function in x, w and a smooth function in $\lambda \in \mathbb{C}P^1 \times \mathbb{C}P^1$ with values in $\mathrm{GL}(n, \mathbb{C})$ satisfying the linear system (2.1^1), where $k = 1, 2, \ldots$. By developing the constructions of Chap. 4, let us consider the $(0, 1)$-form

$$\theta = \theta_1(x, w, \lambda)d\bar{\lambda}_1 + \theta_2(x, w, \lambda)d\bar{\lambda}_2, \tag{7.1}$$

where

$$\theta_j(x, w, \lambda) = \mu^{-1}(x, w, \lambda)\frac{\partial\mu(x, w, \lambda)}{\partial\bar{\lambda}_j}, \; j = 1, 2.$$

The following extensions of (4.8) are implied from (2.1^1) and (7.1):

$$\frac{\partial\theta_2}{\partial\bar{\lambda}_1} - \frac{\partial\theta_1}{\partial\bar{\lambda}_2} + [\theta_1, \theta_2] = 0, \tag{7.2}$$

$$D_j\theta_\nu = O(w^k), \quad \nu = 1, 2; \quad j = 0, 1, 2, 3, \tag{7.3}$$

$$|\theta_j(x, w, \lambda)| = O\left(\frac{1}{1 + |\lambda_j|^2}\right), \tag{7.4}$$

where the $\{D_j\}$ are operators defined on (2.3). The form θ should be considered up to the $\bar{\partial}$-gauge equivalence

$$\theta \sim h^{-1}\bar{\partial}h + h^{-1}\theta h, \tag{7.5}$$

where the matrix function h with values in $\mathrm{GL}(n, \mathbb{C})$ satisfies the equations

$$D_j h = O(w^k).$$

For $w = 0$, the form $\theta(x, 0, \lambda)$ turns into scattering data (the Radon–Penrose transform) of the field $a(x) = a_{x^j}(x, 0)dx^j$. The form θ satisfying (7.2)–(7.5) without loss of generality may be considered to be a polynomial in w of k-degree.

The properties (7.2)–(7.5) mean, in particular, that the form θ is defined on $L^{(k)}(\mathcal{D})$ and $\bar{\partial}\theta + \theta \wedge \theta = 0$ on $L^{(k)}(\mathcal{D})$, where $L^{(k)}(\mathcal{D})$ is the kth infinitesimal neighbourhood of the manifold $L(\mathcal{D}) \subset L_+(\mathcal{D}) \times L_-(\mathcal{D})$. Such forms are in one-to-one correspondence with the topologically trivial n-dimensional holomorphic vector bundles E_θ over the manifold $L^{(k)}(\mathcal{D})$, which are furthermore analytically trivial on each quadric $\mathcal{L}(x)$, $x \in \mathcal{D}$ (see Chaps. 3 and 8).

Theorems 2.1, 2.2, and 4.2 lead to the following interpretation of the Yang–Mills equations in terms of scattering data that develops results of the papers Buchdahl [1985], Khenkin [1980], Isenberg and Yasskin [1979], and Witten [1978].

Theorem 7.1. *Let $\mathcal{D}$ be a convex domain in $\mathbb{C}^4$ (respectively $R^4 \subset \mathbb{C}^4$). Then the following statements are valid:*

(a) the scattering data $\theta(x, \lambda)$ of an arbitrary holomorphic (respectively, smooth) field $a(x)$, $x \in \mathcal{D}$, admit an extension to the form $\theta(x, w, \lambda)$ of the type (7.1)–(7.5), where $k = 2$;

(b) in order for the form $\theta(x, \lambda)$, $x \in \mathcal{D}$, $\lambda \in \mathcal{L}$ to correspond to the holomorphic (respectively, smooth) field $a(x)$, $x \in \mathcal{D}$, satisfying the Yang–Mills vacuum equation (1.11), it is necessary and sufficient that this form admits an extension to the form $\theta(x, w, \lambda)$ of the type (7.1)–(7.5), where $k = 3$;

(c) in order for the form $\theta(x, \lambda)$ to correspond to the holomorphic (respectively, smooth) field $a(x), x \in \mathcal{D}, \lambda \in \mathcal{L}$, satisfying the equations (1.11) and (2.8) (or (2.8^1)), it is necessary and sufficient that the form admits an extension to the form $\theta(x, w, \lambda)$ of the type (7.1)–(7.5), where $k = 4$ (or $k \geq 5$).

Thus, in order to obtain solutions of the homogeneous Yang–Mills equation (1.11), it is sufficient to obtain the solutions $\theta(x, w, \lambda)$ of the system (7.2)–(7.4), where $k = 3$. Having carried out the Radon–Penrose inverse transform from such a form for $w = 0$, we obtain solutions of equation (1.11).

Now we provide a general method of constructing all local solutions of the system (7.2)–(7.4) for $k = 3$ that extends a method of the perturbation theory well known in mathematical physics to scattering data of the form (7.1)–(7.4).

Let us introduce the spaces $C_{0,q}(\mathcal{D}^{(k)} \times \mathcal{L})$ consisting of such $(0, q)$-forms g on the quadric $\mathcal{L} = \mathbb{C}P^1 \times \mathbb{C}P^1$, which are holomorphic in $x \in \mathcal{D}$, are polynomials in w of degree k and have the finite norm of the form

$$\|g\|_{C(\mathcal{D}^{(k)} \times \mathcal{L})} = \sup_{x \in \mathcal{D}, \lambda \in \mathcal{L}, |w| \leq 1} |g(x, w, \lambda)|.$$

Let $\{D_j\} = D$ be differential operators of the form (2.3) depending on the parameter $\lambda = (\lambda_1, \lambda_2) \in \mathbb{C}P^1 \times \mathbb{C}P^1$. Let $\mathcal{D}_\varepsilon$ denote the ε-neighbourhood of the domain $\mathcal{D} \subset \mathbb{C}^4$.

Proposition 7.1. *Let $\mathcal{D}$ be a bounded convex domain $\mathcal{D} \subset \mathbb{C}^4$. For any $k = 1, 2, 3, \ldots,$ and for any $\varepsilon > 0$, there exists the continuous (integral) operator*

$$R : C_{0,2}(\mathcal{D}_\varepsilon^{(k)} \times \mathcal{L}) \to C_{0,1}(\mathcal{D}^{(k)} \times \mathcal{L})$$

with the following properties: For any form g from $C_{0,2}(\mathcal{D}_\varepsilon^{(k)} \times \mathcal{L})$ of the type $g = \bar{\partial}f$, where $Df = O(w^k)$, we have $g = \bar{\partial}Rg$, where $DRg = O(w^k)$ and $||R|| = O(\varepsilon^{-2})$.

The integral operator R possessing necessary properties may be written in an explicit form. Let R^1, R^2 be operators of the form (3.13) inverting the $\bar{\partial}$-operator on functions and on $(0,1)$-forms on the quadric $\mathcal{L}$. Let G be an operator that inverts the differential operator D of the form (2.3), and let $\hat{G}$ be a suitable operator that inverts the differential operator $\hat{D}$ of the form (2.4). Then for the form $g \in C_{0,2}(\mathcal{D}_\varepsilon^{(k)}, \mathcal{L})$, we suppose

$$Rg = R^2g - \bar{\partial}G(R^1DR^2g - \hat{G}\hat{D}R^1DR^2g).$$

The condition $g = \bar{\partial}f$, where $Df = O(w^k)$, makes it possible for a suitable choice of $\hat{G}$ to obtain the compatibility condition

$$\hat{D}(R^1DR^2g - \hat{G}\hat{D}R^1DR^2g) = O(w^k)$$

simultaneously with the condition $\bar{\partial}\hat{G}\hat{D}R^1DR^2g = 0$, where $x + w \in \mathcal{D}$, $x - w \in \mathcal{D}, \lambda \in \mathcal{L}$. Hence it follows that the equality $DRg = O(w^k)$ stated in Proposition 7.1 is valid. The equality $g = \bar{\partial}Rg$ and the estimate $||R|| = O(\varepsilon^{-2})$ immediately follow from the definition of the operator R and from properties taking place in the definition of the operators $R^1, R^2, D, \hat{D}, G, \hat{G}$.

Let us now consider the linearized system (7.2)–(7.4) for $k \geq 3$. We denote the solution of such a system by

$$\theta^0 = \theta_1^0(x, w, \lambda)d\bar{\lambda}_1 + \theta_2^0(x, w, \lambda)d\bar{\lambda}_2.$$

The form θ^0 satisfies the equations (7.3) and (7.4) for $k \geq 3$, and instead of (7.2), it satisfies the linear equation

$$\frac{\partial \theta_2^0}{\partial \bar{\lambda}_1} - \frac{\partial \theta_1^0}{\partial \bar{\lambda}_2} = 0. \tag{7.6}$$

The properties (7.3), (7.4), and (7.6) mean, in particular, that the form θ^0 is defined on $L^{(k)}(\mathcal{D})$ and $\bar{\partial}\theta^0 = 0$ on $L^{(k)}(\mathcal{D})$.

Due to Theorem 7.1, each scalar element of such a matrix form θ^0 is the Radon–Penrose transform of a (holomorphic) solution of the Maxwell homogeneous system in the domain $\mathcal{D} \subset \mathbb{C}^4$.

The following result makes it possible to obtain all solutions of the Yang–Mills equation with sufficiently small norms in the fixed convex domain $\mathcal{D} \subset \mathbb{C}^4$, starting with a solution of the linearized system (7.3), (7.4), (7.6) with help of a suitable integral equation.

Theorem 7.2. *Let $\mathcal{D}$ be a bounded convex domain in $\mathbb{C}^4$. For any $\varepsilon > 0$ and for any form $\theta^0 \in C_{0,1}(\mathcal{D}_\varepsilon^{(k)}, \mathcal{L})$ with a sufficiently small norm and with the properties (7.3), (7.4), and (7.6), where $x \in \mathcal{D}_\varepsilon, k \geq 3$, the integral equation of the form*

$$\theta = \theta^0 + R(\theta \wedge \theta) \tag{7.7}$$

has the (moreover, unique) solution θ with a sufficiently small norm in the space $C_{0,1}(\mathcal{D}^{(k)}, \mathcal{L})$. In addition, the form θ possesses the properties (7.2)–(7.4) for $k = 3$; i.e., in view of Theorem 7.1 (and Proposition 3.3), θ is the Radon–Penrose transform of the holomorphic field $a(x)$, $x \in \mathcal{D}$, satisfying the Yang–Mills equation.

Remark. The integral equation (7.7) may be also applied to the solution of the Cauchy problem for the Yang–Mills equation on the real Minkowski space. We assume that in this way, we shall manage to develop interesting results for recent years about the existence of global solutions to the Cauchy problem for the Yang–Mills equations (Choquet-Bruhat and Christodoulou [1981]; Eardley and Moncrief [1982]; Segal [1979]).

The first statement of Theorem 7.2 is being proved due to the use of the estimate of the norm of the operator R given in Proposition 7.1. In addition, a solution of the integral equation (7.7) is obtained by the following iterative procedure

$$\theta = \theta_0 + \theta_1 + \theta_2 + \cdots ,$$

where

$$\theta_0 = \theta^0,$$
$$\theta_1 = R(\theta_0 \wedge \theta_0),$$
$$\theta_2 = R(\theta_0 \wedge \theta_1 + \theta_1 \wedge \theta_0),$$
$$\cdots$$
$$\theta_{j+1} = R(\theta_0 \wedge \theta_j + \theta_1 \wedge \theta_{j-1} + \cdots + \theta_j \wedge \theta_0).$$

The most meaningful part of the proof consists of verification of the properties (7.2)–(7.4) for the form θ for $k = 3$ if the form θ^0 possesses the properties (7.3), (7.4), and (7.6) for $k \geq 3$. This verification is based on the joint use of Proposition 7.1 and the following proposition.

Proposition 7.2. *For any $j = 0, 1, 2, \ldots$ and for any $(0,1)$-forms $\theta_0, \theta_1,$ $\ldots, \theta_j$ on the manifold $L^{(3)}(\mathcal{D}) \subset L_+(\mathcal{D}) \times L_-(\mathcal{D})$ with smooth $(n \times n)$-matrix coefficients satisfying on $L^{(3)}(\mathcal{D})$ the system of $\bar{\partial}$-equations of the form*

$$\bar{\partial}\theta_0 = 0,$$
$$\bar{\partial}\theta_1 = \theta_0 \wedge \theta_0,$$
$$\cdots$$
$$\bar{\partial}\theta_j = \theta_0 \wedge \theta_{j-1} + \theta_1 \wedge \theta_{j-2} + \cdots + \theta_j \wedge \theta_0,$$

there exists the $(0,1)$-form θ_{j+1} with smooth $(n \times n)$-matrix coefficients satisfying on $L^{(3)}(\mathcal{D})$ the equation of the form

$$\bar{\partial}\theta_{j+1} = \theta_0 \wedge \theta_j + \theta_1 \wedge \theta_{j-1} + \cdots + \theta_j \wedge \theta_0.$$

This statement is deduced from Proposition 6.6. Suppose, we have the $(0,1)$-form θ_0 with the property

$$\bar{\partial}\theta_0 = 0 \quad \text{on } L^{(3)}(\mathcal{D}). \tag{7.8}$$

Then the equality $\bar{\partial}(\theta_0 \wedge \theta_0) = 0$ on $L^{(3)}(\mathcal{D})$ is also valid. Due to Proposition 4.1, the form θ_0 is $\bar{\partial}$-exact on any manifold $\mathcal{L}^{(1)}(x)$, $x \in \mathcal{D}$. Hence it follows that the form $\theta_0 \wedge \theta_0$ is $\bar{\partial}$-exact on any manifold $\mathcal{L}^{(3)}(x)$, $x \in \mathcal{D}$. Due to Proposition 6.6, the form $\theta_0 \wedge \theta_0$ is $\bar{\partial}$-exact on $L^{(3)}(\mathcal{D})$. So, there exists the $(0,1)$-form θ_1 with smooth coefficients satisfying the equation

$$\theta_0 \wedge \theta_0 = \bar{\partial}\theta_1 \quad \text{on } L^{(3)}(\mathcal{D}). \tag{7.9}$$

Let us further consider the form $\theta_0 \wedge \theta_1 + \theta_1 \wedge \theta_0$. This form is $\bar{\partial}$-closed on $L^{(3)}(\mathcal{D})$, since from (7.9) and (7.8), the following equality is valid on $L^{(3)}(\mathcal{D})$:

$$\bar{\partial}(\theta_0 \wedge \theta_1 + \theta_1 \wedge \theta_0) = -\theta_0 \wedge \theta_0 \wedge \theta_0 + \theta_0 \wedge \theta_0 \wedge \theta_0 = 0.$$

The exactness of the form $\theta_0 \wedge \theta_1 + \theta_1 \wedge \theta_0$ does not depend on $\bar{\partial}$-calibration of the form θ_0 and on the consequent choice of the solution θ_1 of equation (7.9). Thus, for any fixed $x \in \mathcal{D}$, we choose a $\bar{\partial}$-calibration of the form θ_0 such that $\theta_0 \wedge \theta_0 = 0$ on $\mathcal{L}^{(3)}(x)$. For such a calibration, it follows from (7.9) that $\bar{\partial}\theta_1 = 0$ on $\mathcal{L}^{(3)}(x)$. Due to Proposition 4.1, it is possible to choose a solution (7.9) on $L^{(3)}(\mathcal{D})$ such that $\theta_1 = 0$ on $\mathcal{L}^{(1)}(x)$. For such a calibration of the forms θ_0 and θ_1, the equality $\theta_0 \wedge \theta_1 + \theta_1 \wedge \theta_0 = 0$ is valid on $\mathcal{L}^{(3)}(x)$. Thus, for any $x \in \mathcal{D}$, the form $\theta_0 \wedge \theta_1 + \theta_1 \wedge \theta_0$ is $\bar{\partial}$-exact on $\mathcal{L}^{(3)}(x)$. Due to Proposition 6.6, this form is also $\bar{\partial}$-exact on $L^{(3)}(\mathcal{D})$, i.e., there exists the $(0,1)$-form θ_2 satisfying the equation $\bar{\partial}\theta_2 = \theta_0 \wedge \theta_1 + \theta_1 \wedge \theta_0$ on $L^{(3)}(\mathcal{D})$, and so on.

The Radon–Penrose transform reduces a sophisticated nonlinear system of the Yang–Mills equations to a related, less sophisticated one, but unfortunately also to a nonlinear system (7.2)–(7.4) for scattering data. The following result provides additional conditions under which this transform leads to a linearization of the system of the Yang–Mills equations.

Theorem 7.3. *Let $\mathcal{D}$ be a convex domain in $\mathbb{C}^4$ (respectively, R^4). In order for a holomorphic (respectively, smooth) connection $a(x)$, $x \in \mathcal{D}$ to satisfy the Yang–Mills equations (1.11) and the additional commutation equalities (2.8) and (2.8^1) for any k, it is necessary and sufficient that scattering data for the field $\theta(x,\lambda)$ are gauge-equivalent to the form $\theta(x,w,\lambda)$ of the type (7.1) with the following properties:*

$$\frac{\partial \theta_2}{\partial \bar{\lambda}_1} - \frac{\partial \theta_1}{\partial \bar{\lambda}_2} = 0; \qquad [\theta_1, \theta_2] = 0;$$

$$D_\nu \theta_j = 0, \qquad \theta_j = O\left(\frac{1}{1 + |\lambda_j|^2}\right), \tag{7.10}$$

where

$$\lambda \in \mathcal{L}; \quad (x + w) \in \mathcal{D}, \quad (x - w) \in \mathcal{D}; \quad j = 1, 2; \quad \nu = 0, 1, 2, 3.$$

In addition, self-dual fields correspond to those forms θ in which the component $\theta_2 = 0$, and anti-self-dual fields correspond to forms in which $\theta_1 = 0$.

Supplement 1. Theorem 7.3 is valid for the case of any matrix group. Due to Theorem 2.3 for the group $GL(2, \mathbb{C})$, i.e., for the connections a with values in $gl(2, \mathbb{C})$, the assertion of Theorem 7.3 is valid if one ignores the additional condition (2.8^1) in this assertion, leaving only the commutation conditions (2.8) of the self-dual and anti-self-dual components of the curvature of the connection a.

Supplement 2. Theorem 7.3 applied to the Minkowski space

$$\mathcal{M}_0 = \{x \in \mathbb{C}^4 : \operatorname{Im} x_0 = 0, \operatorname{Re} x_\nu = 0, \nu = 1, 2, 3\},$$

in particular, provides a linearization of the Cauchy problem for the Yang–Mills equation (1.11) complemented with the commutation relations (2.8) and (2.8^1). In fact, due to Theorem 7.3 and the Cauchy data of this system on the plane $Q_0 = \{x \in \mathcal{M}_0 : x_0 = 0\}$, it is possible to construct the corresponding scattering data

$$\theta(x, w, \lambda), \ (x + w) \in Q_0, \ (x - w) \in Q_0, \ \lambda \in \mathcal{L}$$

with properties (7.10). In twistor coordinates, these scattering data define a form of the $\bar{\partial}$-connection $\theta(\zeta, \eta)$ on the CR-manifold $L_+(\mathcal{M}_0) \times L_-(\mathcal{M}_0) \supset L(\mathcal{M}_0)$. The Radon–Penrose inverse transform turns the form $\theta(\zeta, \eta)$ into a solution of the Cauchy problem for (1.11) and (2.8) on the space $\mathcal{M}_0$. The aforementioned procedure extends a well-known twistor procedure for the solution to the Cauchy problem for the Maxwell equations (see Gindikin and Khenkin [1981]; Woodhouse [1985]) to the Yang–Mills equations.

The sufficiency condition in Theorem 7.3 is an immediate consequence of Theorem 7.1(c). It is possible to reduce the necessity condition to the case of the holomorphic fields $a(x), x \in \mathcal{D}$, with a sufficiently small norm. For this case, due to Theorem 7.2, the scattering data θ corresponding to the field a may be obtained from the form θ^0 with the properties (7.3), (7.4), and (7.6) by solving the integral equation (7.7). By the Radon–Penrose inverse transform, the elements of the matrix form θ^0 due to Theorem 7.1(b) are turned into the holomorphic fields a_{ij}^0 satisfying the Maxwell equations. From (7.7) and from the commutation relations (2.8) and (2.8^1), it follows that for the matrix field $a^0 = \{a_{ij}^0\}$, the following commutation relations hold: For any two points x' and x'' from $\mathcal{D}$,

$$[f_{\alpha\beta}(x'), f_{\alpha'\beta'}(x'')] = 0, \tag{7.11}$$

where $f_{\alpha\beta}$ and $f_{\alpha'\beta'}$, respectively, are components of the self-dual and of the anti-self-dual parts

$$f^+ = f_{\alpha\beta} dx^{\alpha 0'} \wedge dx^{\beta 1'}, \qquad f^- = f_{\alpha'\beta'} dx^{0\alpha'} \wedge dx^{1\beta'}$$

of the two-form $da^0 = f^+ + f^-$. Furthermore, from the Penrose–Ward scalar transforms (cf. (2.1)) of the matrix elements of the forms $f^\pm$ and due to (7.11), we obtain the form that is gauge-equivalent to the original one,

$$\theta^0(x, w, \lambda_1, \lambda_2) = \theta_1^0(x + w, \lambda_1) d\bar{\lambda}_1 + \theta_2^0(x - w, \lambda_2) d\bar{\lambda}_2, \tag{7.12}$$

where

$$[\theta_1^0, \theta_2^0] = 0, \qquad D_\nu \theta_j^0 = 0, \qquad \theta_j^0 = O\left(\frac{1}{1 + |\lambda_j|^2}\right), \qquad j = 1, 2.$$

Let us now consider the nonlinear equation (7.7) with the form θ^0 of the type(7.12). Due to Theorem 7.2 and the conditions (7.12), equation (7.7) has a unique solution (with a small norm) of the form $\theta = \theta^0$. Thus, the scattering data θ of the original field a turn out to be gauge-equivalent to the form θ^0 with the properties (7.12) and consequently with (7.10).

The Yang–Mills equation also admits a very meaningful interpretation in terms of the partial scattering data introduced in Chap. 4.

Theorem 7.4. *Let $\mathcal{D}$ be a convex domain in $\mathbb{C}^4$. The form $\theta'(x, \lambda)d\bar\lambda$ (or $\theta''(x, \lambda)d\bar\lambda$), $x \in \mathcal{D}$, $\lambda \in \mathbb{C}$, with the properties (4.10^1) and (4.12^1) (respectively, (4.10^{11}) and (4.12^{11})) is turned by (4.6) into the holomorphic field $a(x)$ satisfying the Yang–Mills equation (1.11) in the domain $\mathcal{D}$ iff there exists the form $\theta(x, w, \lambda)d\bar\lambda$ with the following properties:*

(a) $\theta(x, 0, \lambda) = \theta'(x, \lambda)$ (respectively, $\theta''(x, \lambda)$);

(b) $L_\nu \theta(x, w, \lambda) = O(w^3)$, $\nu = 0, 1, 2, 3$,
where L_ν are operators of the form (2.3), where $\lambda_1 = \lambda + \frac{1}{\lambda}, \lambda_2 = \lambda - \frac{1}{\lambda}$ (respectively, $\lambda_1 = \lambda^2 + \lambda, \lambda_2 = \lambda^2 - \lambda$);

(c) $\theta(x, w, \lambda) = O(|\lambda|^{-2})$ (respectively, $O(|\lambda|^{-3})$);

(d) $\theta(x, w, \lambda) = \mu^{-1}(x, w, \lambda)\bar\partial_\lambda \mu(x, w, \lambda)$,
where μ and μ^{-1} are smooth in λ, holomorphic in x, w matrix functions with the property

$$\mu(x, w, 0) = \mu(x, w, \infty) + O(w^4)$$

(respectively, $\lambda^2 \frac{\partial}{\partial \lambda}\mu(x, w, \lambda)|_{\lambda=\infty} = O(w^4)$, $\lambda^2 \frac{\partial}{\partial \lambda}\mu(x, w, \lambda)|_{\lambda=\infty} = O(w^4)$).

Such forms $\theta(x, w, \lambda)d\bar\lambda$ are in one-to-one correspondence with the topologically trivial n-dimensional holomorphic vector bundles E_θ over the manifold

$$L^{(3)}(\mathcal{D}) \cap (\mathbb{C}P_+^3 \times \mathbb{C}P_-^3)' \quad (\text{or } L^{(3)}(\mathcal{D}) \cap (\mathbb{C}P_+^3 \times \mathbb{C}P_-^3)''),$$

which are, furthermore, analytically trivial on each degenerated torus $\mathcal{L}'(x, w)$ (or $\mathcal{L}''(x, w)$) lying in this manifold.

Now we provide a curious interpretation of the Yang–Mills equations in terms of the Cauchy–Riemann equations. Let us again consider the holomorphic (respectively, smooth) fields $a_{yj}(y, z)$ and $a_{zj}(y, z)$ with values in $gl(n, \mathbb{C})$ satisfying the compatibility equations (2.2) for $k = 3$ in the convex domain $y \in \mathcal{D}$, $z \in \mathcal{D}$. Taking into account Proposition 2.3, we substitute $\lambda_1 = \lambda$, $\lambda_2 = \bar\lambda$ in the linear system (2.1) for $k = 3$ and consider the one-form

$$\Omega = \Omega_1(y, z, \lambda)d\lambda + \Omega_2(y, z, \lambda)d\bar\lambda, \tag{7.13}$$

where

$$\Omega_1 = \eta^{-1}(y, z, \lambda)\frac{\partial}{\partial \lambda}\eta(y, z, \lambda); \qquad \Omega_2 = \eta^{-1}(y, z, \lambda)\frac{\partial}{\partial \bar\lambda}\eta(y, z, \lambda).$$

From (2.1), where $k = 3$, $\lambda_1 = \bar{\lambda}_2$, and from (7.13), it follows that

$$\frac{\partial \Omega_2}{\partial \lambda} - \frac{\partial \Omega_1}{\partial \bar{\lambda}} + [\Omega_1, \Omega_2] = 0, \tag{7.14}$$

$$\left(\alpha_\nu \frac{\partial}{\partial y}\right) \Omega_2(y, z, \lambda) = O((y - z)^3),$$

$$\left(\beta_\nu \frac{\partial}{\partial z}\right) \Omega_1(y, z, \lambda) = O((y - z)^3), \quad \nu = 1, 2, \tag{7.15}$$

$$|\Omega_j(y, z, \lambda)| = O\left(\frac{1}{1 + |\lambda|^2}\right), \quad j = 1, 2, \tag{7.16}$$

where the vectors $\alpha_1, \alpha_2, \beta_1, \beta_2$ are defined in (2.1). In addition, the form Ω is defined up to the gauge equivalence

$$\Omega \sim h^{-1} dh + h^{-1} \Omega h, \tag{7.17}$$

where

$$\left(\alpha_\nu \frac{\partial}{\partial y}\right) h = O((y - z)^3), \qquad \left(\beta_\nu \frac{\partial}{\partial z}\right) h = O((y - z)^3).$$

Conversely, let the form Ω have the properties (7.14)–(7.17). The existence of the smooth function $\eta(y, z, \lambda)$, $y, z \in \mathcal{D}$, $\lambda \in \mathbb{C}P^1$ with values in $\mathrm{GL}(n, \mathbb{C})$, which satisfies (7.13), follows from (7.14) and (7.16). The equalities (2.1), (7.15), and Proposition 2.3 make it possible to obtain the fields $a_{y^j}(y, z)$ and $a_{z^j}(y, z)$ satisfying the compatibility equations (2.2). Hence and from Theorem 2.1, we obtain the following statement.

Proposition 7.3. *The transform $a \mapsto \Omega$ of the form (7.13) realizes a one-to-one correspondence between holomorphic (respectively, smooth) solutions of the Yang–Mills equation in the convex domain $\mathcal{D} \subset \mathbb{C}^4$ (respectively, R^4) and the forms Ω with the properties (7.14)–(7.17).*

Chapter 8
The Maxwell–Yang–Mills, the Weyl–Dirac, and the Klein–Gordon Equations as the Cauchy–Riemann Equations on the Twistor Space

This chapter is concerned with the results of Eastwood, Penrose and Wells [1981], Gindikin and Khenkin [1981], Khenkin [1984; 1980], Khenkin and Manin [1982; 1980], Khenkin and Polyakov [1990], Isenberg and Yasskin [1979], LeBrun [1982], Ward [1977], Witten [1978], and Woodhouse [1985], which show that the Maxwell–Yang–Mills, Weyl–Dirac and Klein–Gordon nonhomogeneous equations are turned into the Cauchy–Riemann nonhomogeneous equations on the twistor space due to the Radon–Penrose transform.

First of all, we present the following important result in order to further supplement Theorem 7.1(a).

Theorem 8.1 (Khenkin [1980]; Khenkin and Manin [1982]). *Let $\mathcal{D}$ be a domain of holomorphy in $\mathbb{C}M$ (or a domain in $\mathcal{M}$ or $\mathcal{E}$), the intersections of which with all complex null lines are contractible. Then any element $\theta \in \tilde{H}^{0,1}(L(\mathcal{D}), \mathrm{GL}(n, \mathbb{C}))$ may be extended (not in a unique way) to the element $\tilde{\theta}$ of the space $\tilde{H}^{0,1}(L^{(1)}(\mathcal{D}), \mathrm{GL}(n, \mathbb{C}))$. However, there exists a unique extension of θ to the element of the space $\tilde{H}^{0,1}(L^{(2)}(\mathcal{D}), \mathrm{GL}(n, \mathbb{C}))$, which is represented by a $\mathrm{gl}(n, \mathbb{C})$-valued $(0,1)$-form on $L_+(\mathcal{D}) \times L_-(\mathcal{D})$ satisfying the $\bar{\partial}$-equation*

$$\bar{\partial}\theta + \theta \wedge \theta = \mathcal{J}\langle \zeta \cdot \eta \rangle^3, \tag{8.1}$$

where $\mathcal{J} \in H^2(L(\mathcal{D}), \mathrm{End}\, E_\theta(-3, -3))$. All extensions of $\theta|_{L(\mathcal{D})}$ to the element of the space $\tilde{H}^{0,1}(L^{(1)}(\mathcal{D}), \mathrm{GL}(n, \mathbb{C}))$ are described by the forms

$$\tilde{\theta} = \theta + \Omega\langle \zeta \cdot \eta \rangle, \tag{8.2}$$

where $\Omega \in H^1(L(\mathcal{D}), \mathrm{End}\, E_\theta(-1, -1))$.

The assertion of Theorem 7.1(a) is a specialization of Theorem 8.1 for the case when $\mathcal{D}$ is a convex domain in CM_0.

Due to Proposition 6.1, we have $H^2(L(\mathcal{D}), \mathrm{End}\, E_\theta(-1, -1)) = 0$. Hence it follows (see Proposition 3.2) that any element θ from $\tilde{H}^{0,1}(L(\mathcal{D}), \mathrm{GL}(n, \mathbb{C}))$ is being extended without any obstructions to the element $\tilde{\theta}$ from $\tilde{H}^{0,1}(L^{(1)}(\mathcal{D}), \mathrm{GL}(n, \mathbb{C}))$. Such an extension is not unique, since the group $H^1(L(\mathcal{D}), \mathrm{End}\, E_\theta(-1, -1))$, which is different from zero (due to Theorem 6.1), acts effectively and transitively on the set of extensions of the bundle E_θ to the bundle $E_{\tilde{\theta}}$. The existence of the (moreover, unique) extension θ to the element $\tilde{H}^{(0,1)}(L^{(2)}(\mathcal{D}), \mathrm{GL}(n, \mathbb{C}))$ is based on the following assertion.

Proposition 8.1 (Khenkin and Manin [1982; 1980]). *Suppose the $(0,1)$-form Ω with values in $\mathrm{End}\, E(-1, -1)$ is given on $L_+(\mathcal{D}) \times L_-(\mathcal{D})$ and*

$$\bar{\partial}\Omega + [\tilde{\theta}, \Omega] = \Gamma\langle \zeta \cdot \eta \rangle \quad on \ L^{(1)}(\mathcal{D}), \tag{8.3}$$

where

$$\tilde{\theta} \in \tilde{H}^{0,1}\big(L^{(1)}(\mathcal{D}), \mathrm{GL}(n, \mathbb{C})\big), \qquad \Gamma \in H^2\big(L(\mathcal{D}), \mathrm{End}\, E_\theta(-2, -2)\big).$$

Then

(a) The $(0,2)$-form $\Omega \wedge \Omega$ gives a zero element in $H^2(L(\mathcal{D}), \mathrm{End}\, E_\theta(-2, -2))$;

(b) the Radon–Penrose transforms $\omega_\pm$ and γ of the forms Ω and Γ of the type (6.2) and (6.5) are connected by the equality

$$\omega_\pm(x) = -\frac{1}{2\pi i}\gamma(x), \quad x \in \mathcal{D}. \tag{8.4}$$

Let us deduce Theorem 8.1 from Propositions 8.1 and 6.2, and Theorem 6.1. Any element $\tilde{\theta} \in \tilde{H}^{0,1}(L^{(1)}(\mathcal{D}), \mathrm{GL}(n, \mathbb{C}))$ satisfies the relation

$$\bar{\partial}\tilde{\theta} + \tilde{\theta} \wedge \tilde{\theta} = \tilde{\Gamma}\langle \zeta \cdot \eta \rangle^2,$$

where $\tilde{\Gamma} \in H^2(L(\mathcal{D}), \mathrm{End}\, E_\theta(-2, -2))$. Due to Theorem 6.1, there will be an element $\Omega \in H^1(L(\mathcal{D}), \mathrm{End}\, E_{\tilde{\theta}}(-1, -1))$ such that

$$\omega(x) = -\frac{1}{4\pi^2}\tilde{\gamma}(x), \quad x \in \mathcal{D},$$

where $\omega(x)$ and $\tilde{\gamma}(x)$ are transforms of the form (6.2^1) and (6.5). Due to Proposition 3.1, the element Ω may be represented by the from on $L_+(\mathcal{D}) \times L_-(\mathcal{D})$ satisfying the relation

$$\bar{\partial}\Omega + [\tilde{\theta}, \Omega] = \Gamma\langle \zeta \cdot \eta \rangle,$$

where $\Gamma \in H^2(L(\mathcal{D}), \mathrm{End}\, E_{\tilde{\theta}}(-2, -2))$.

From these relations, Propositions 8.1 and 6.2, and Theorem 6.1, it follows that the forms Γ and $\tilde{\Gamma}$ are equivalent in

$$H^2\big(L(\mathcal{D}), \mathrm{End}\, E_{\tilde{\theta}}(-2, -2)\big).$$

Let us now consider another extension of the $\bar{\partial}$-connection $\tilde{\theta}$ from $L(\mathcal{D})$ into $L^{(1)}(\mathcal{D})$ given by the form $\tilde{\tilde{\theta}} = \tilde{\theta} - \Omega\langle \zeta \cdot \eta \rangle$. We have

$$\bar{\partial}\tilde{\tilde{\theta}} + \tilde{\tilde{\theta}} \wedge \tilde{\tilde{\theta}} = \tilde{\Gamma}\langle \zeta \cdot \eta \rangle^2 - (\bar{\partial}\Omega + [\tilde{\theta}, \Omega])\langle \zeta \cdot \eta \rangle$$
$$+ \Omega \wedge \Omega\langle \zeta \cdot \eta \rangle^2 = (\tilde{\Gamma} - \Gamma + \Omega \wedge \Omega)\langle \zeta \cdot \eta \rangle^2 = \tilde{\tilde{\Gamma}}\langle \zeta \cdot \eta \rangle^2.$$

From Proposition 8.1 and Theorem 6.1, it follows that the form $\tilde{\tilde{\Gamma}}$ gives a zero element in $H^2(L(\mathcal{D}), \mathrm{End}\, E_{\tilde{\theta}}(-2, -2))$, i.e.,

$$\tilde{\tilde{\Gamma}} = \bar{\partial}\tilde{\Omega} + [\tilde{\tilde{\theta}}, \tilde{\Omega}] + O(\langle \zeta \cdot \eta \rangle) \quad \text{on } L^{(1)}(\mathcal{D}).$$

Now put $\theta = \tilde{\tilde{\theta}} - \tilde{\Omega}\langle \zeta \cdot \eta \rangle^2$. From the last two equalities, we have

$$\bar{\partial}\theta + \theta \wedge \theta = (\tilde{\tilde{\Gamma}} - \bar{\partial}\tilde{\Omega} - [\tilde{\tilde{\theta}}, \tilde{\Omega}])\langle \zeta \cdot \eta \rangle^2 + \tilde{\Omega} \wedge \tilde{\Omega}\langle \zeta \cdot \eta \rangle^4 = \mathcal{J}\langle \zeta \cdot \eta \rangle^3,$$

where $\mathcal{J} \in H^2(L(\mathcal{D}), \mathrm{End}\, E_\theta(-3, -3))$.

Thus, a desired extension of the bundle E_θ from $L(\mathcal{D})$ into $L^{(2)}(\mathcal{D})$ is constructed.

Due to Proposition 8.1 and Theorem 6.1, there is only one among the extensions E_θ from $L(\mathcal{D})$ into $L^{(1)}(\mathcal{D})$ that extends further from $L^{(1)}(\mathcal{D})$ into $L^{(2)}(\mathcal{D})$. All other extensions E_θ into $L^{(1)}(\mathcal{D})$ are described by the forms of the type (8.2). At last, the set of extensions of the bundle E_θ from $L^{(1)}(\mathcal{D})$ into $L^{(2)}(\mathcal{D})$ consists of one element, since a zero group $H^1(L(\mathcal{D}), \mathrm{End}\, E_\theta(-2, -2))$ acts on the set of such extensions (see Propositions 6.1 and 3.2).

Let us further fix the form θ with the property (8.1) and the form $\tilde\theta$ of the type (8.2). Due to Propositions 6.3 and 3.1, the elements $\Psi_\pm, \Phi_\pm$ of the spaces

$$H^1(L(\mathcal{D}), \operatorname{End} E_\theta[(-1,0), \quad \text{or} \quad (0,-1)]),$$
$$H^1(L(\mathcal{D}), \operatorname{End} E_\theta[(-2,0), \quad \text{or} \quad (0,-2)])$$

are represented by the forms with the properties

$$\bar\partial \Psi_\pm + [\theta, \Psi_\pm] = G_\pm \langle \zeta \cdot \eta \rangle^2 \quad \text{on } L^{(2)}(\mathcal{D}), \tag{8.5}$$
$$\bar\partial \Phi_\pm + [\tilde\theta, \Phi_\pm] = F_\pm \langle \zeta \cdot \eta \rangle \quad \text{on } L^{(1)}(\mathcal{D}), \tag{8.6}$$

where

$$G_\pm \in H^2(L(\mathcal{D}), \operatorname{End} E_\theta[(-3,-2), \quad \text{or} \quad (-2,-3)]);$$
$$F_\pm \in H^2(L(\mathcal{D}), \operatorname{End} E_\theta[(-3,-1), \quad \text{or} \quad (-1,-3)]).$$

It turned out that $\bar\partial$-equations of the form (8.1), (8.5), and (8.6) are equivalent, respectively, to the Maxwell–Yang–Mills, Weyl–Dirac, and Klein–Gordon nonhomogeneous equations.

Theorem 8.2 (Khenkin [1980]; Khenkin and Manin [1982; 1980]). *In order that (under the conditions of Theorem 8.1) the forms θ and J; $\Psi_\pm$, θ and $G_\pm$; $\Phi_\pm$, $\tilde\theta$ and $F_\pm$, respectively, satisfy the equations (8.1), (8.5), and (8.6), it is necessary and sufficient that the Penrose transforms of these forms of the type (4.19), (6.2^1), (6.3), (6.4), (6.6^1), (6.7), and (6.8) satisfy the equations*

$$\text{(a)} \quad [\nabla^{CA'}, f_{AC}] = \frac{3}{2\pi^2} j_A^{A'}; \quad [\nabla_{AC'}, f^{A'C'}] = \frac{3}{2\pi^2} j_A^{A'}; \tag{8.7}$$

$$\text{(b)} \quad [\nabla_{AA'}, \psi^{A'}] = \frac{1}{2\pi^2} g_A; \quad [\nabla_{AA'}, \psi^{A}] = \frac{1}{2\pi^2} g_{A'}; \tag{8.8}$$

$$\text{(c)} \quad \Box \varphi_\pm + [\omega, \varphi_\pm] = \frac{1}{\pi i} f_\pm, \tag{8.9}$$

where

$$f_{AC} = [\nabla_{CC'}, \quad \nabla_A^{C'}], f^{A'C'} = [\nabla^{CC'}, \nabla_C^{A'}],$$
$$\nabla_{AB'} = \frac{\partial}{\partial x^{AB'}} + a_{AB'}, \quad \Box = [\nabla^{AB'}, [\nabla_{AB'}, \cdot]].$$

From Theorem 8.2, and due to Proposition 6.1 and Theorems 6.1 and 8.1, the most important consequence follows:

Theorem 8.3 (Isenberg and Yasskin [1979]; Khenkin and Manin [1980]; Witten [1978]). *Under the conditions of* Theorem 8.1, *the Penrose transforms $\theta \mapsto a_A^{B'}, \Omega \mapsto \omega, \Psi_\pm \mapsto \psi_A, \psi_{A'}; \Phi_\pm \mapsto \varphi_\pm$ of the form (4.19), (6.3), (6.4), and (6.2^1) determine the isomorphisms:*

(a) *between a space of the topologically trivial holomorphic (respectively, CR-) bundles E_θ of rank n over $L^{(3)}(\mathcal{D})$, analytically trivial on all quadrics $\mathcal{L}(x)$, $x \in \mathcal{D}$, and a space of the holomorphic (respectively, smooth) $\mathrm{GL}(n,\mathbb{C})$ connections $\nabla_{AB'}$ in the trivial n-dimensional bundle over $\mathcal{D}$ that satisfy the Maxwell–Yang–Mills homogeneous equation*

$$[\nabla^{CA'}, f_{AC}] = [\nabla_{AC'}, f^{A'C'}] = 0; \tag{8.10}$$

(b) *between a space of the $\bar{\partial}$-cohomologies $H^1(L^{(2)}(\mathcal{D}), \mathrm{End}\, E_\theta[(-1,0)$ or $(0,-1)])$, where $\theta \in \tilde{H}^{0,1}(L^{(2)}(\mathcal{D}), \mathrm{GL}(n,\mathbb{C}))$, and a space of the holomorphic (respectively, smooth) sections $\psi_A, \psi_{B'}$ of the spinor bundles $\mathrm{End}\, E \otimes S_\mp \otimes \bigwedge^2 S_\pm$ over $\mathcal{D}$ that satisfy the Weyl–Dirac equations*

$$[\nabla_{AB'}, \psi^{B'}] = 0, \quad [\nabla_{AB'}, \psi^A] = 0;$$

(c) *between spaces of the $\bar{\partial}$-cohomologies $H^1(L^{(1)}(\mathcal{D}), \mathrm{End}\, E_{\tilde\theta}[(-2,0)$ or $(0,-2)])$, where $\tilde\theta \in \tilde{H}^{0,1}(L^{(1)}(\mathcal{D}), \mathrm{GL}(n,\mathbb{C}))$, and spaces of the holomorphic (respectively, smooth) sections $\varphi_\pm$ of the bundles $\mathrm{End}\, E \otimes \bigwedge^2 S_\pm$ over $\mathcal{D}$ that satisfy the Klein–Gordon equation*

$$\Box\varphi_\pm + [\omega, \varphi_\pm] = 0.$$

The assertion of Theorem 7.1(b) is a specialization of Theorem 8.3(a).

Theorems 8.3(a), 8.3(b), and 8.3(c) for the self-dual and anti-self-dual cases when

$$\theta \in \tilde{H}^{0,1}\big(L_\pm(\mathcal{D}), \mathrm{GL}(n,\mathbb{C})\big),$$
$$\Psi_\pm \in H^1\big(L_\pm(\mathcal{D}), \mathrm{End}\, E_\theta(-1)\big) \qquad \Psi_\pm \in H^1\big(L_\pm(\mathcal{D}), \mathrm{End}\, E_\theta(-2)\big)$$

were first obtained for holomorphic fields and in terms of Chech cohomologies in Penrose [1969] and Ward [1977]; see also Eastwood, Penrose and Wells [1981], Gindikin and Khenkin [1981], and Hitchin [1980]. Extensions of these results to the case of nonanalytical self-dual fields and $\bar{\partial}$-cohomologies were obtained in Gindikin and Khenkin [1981], Khenkin and Polyakov [1990], Wells [1981], and Woodhouse [1985]. In full generality, Theorem 8.3(a) for holomorphic fields was first obtained in Witten [1978] and Isenberg, Yasskin and Green [1978]; see also proofs in Buchdahl [1985], Khenkin [1980], Khenkin and Manin [1982], Manin [1984], and Pool [1984; 1986]. Theorems 8.3(b) and 8.3(c) for holomorphic fields were obtained in Khenkin and Manin [1982; 1980]. The extensions of these results to nonanalytical fields were obtained in Khenkin [1984; 1980; 1981].

The assertion equivalent to Theorem 8.2(a) with an undefined factor on the right-hand side of (8.7) was first obtained in Manin [1981] in terms of the obstruction theory (Griffiths [1966]) by using the results of Isenberg and Yasskin [1979], Witten [1978], and group-theoretic arguments. Theorem 8.2(a) was obtained in Khenkin [1980] not so much for the sake of refinement of the constant on the right-hand side of (8.7) as for finding an elementary way to this result that does not depend on (not completely detailed) constructions from Isenberg and Yasskin [1979] and Witten [1978]. Theorems 8.2(b) and 8.2(c) are obtained from Khenkin and Manin [1982; 1980].

The proof of Theorem 8.2 in Khenkin and Manin [1982] is pointwise and for each fixed point $x^* \in \mathcal{D}$ uses special calibrations of the corresponding forms of $\bar{\partial}$-connections θ on $L(\mathcal{D})$ and of a usual connection a on $\mathcal{D}$. Let $\mu_{x^*}(\zeta, \eta)$ be a smooth function on $L(\mathcal{D})$ with values in $\mathrm{GL}(n, \mathbb{C})$ and with the property (6.1), and let this function depend on the parameter $x^* \in \mathcal{D}$. For the form θ satisfying (8.1), suppose

$$\tilde{\theta} = \mu_{x^*} \theta \mu_{x^*}^{-1} - (\bar{\partial}\mu_{x^*})\mu_{x^*}^{-1}.$$

The form $\tilde{\theta}$ is equal to zero on $\mathcal{L}^{(1)}(x^*)$, i.e., this form is equal to zero on $\mathcal{L}(x^*)$ together with the first derivatives. We call such a calibration $\tilde{\theta}$ of the form θ an x^*-calibration.

Suppose further that

$$\tilde{f}^{A'B'}(x^*, \eta^{C'}) = \int_{\zeta \in \mathcal{L}_+(x^*)} \frac{\partial^2 \tilde{\theta}}{\partial \zeta_{A'} \partial \zeta_{B'}} \zeta^C d\zeta_C,$$

$$\tilde{f}^{AB}(x^*, \zeta^C) = \int_{\eta \in \mathcal{L}_-(x^*)} \frac{\partial^2 \tilde{\theta}}{\partial \eta_A \partial \eta_B} \eta^{C'} d\eta_{C'}.$$

(8.11)

In view of the $\bar{\partial}$-closure of forms under the sign of the integrals (8.11), these integrals are holomorphic in $\eta \in \mathcal{L}_-(x^*)$ and in $\zeta \in \mathcal{L}_+(x^*)$, respectively. Due to Liouville's theorem, the integrals (8.11) do not depend on the variables η and ζ. Due to the equality $\theta|_{\mathcal{L}^{(1)}(x^*)} = 0$, there exists the function $\tilde{\mu}$ on $L(\mathcal{D})$ with values in $\mathrm{GL}(n, \mathbb{C})$, which depends on the parameter $x^* \in \mathcal{D}$ and has the property (4.20) and $\tilde{\mu} = I + O(|x - x^*|^2)$. Let us determine the field $\tilde{a}^{AA'}$ by the equality (4.19), where $\mu = \tilde{\mu}$. We have $\tilde{a}^{AA'}(x^*) = 0$.

The following proposition plays a very important role in the proof of Theorem 8.2(a).

Proposition 8.2. *The transforms*

$$\tilde{\theta} \mapsto \tilde{a}^{AA'}, \qquad \tilde{\theta} \mapsto \tilde{f}^{A'B'}, \qquad \tilde{\theta} \mapsto \tilde{f}^{AB}$$

are connected by the equalities

$$\nabla_A^{B'} \tilde{a}^{AA'}(x^*) = \frac{\partial \tilde{a}^{AA'}}{\partial x_{B'}^A}(x^*) = -\frac{1}{2\pi i} \tilde{f}^{A'B'}(x^*),$$

(8.12)

$$\nabla_{A'}^B \tilde{a}^{AA'}(x^*) = \frac{\partial \tilde{a}^{AA'}}{\partial x_B^{A'}}(x^*) = -\frac{1}{2\pi i} \tilde{f}^{AB}(x^*).$$

This proposition was obtained in Gindikin and Khenkin [1981] for the self-dual case and in Khenkin [1980] for the general case by direct computation. An equivalent statement with an indeterminate factor on the right-hand side of (8.12) was also obtained in Manin [1981], where the corresponding transform was defined as an image of the element of the group $H^1(\mathcal{L}(x), S^2 N^*)$ corresponding to the restriction of the bundle E_θ to $\mathcal{L}^{(2)}(x^*)$; N^* is a conormal bundle for the quadric $\mathcal{L}(x^*) \subset L(\mathcal{D})$.

Now let us deduce Theorem 8.2(a) by using Proposition 8.2. Let us first prove the equality (8.7) at a fixed point $x^* \in \mathcal{D}$ for $\theta = \tilde{\theta}$ and $\tilde{a}$ in the x^*-calibration. Due to Statement 8.2, it is sufficient to prove the equalities

$$\nabla^{BB'} \tilde{f}_{AB}(x^*) = \nabla_{AA'} \tilde{f}^{A'B'}(x^*) = \frac{3}{\pi i} j_A^{B'}(x^*), \qquad (8.13)$$

where the value $j_A^{B'}$ is defined by the transform (6.8) in which $\mu = \tilde{\mu}$, $\nabla^{BB'} = \frac{\partial}{\partial x_{BB'}} + \tilde{a}^{BB'}$.

The following equalities follow from using (4.17) and (8.11),

$$\nabla_{AA'} \tilde{f}^{A'B'}(x^*) = \int_{\mathcal{L}_+(x^*)} \left(\zeta_A \frac{\partial}{\partial \zeta^{A'}} + \eta_{A'} \frac{\partial}{\partial \eta^A} \right) \frac{\partial^2 \tilde{\theta}}{\partial \zeta_{A'} \partial \zeta_{B'}} \zeta^C d\zeta_C$$

$$= \int_{\mathcal{L}_+(x^*)} \eta_{A'} \frac{\partial}{\partial \eta^A} \frac{\partial^2 \tilde{\theta}}{\partial \zeta_{A'} \partial \zeta_{B'}} \zeta^C d\zeta_C$$

(by the Cauchy–Green formula)

$$= \frac{1}{2\pi i} \iint_{\mathcal{L}(x^*)} \eta_{0'} \frac{\partial}{\partial \eta^A} \frac{\partial^2 \bar{\partial} \tilde{\theta}}{\partial \zeta_{0'} \partial \zeta_{B'}} \zeta^C d\zeta_C \frac{\eta^{C'} d\eta_{C'}}{\eta^{1'} \eta^{0'}}$$

(due to (8.1))

$$= \frac{1}{2\pi i} \iint_{\mathcal{L}(x^*)} \eta_{0'} 6 \mathcal{J} \zeta_A \eta^{B'} \eta^{0'} \zeta^C d\zeta_C \wedge \frac{\eta^{C'} d\eta_{C'}}{\eta^{1'} \eta^{0'}}$$

$$= \frac{3}{\pi i} \iint_{\mathcal{L}(x^*)} \mathcal{J} \zeta_A \eta^{B'} \wedge \zeta^C d\zeta_C \wedge \eta^{C'} d\eta_{C'} = \frac{3}{\pi i} j_A^{B'}(x^*).$$

If the equality (8.13) is valid for the x^*-calibrations $\theta = \tilde{\theta}$ and $a = \tilde{a}$, then it is also valid for any calibrations, since the left-hand and the right-hand side of (8.13) are being consistently transformed under gauge transforms. Conversely, if the equality (8.13) is valid, then due to Theorem 8.1 and to Propositions 3.1 and 3.2 and to Theorems 4.2 and 6.6 corresponding to the fields $a_{AB'}$ and $j_{AB'}$, the elements θ and $\mathcal{J}$ of the spaces

$$\tilde{H}^{0,1}\big(L(\mathcal{D}), \mathrm{GL}(n, \mathbb{C})\big) \quad \text{and} \quad H^2\big(L(\mathcal{D}), \mathrm{End}\, E_\theta(-3, -3)\big)$$

have presentations connected by the $\bar{\partial}$-equations (8.1).

The proofs of Theorems 8.2(b) and 8.2(c) are similar (see Khenkin and Manin [1982]).

Under certain conditions, the following result provides an extension of the bundle E_θ from $L^{(2)}(\mathcal{D})$ into $L^{(2+k)}(\mathcal{D})$ and at the same time computes the obstruction to the extension E_θ into $L^{(2+k)}(\mathcal{D})$, $k \geq 2$.

Theorem 8.4. *In order for the representations of elements*

$$\theta \in \tilde{H}^{0,1}\big(L^{(2)}(\mathcal{D}), \mathrm{GL}(n, \mathbb{C})\big)$$

and

$$\Omega \in H^2\big(L(\mathcal{D}), \mathrm{End}\, E_\theta(-2 - k, -2 - k)\big)$$

to be connected by the Cauchy–Riemann equations of the form

$$\bar{\partial}\theta + \theta \wedge \theta = \Omega \langle \zeta \cdot \eta \rangle^{2+k} \quad on \ L^{(2+k)}(\mathcal{D}), \tag{8.14}$$

it is necessary and sufficient that their Radon–Penrose transforms

$$a_A^{B'} \quad and \quad \omega_{A_1 \cdots A_k}^{B'_1 \cdots B'_k}, \ k \geq 2$$

satisfy the Yang–Mills equations and the following additional commutation relations

$$\sum \big[[\nabla^{A_{i_\ell} B'_{j_\ell}}[\dots [\nabla^{A_{i_3} B'_{j_3}} f_{A_{i_1} A_{i_2}}]\dots]], f^{B'_{j_1} B'_{j_2}}\big]$$
$$= C(k)\begin{cases} 0, & if \ k - 2 \leq \ell < k \\ \omega_{A_1 \cdots A_k}^{B'_1 \cdots B'_k}, & if \ \ell = k, \end{cases} \tag{8.15}$$

where the sum is being taken over all transpositions $(i_1, \dots, i_\ell)$ and $(j_1, \dots, j_\ell)$ of the sequence $(1, 2, \dots, \ell)$.

For $k = 2$, Theorem 8.4 yields Theorem 7.1(c).

In terms of bundles, a similar (less exact) assertion is stated without a proof in Buchdahl [1985] and as a hypothesis in Yasskin [1987].

We provide a proof of Theorem 8.4 for the most interesting case $k = 2$. For this case, due to Theorem 8.3(a), the equalities (8.10) are necessary and sufficient in order that the form θ be gauge equivalent to the form (8.14), where $k = 2$. Let us prove that the equality

$$\bar{\partial}\theta + \theta \wedge \theta = \Omega \langle \zeta \cdot \eta \rangle^4, \quad (\zeta \cdot \eta) \in L(\mathcal{D}) \tag{8.16}$$

is equivalent to the equalities (8.10) and to the following:

$$- f^{B'_1 B'_2}(x) f_{A_1 A_2}(x) + f_{A_1 A_2}(x) f^{B'_1 B'_2}(x) = 24 \omega_{A_1 A_2}^{B'_1 B'_2}(x). \tag{8.17}$$

180   G.M. Khenkin, R.G. Novikov

Suppose the equality (8.16) is valid. Then (8.10) follows from Theorem 8.3(a). Let us prove (8.17) for a fixed point $x^* \in \mathcal{D}$, using x^*-calibration, when $a(x^*) = 0$, $\theta|_{\mathcal{L}^{(1)}(x^*)} = 0$.

Since

$$\frac{\partial^4 \langle \zeta \cdot \eta \rangle^4}{\partial \eta_{A_1} \partial \eta_{A_2} \partial \zeta_{B_1'} \partial \zeta_{B_2'}}\bigg|_{L(\mathcal{D})} = 24 \eta^{B_1'} \eta^{B_2'} \zeta^{A_1} \zeta^{A_2},$$

then due to definition (6.9), we have

$$\omega^{B_1' B_2' A_1 A_2}(x^*) = \frac{1}{24} \int_{(\zeta,\eta) \in \mathcal{L}(x^*)} \frac{\partial^4 \Omega \langle \zeta \eta \rangle^4}{\partial \eta_{A_1} \partial \eta_{A_2} \partial \zeta_{B_1'} \partial \zeta_{B_2'}} \wedge \zeta^C d\zeta_C \wedge \eta^{C'} d\eta_{C'}.$$

After substitution of (8.16) into this integral and due to the use of the vanishing of the cohomologies

$$H^1(\mathcal{L}, \mathcal{O}), H^1(\mathcal{L}, \mathcal{O}(-1,0)), \qquad H^1(\mathcal{L}, \mathcal{O}(-1,-1)),$$

we have

$$\omega^{B_1' B_2' A_1 A_2}(x^*)$$

$$= \frac{1}{24} \int_{\mathcal{L}(x^*)} \left(\frac{\partial^2 \theta}{\partial \eta_{A_1} \partial \eta_{A_2}} \wedge \frac{\partial^2 \theta}{\partial \zeta_{B_1'} \partial \zeta_{B_2'}} + \frac{\partial^2 \theta}{\partial \zeta_{B_1'} \partial \zeta_{B_2'}} \wedge \frac{\partial^2 \theta}{\partial \eta_{A_1} \partial \eta_{A_2}} \right)$$

$$\wedge \zeta^C d\zeta_C \wedge \eta^{C'} d\eta_{C'}.$$

Due to the use of the fact that all $(0,1)$-forms in the last integral are cohomologous on $\mathcal{L}(x^*)$ to the forms that depend either only on η or only on ζ, we obtain

$$\omega^{B_1' B_2' A_1 A_2}(x^*)$$

$$= -\frac{1}{24} \int_{\zeta \in \mathcal{L}_+(x^*)} \frac{\partial^2 \theta}{\partial \zeta_{B_1'} \partial \zeta_{B_2'}} \wedge \zeta^C d\zeta_C \cdot \int_{\eta \in \mathcal{L}_-(x^*)} \frac{\partial^2 \theta}{\partial \eta_{A_1} \partial \eta_{A_2}} \wedge \eta^{C'} d\eta_{C'}$$

$$+ \frac{1}{24} \int_{\eta \in \mathcal{L}_-(x^*)} \frac{\partial^2 \theta}{\partial \eta_{A_1} \partial \eta_{A_2}} \wedge \eta^{C'} d\eta_{C'} \cdot \int_{\zeta \in \mathcal{L}_+(x^*)} \frac{\partial^2 \theta}{\partial \zeta_{B_1'} \partial \zeta_{B_2'}} \wedge \zeta^C d\zeta_C$$

$$= \frac{1}{24} (f^{A_1 A_2} \cdot f^{B_1' B_2'} - f^{B_1' B_2'} \cdot f^{A_1 A_2}).$$

Conversely, if the relations (8.10) are valid, then due to Theorem 8.3(a) and Proposition 3.1, the form θ is gauge-equivalent to the form with the property

$$\bar{\partial}\theta + \theta \wedge \theta = \tilde{\Omega} \langle \zeta \cdot \eta \rangle^4.$$

If (8.17) is valid, then, as above, we obtain the equalities

$$\int_{(\zeta,\eta) \in \mathcal{L}(x)} \zeta^{A_1} \zeta^{A_2} \eta_{B_1'} \eta_{B_2'} \mu_x \tilde{\Omega} \mu_x^{-1} \wedge \zeta^C d\zeta_C \wedge \eta^{C'} d\eta_{C'} = \omega^{A_1 A_2}_{B_1' B_2'}(x), \quad x \in \mathcal{D}.$$

Hence and from Proposition 6.6, it follows that the form $\tilde{\Omega}$ is $\bar{\partial} + \theta$-cohomologous to the form Ω. Theorem 8.4 is proved for $k = 2$.

Chapter 9
The Yang–Mills–Higgs–Dirac Equations as Holomorphic Vector Bundles

Here we provide the results developed in Khenkin [1981; 1982] that show that the Yang–Mills fields, interacting in a determinate way with scalar and spinor fields, may be also described in terms of holomorphic vector bundles over a space of complex light rays.

Let $\nabla^{\pm}_{AB'} = \frac{\partial}{\partial x^{AB'}} + a^{\pm}_{AB'}$ be holomorphic or smooth $\mathrm{GL}(n_{\pm}, \mathbb{C})$-connections, respectively, in fixed (for simplicity, trivial) $n_{\pm}$-dimensional bundles over a holomorphy domain $\mathcal{D} \subset \mathbb{C}M$ (respectively, $\mathcal{M}$ or $\mathcal{E}$). As before, we suppose that the intersections $\mathcal{D}$ with complex null lines are contractible.

Due to results stated in Chap. 8, the holomorphic (respectively, CR-) bundles $E_{\pm}$ topologically equivalent to the $n_{\pm}$-dimensional trivial bundles $E^0_{\pm}$ and analytically equivalent to $E^0_{\pm}$ on all quadrics $\mathcal{L}(x)$, $x \in \mathcal{D}$ correspond on $L^{(2)}(\mathcal{D})$ to the gauge fields $a_{\pm}$. It turned out that to gauge fields $a_{\pm}$, interacting with the Higgs scalar fields $\varphi_{\pm}$ in the form of (1.18), correspond such holomorphic (respectively, CR-) bundles $\mathbb{E}$ on $L^{(2)}(\mathcal{D})$ that under restriction to $L(\mathcal{D})$ have the following form

$$\mathbb{E}|_{L(\mathcal{D})} = E_{+}(-1, 0) \oplus E_{-}(0, -1)|_{L(\mathcal{D})}. \tag{9.1}$$

Due to the results of Chaps. 3, 4, and 8 (Theorem 8.1 and Propositions 3.1 and 4.1), the bundles $E_{\pm}$ may be coded by the smooth $\mathrm{gl}(n_{\pm}, \mathbb{C})$-valued $(0, 1)$-forms $\theta_{\pm}$ on $L_{+}(\mathcal{D}) \times L_{-}(\mathcal{D})$ satisfying the relations

$$\begin{aligned}
\bar{\partial}\theta_{\pm} + \theta_{\pm} \wedge \theta_{\pm} &= 0 \quad \text{on } L^{(2)}(\mathcal{D}), \\
\theta_{\pm}|_{\mathcal{L}^{(1)}(x)} &= \mu_{\pm}^{-1}\bar{\partial}\mu_{\pm} \quad \text{on } \mathcal{L}^{(1)}(x), \ x \in \mathcal{D},
\end{aligned} \tag{9.2}$$

where the $\mu_{\pm}$ are functions on $\mathcal{L}^{(1)}(x)$ with values in $\mathrm{GL}(n_{\pm}, \mathbb{C})$ smoothly depending on the parameter $x \in \mathcal{D}$.

Suppose $E^0 = E^0_{+}(-1, 0) \oplus E^0_{-}(0, -1)$.

Proposition 9.1 (Khenkin [1981]). *Any holomorphic vector bundle E over $L^{(2)}(\mathcal{D})$ that under restriction to $L(\mathcal{D})$ has the form (9.1) is coded, up to the $\bar{\partial}$-gauge equivalence on $L^{(2)}(\mathcal{D})$, by the $(0, 1)$-form θ on $L_{+}(\mathcal{D}) \times L_{-}(\mathcal{D})$ with values in $\mathrm{End}\, E^0$,*

$$\theta = \begin{pmatrix} \theta_{+} + \Omega_{+}\langle \zeta \cdot \eta \rangle & \Phi_{+}\langle \zeta \cdot \eta \rangle \\ \Phi_{-}\langle \cdot \eta \rangle & \theta_{-} + \Omega_{-}\langle \zeta \cdot \eta \rangle \end{pmatrix}, \tag{9.3}$$

where the $(0, 1)$-forms $\Omega_{\pm}$ with values in $\mathrm{End}\, E^0_{\pm}$ and $\Phi_{\pm}$ with values on $\mathrm{Hom}(E_{-}, E_{+}(-2, 0))$ (or $\mathrm{Hom}(E_{+}, E_{-}(0, -2)))$ satisfy the equations

$$\bar{\partial}\Phi_{\pm} + \theta_{\pm} \wedge \Phi_{\pm} + \Phi_{\pm} \wedge \theta_{\mp} + (\Omega_{\pm} \wedge \Phi_{\pm} + \Phi_{\pm} \wedge \Omega_{\mp})\langle \zeta \cdot \eta \rangle = 0, \tag{9.4}$$

$$\bar{\partial}\Omega_{\pm} + \theta_{\pm} \wedge \Omega_{\pm} + \Omega_{\pm} \wedge \theta_{\pm} + \Phi_{\pm} \wedge \Phi_{\mp}\langle \zeta \cdot \eta \rangle + \Omega_{\pm} \wedge \Omega_{\pm}\langle \zeta \cdot \eta \rangle = 0, \quad (9.5)$$

on $L^{(1)}(\mathcal{D})$.

In addition, the elements of the space $H^1(L^{(2)}(\mathcal{D}), E_\theta)$ are described, up to the $\bar{\partial}$-gauge equivalence, by the $(0,1)$-forms $\Psi_{\pm}$ with values in $E_{\pm}^0$ satisfying the equations

$$\begin{aligned}
\bar{\partial}\Psi_{+} + \theta_{+} \wedge \Psi_{+} + (\Omega_{+} \wedge \Psi_{+} + \Phi_{+} \wedge \Psi_{-})\langle \zeta \cdot \eta \rangle = 0, \\
\bar{\partial}\Psi_{-} + \theta_{-} \wedge \Psi_{-} + (\Omega_{+} \wedge \Psi_{-} + \Phi_{-} \wedge \Psi_{+})\langle \zeta \cdot \eta \rangle = 0
\end{aligned} \qquad (9.6)$$

on $L^{(2)}(\mathcal{D})$.

To prove Proposition 9.1, it is important to note that the $(0,1)$-form θ provides a new complex structure in the bundle E^0 over $L^{(2)}(\mathcal{D})$ iff the Cauchy–Riemann–Cartan integrability condition $\bar{\partial}\theta + \theta \wedge \theta = 0$ is valid on $L^{(2)}(\mathcal{D})$. In the detailed record (and under Condition (9.1)), this integrability condition is equivalent to the relations (9.2) and (9.4). Similarly, the equations (9.5) are the detailed record of the Cauchy–Riemann equations of the form

$$\bar{\partial}\Psi + \theta \wedge \Psi = 0 \quad \text{on } L^{(2)}(\mathcal{D}), \quad \text{where } \Psi = \begin{pmatrix} \Psi_{+} \\ \Psi_{-} \end{pmatrix}.$$

Generalizing the definitions (6.2), (6.3), and (6.4), let us introduce the Penrose transforms of the forms $\Omega_{\pm}, \Phi_{\pm}, \Psi_{\pm}$,

$$\omega_{\pm}(x) = \frac{1}{\pi i} \int_{\mathcal{L}_{+}(x)} \sum_{A'} \eta^{A'} \frac{\partial}{\partial \zeta^{A'}}(\mu_{\pm}\Omega_{\pm}\mu_{\pm}^{-1}) \wedge \zeta^C d\zeta_C,$$

$$\varphi_{+}(x) = \frac{1}{\pi i} \int_{\mathcal{L}_{+}(x)} \mu_{+}\Phi_{+}\mu_{-}^{-1} \wedge \zeta^C d\zeta_C \qquad (9.7)$$

$$\varphi_{-}(x) = \frac{1}{\pi i} \int_{\mathcal{L}_{-}(x)} \mu_{-}\Phi_{-}\mu_{+}^{-1} \wedge \eta^{C'} d\eta_{C'}$$

$$\psi^{A'}(x) = \frac{1}{\pi i} \int_{\mathcal{L}_{+}(x)} \frac{\partial}{\partial \zeta_{A'}}\mu_{+}\Psi_{+} \wedge \zeta^C d\zeta_C,$$

$$\psi^{A}(x) = \frac{1}{\pi i} \int_{\mathcal{L}_{-}(x)} \frac{\partial}{\partial \eta_A}\mu_{-}\Psi_{-} \wedge \eta^{C'} d\eta_{C'}. \qquad (9.8)$$

Proposition 9.2 (Khenkin [1981]). *In order that the $(0,1)$-forms $\Omega_{\pm}$, $\Phi_{\pm}, \Psi_{\pm}$ representing the corresponding classes of cohomologies*

$$H^1\big(L(\mathcal{D}), \operatorname{End} E_\theta^{\pm}(-1,-1)\big); \qquad H^1\big(L(\mathcal{D}), \operatorname{Hom}\big(E_{\theta_{\mp}}, E_{\theta_{\pm}}((-2,0)(0,-2))\big)\big);$$

$$H^1\big(L(\mathcal{D}), E_{\theta_{\pm}}((-1,0) \text{ or } (0,-1))\big)$$

satisfy the $\bar{\partial}$-equations (9.4)–(9.6), it is necessary and sufficient that their

Penrose transforms $\omega_\pm$, $\varphi_\pm$, $\psi^{A'}$ and ψ^A satisfy the relations

$$\omega_+ = -\tfrac{1}{2}\varphi_+\varphi_-; \quad \omega_- = \tfrac{1}{2}\varphi_-\varphi_+; \tag{9.9}$$

$$\Box_{+-}\varphi_+ = -\omega_+\varphi_+ + \varphi_+\omega_-; \quad \Box_{-+}\varphi_- = \omega_-\varphi_- - \varphi_-\omega_+; \tag{9.10}$$

$$\nabla^+_{AA'}\psi^{A'} + \varphi_+\psi_A = 0;$$
$$\nabla^-_{AA'}\psi^A + \varphi_-\psi_{A'} = 0, \tag{9.11}$$

where

$$\nabla^+_{AA'}\psi^{A'} = \frac{\partial \psi^{A'}}{\partial x^{AA'}} + a^+_{AA'}\psi^{A'}; \qquad \Box_{\pm\mp}\varphi_\pm = \left[\nabla^{AA'}_{\mp\pm}, [\nabla^{\pm\mp}_{AA'}, \varphi_\pm]\right];$$

$$[\nabla^{\pm\mp}_{AA'}, \varphi_\pm] = \frac{\partial \varphi_\pm}{\partial x^{AA'}} + a^\pm_{AA'}\varphi_\pm - \varphi_\pm a^\mp_{AA'}.$$

From Proposition 8.1, it follows that after the transforms $\Phi_\pm \mapsto \varphi_\pm, \Omega_\pm \mapsto \omega_\pm$, the $\bar{\partial}$-equations (9.5) go into the equations (9.9). In order to prove that the equations (9.4) go into the equations (9.10) after the same transforms, it is sufficient due to Theorem 8.2(c) to verify that the Penrose transforms $f_\pm$ of the $(0,2)$-forms $F_\pm = \Omega_\pm \wedge \Phi_\pm + \Phi_\pm \wedge \Omega_\mp$ are equal to

$$f_+ = (\omega_+\varphi_+ - \varphi_+\omega_-)\pi i,$$
$$f_- = (\varphi_-\omega_+ - \omega_-\varphi_-)\pi i, \tag{9.12}$$

respectively.

Generalizing the definitions 6.6, we have

$$f_+ = \frac{1}{\pi i}\iint\limits_{\mathcal{L}(x)} \sum_A \zeta^A \frac{\partial}{\partial \eta^A}(\mu_+ F_+ \mu_-^{-1}) \wedge \zeta^C d\zeta_C \wedge \eta^{C'} d\eta_{C'}$$

$$= \frac{1}{\pi i}\iint\limits_{\mathcal{L}(x)} \left[\sum_A \zeta^A \frac{\partial}{\partial \eta^A}(\mu_+ \Omega_+ \mu_+^{-1}) \wedge (\mu_+ \Phi_+ \mu_-^{-1})\right.$$

$$\left. + (\mu_+ \Phi_+ \mu_-^{-1}) \wedge \sum_A \zeta^A \frac{\partial}{\partial \eta^A}(\mu_- \Omega_- \mu_-^{-1})\right] \wedge \zeta^C d\zeta_C \wedge \eta^{C'} d\eta_{C'}$$

$$+ \frac{1}{\pi i}\iint\limits_{\mathcal{L}(x)} \left[\mu_+ \Omega_+ \mu_+^{-1} \wedge \sum_A \zeta^A \frac{\partial}{\partial \eta^A}(\mu_+ \Phi_+ \mu_-^{-1})\right.$$

$$\left. + \sum_A \zeta^A \frac{\partial}{\partial \eta^A}(\mu_+ \Phi_+ \mu_-^{-1}) \wedge (\mu_- \Omega_- \mu_-^{-1})\right] \wedge \zeta^C d\zeta_C \wedge \eta^{C'} d\eta_{C'}.$$

Since the forms $\mu_\pm \Omega_\pm \mu_\pm^{-1}$ are $\bar{\partial}$-exact on $\mathcal{L}(x)$, the second integral on the right-hand side of the last equality is equal to zero. According to the definitions (9.7) and (9.8) and to the fact that in the first term of the right-hand side the forms

$$\sum_A \zeta^A \frac{\partial}{\partial \eta^A}(\mu_\pm \Omega_\pm \mu_\pm^{-1}) \quad \text{and} \quad (\mu_\pm \Phi_\pm \mu_\mp^{-1})$$

are cohomologous to the forms depending only on ζ or only on η, we obtain (9.12).

Similarly, due to Theorem 8.2(b), the correspondence under the Radon–Penrose transform of the equations (9.6) and (9.11) is proved.

The following result for the presentation of the solutions φ^4 of the equation in $\mathcal{D}$ in the form of holomorphic bundles over $L^{(2)}(\mathcal{D})$ is obtained from Propositions 9.1 and 9.2.

Theorem 9.1 (Khenkin [1981]). *Suppose we have a fixed domain of holomorphy $\mathcal{D}$ in $\mathbb{C}M$ (or $\mathcal{M}$ or $\mathcal{E}$) whose intersections with any complex light ray is contractible, holomorphic (respectively, smooth) connections $\nabla_{\pm}$ in $n_{\pm}$-dimensional bundles over $\mathcal{D}$, and, corresponding to $\nabla_{\pm}$, the bundles $E_{\pm}$ over $L^{(2)}(\mathcal{D})$ with $(0,1)$-forms of the $\bar{\partial}$-connection $\theta_{\pm}$, then the Radon–Penrose transform $\mathcal{P}\Phi_{\pm} = \varphi_{\pm}$ of the type (9.7) realizes an isomorphism between a space of the bundles E_{θ}, $\theta = \theta(\Phi_{\pm})$ over $L^{(2)}(\mathcal{D})$ coinciding with the bundle $E_{+}(-1,0)+E_{-}(0,-1)$ on $L(\mathcal{D})$ and a set of holomorphic (respectively, smooth) solutions φ_{+} of the system of the equations*

$$\square_{+-}\varphi_{+} = \varphi_{+}\varphi_{-}\varphi_{+}; \qquad \square_{-+}\varphi_{-} = \varphi_{-}\varphi_{+}\varphi_{-}. \tag{9.13}$$

In addition, for fixed matrix-functions $\varphi_{\pm} = \mathcal{P}\Phi_{\pm}$ satisfying (9.13), the Penrose transform of the type (9.8) realizes an isomorphism between the space of cohomologies $H^1(L^{(2)}(\mathcal{D}), E_{\theta})$ and the space of holomorphic (respectively, smooth) solutions of the Dirac system (9.11) on $\mathcal{D}$.

Let us now describe, in terms of bundles, the solutions of the Yang–Mills–Higgs system that consists of the equations (9.13) and is supplemented with the equations

$$\nabla_{AB'}^{++} f_{+A'}^{B'} = \tfrac{1}{2}\left([\nabla_{AA'}^{+-},\varphi_{+}]\varphi_{-} - \varphi_{+}[\nabla_{AA'}^{-+},\varphi_{-}]\right),$$
$$\nabla_{AB'}^{--} f_{-A'}^{B'} = \tfrac{1}{2}\left([\nabla_{AA'}^{-+},\varphi_{-}]\varphi_{+} - \varphi_{-}[\nabla_{AA'}^{+-},\varphi_{+}]\right), \tag{9.14}$$

where $f_{\pm} = da_{\pm} + a_{\pm} \wedge a_{\pm}$ are forms of the curvature of the connections $\nabla_{\pm}$. Let θ be a form of the type (9.3) satisfying the relation

$$\bar{\partial}\theta + \theta \wedge \theta = O\begin{pmatrix} \langle\zeta\cdot\eta\rangle^4 & \langle\zeta\cdot\eta\rangle^3 \\ \langle\zeta\cdot\eta\rangle^3 & \langle\zeta\cdot\eta\rangle^4 \end{pmatrix} \tag{9.15}$$

on $L_{+}(\mathcal{D}) \times L_{-}(\mathcal{D})$.

In the detailed record, this relation means that the forms $\Phi_{\pm}$, $\theta_{\pm}$, and $\Omega_{\pm}$ satisfy, along with (9.4) and (9.5), the following equations on $L^{(3)}(\mathcal{D})$:

$$\bar{\partial}\theta_{\pm} + \theta_{\pm} \wedge \theta_{\pm} + (\bar{\partial}\Omega_{\pm} + \Omega_{\pm} \wedge \theta_{\pm} \wedge \Omega_{\pm})\langle\zeta\cdot\eta\rangle$$
$$+ \Omega_{\pm} \wedge \Omega_{\pm}\langle\zeta\cdot\eta\rangle^2 + \Phi_{\pm} \wedge \Phi_{\mp}\langle\zeta\cdot\eta\rangle^2 = 0.$$

By developing Proposition 9.2, we obtain the following proposition.

Proposition 9.3. *In order for the $(0,1)$-forms $\theta_\pm, \Omega_\pm, \Phi_\pm$ to satisfy the $\bar\partial$-equations (9.4), (9.5), and (9.16), it is necessary and sufficient that their Radon–Penrose transforms $a_\pm, \omega_\pm, \varphi_\pm$ satisfy the equations (9.9), (9.10), and (9.14).*

In order to show that the equations (9.16) go into the equation (9.14) after the Radon–Penrose transforms, we verify, due to (9.5) and to Theorem 8.2(a), that the following equality is valid:

$$\frac{1}{2}\left([\nabla_{AA'}^{+-},\varphi_+]\varphi_- - \varphi_+[\nabla_{AA'}^{-+},\varphi_-]\right) = \left(-\frac{1}{2\pi i}\right)\left(-\frac{2}{2\pi i}\right)$$
$$\times \iint\limits_{\mathcal{L}(x)} \frac{\partial}{\partial\eta^A}\frac{\partial}{\partial\zeta^{A'}}(\mu_+\mathcal{I}_+\mu_+^{-1}) \wedge \zeta^C d\zeta_C \wedge \eta^{C'} d\zeta_{C'},$$

where

$$\mathcal{I}_+ = -(\bar\partial\Omega_+ + \Omega_+ \wedge \theta_+ + \theta_+ \wedge \Omega_+) - \Omega_+ \wedge \Omega_+\langle\zeta\cdot\eta\rangle - \Phi_+ \wedge \Phi_-\langle\zeta\cdot\eta\rangle.$$

Let us prove this equality by using the "x"-calibration of the forms $\theta_+, \Omega_+, \Phi_\pm$. We have

$$-\frac{1}{2}\left(\frac{1}{\pi i}\right)^2 \iint\limits_{\mathcal{L}(x)} \frac{\partial}{\partial\eta^A\partial\zeta^{A'}}\mathcal{I}_+ \wedge \zeta^C d\zeta_C \wedge \eta^{C'} d\eta_{C'}$$

(due to Proposition 8.1(a))

$$= -\frac{1}{2}\left(\frac{1}{\pi i}\right)^2 \iint\limits_{\mathcal{L}(x)} \left(\zeta_A\frac{\partial}{\partial\zeta^{A'}}(\Phi_+ \wedge \Phi_-) - \eta_{A'}\frac{\partial}{\partial\eta^A}(\Phi_+ \wedge \Phi_-)\right)$$
$$\wedge \zeta^C d\zeta_C \wedge \eta^{C'} d\eta_{C'}$$

(due to the definition (9.7) and to the equality (4.17))

$$= \frac{1}{2}\left([\nabla_{AA'}^{+-},\varphi_+]\varphi_- - \varphi_+[\nabla_{AA'}^{-+},\varphi_-]\right).$$

We denote by $\tilde{H}^{0,1}(L^{(2,5)}(\mathcal{D}),\mathrm{GL}(E^0))$ the space of smooth $(0,1)$-forms θ on $L_+(\mathcal{D}) \times L_-(\mathcal{D})$ with values in $\mathrm{End}\,E^0$ that are considered first up to the gauge equivalence

$$\tilde\theta = B^{-1}\bar\partial B + B^{-1}\theta B + O\begin{pmatrix} \langle\zeta\cdot\eta\rangle^4 & \langle\zeta\cdot\eta\rangle^3 \\ \langle\zeta\cdot\eta\rangle^3 & \langle\zeta\cdot\eta\rangle^4 \end{pmatrix},$$

where B is a smooth function with values in the nondegenerated endomorphisms E^0; second, these forms satisfy equation (9.15), and, third, a restriction of the form θ to any quadric $\mathcal{L}^{(1)}(x) \cap L(\mathcal{D})$ is gauge-trivial. To any form

$\theta \in \tilde{H}^{0,1}(L^{(2,5)}(\mathcal{D}), \mathrm{GL}(E^0))$, we associate the holomorphic (respectively, CR-) bundle E_θ on $L^{(2,5)}(\mathcal{D})$, according to the rule: the holomorphic (respectively, CR-) sections E_θ are those smooth sections H of the bundle E^0 that satisfy the Cauchy–Riemann equations of the form

$$\bar{\partial}H + \theta H = O\begin{pmatrix} \langle \zeta \cdot \eta \rangle^4 & \langle \zeta \cdot \eta \rangle^3 \\ \langle \zeta \cdot \eta \rangle^3 & \langle \zeta \cdot \eta \rangle^4 \end{pmatrix}.$$

It is possible to show that any element of the space $\tilde{H}^{0,1}(L^{(2,5)}(\mathcal{D}), \mathrm{GL}(E^0))$ may be given by the form θ of the type (9.3).

Theorem 9.2 (Khenkin [1982]). *The Radon–Penrose transform* $\theta \mapsto (a_\pm, \varphi_\pm)$ *realizes an isomorphism between the space of holomorphic (respectively, CR-) bundles E_θ on $L^{(2,5)}(\mathcal{D})$ topologically equivalent to E^0 and analytically equivalent to E^0 on each quadric $\mathcal{L}^{(1)}(x) \cap L(\mathcal{D})$ and a space of holomorphic (respectively, smooth) solutions of the Yang–Mills–Higgs system on $\mathcal{D}$ of the form (9.13) and (9.14).*

Remark. To classical solutions of the equations (9.13), (9.14) in the domain $\mathcal{D}$ of the real Minkowski space $\mathcal{M}$ when $\varphi_+ = \varphi_- = \varphi_+^*$, $a_+ = a_- = -a_+^*$ and with the gauge group $U(n)$ correspond the bundles E_θ on $L^{(2,5)}(\mathcal{D})$ given by the forms θ of the type (9.3), where

$$-\theta_-(w, z) = \theta_+(w, z) = -{}^t\theta_+(z, w), \quad \Phi_-(w, z) = {}^t\Phi_+(z, w) = \varphi_+(w, z),$$

and with the gauge function b of the form $\begin{pmatrix} b & 0 \\ 0 & b \end{pmatrix}$, where b has values in $U(n)$; ${}^t(\cdot)$ is a sign of transposition.

In analogy to Theorem 7.2, the form θ of the type (9.3) satisfying the nonlinear $\bar{\partial}$-equation (9.15) may be obtained from the integral equation

$$\theta = \theta^0 + R(\theta \wedge \theta),$$

where the form

$$\theta^0 = \begin{pmatrix} \theta_+^0 & \Phi_+^0 \langle \zeta \cdot \eta \rangle \\ \Phi_-^0 \langle \zeta \cdot \eta \rangle & \theta_-^0 \end{pmatrix}$$

satisfies the linear $\bar{\partial}$-equation

$$\bar{\partial}\theta^0 = \begin{pmatrix} \langle \zeta \cdot \eta \rangle^4 & \langle \zeta \cdot \eta \rangle^3 \\ \langle \zeta \cdot \eta \rangle^3 & \langle \zeta \cdot \eta \rangle^4 \end{pmatrix}; \quad R\theta \wedge \theta = \begin{pmatrix} R_3(\theta \wedge \theta)_{11} & R_2(\theta \wedge \theta)_{12} \\ R_2(\theta \wedge \theta)_{21} & R_3(\theta \wedge \theta)_{22} \end{pmatrix},$$

with R_j being operators inverting the $\bar{\partial}$-operator on the $(0,1)$-forms on $L^{(j)}(\mathcal{D})$, $j = 2, 3$ (cf. Proposition 7.1). Hence and from Theorems 8.3 and 9.2, it follows that all solutions of the Yang–Mills–Higgs equation of the form (9.13), (9.14) in the domain $\mathcal{D} \subset \mathbb{C}M$ with sufficiently small norm may be obtained starting with solutions of the linear Maxwell equations and of the wave equation. Thus, the system (9.13), (9.14) is linearized in the sense of Flato and Simon [1980].

It turned out that the bundles E_θ on $L^{(3)}(\mathcal{D})$ code solutions of the Yang–Mills–Higgs system supplemented with another two nonliniear equations.

Proposition 9.4. *Any bundle E_θ, where $\theta \in \tilde{H}^{0,1}(L^{(2,5)}(\mathcal{D}), \mathrm{GL}(E^0))$ can be extended, and moreover, in a unique way, to a bundle on $L^{(3)}(\mathcal{D})$ iff the corresponding Radon–Penrose transforms $\theta_\pm \mapsto a_\pm$ and $\Phi_\pm \mapsto \varphi_\pm$ satisfy, along with the Yang–Mills–Higgs equations (9.13) and (9.14), the following additional equations,*

$$
\begin{aligned}
f^+_{AB}(x)\varphi_+(x) - \varphi_+(x) \cdot f^-_{AB}(x) = 0, \\
f^-_{A'B'}(x)\varphi_-(x) - \varphi_-(x) \cdot f^+_{A'B'}(x) = 0.
\end{aligned}
\tag{9.17}
$$

From Theorem 9.2 and Proposition 9.4, the next interesting consequence follows.

Theorem 9.3. *Let $n_\pm = 1$, that is, $E^0 = \theta(-1,0) \oplus \mathcal{O}(0,-1)$. The Penrose transform realizes the isomorphism between a space of holomorphic vector bundles on $L^{(3)}(\mathcal{D})$ topologically equivalent to E^0 and analytically equivalent to E^0 on each quadric $\mathcal{L}^{(1)}(x) \cap L(\mathcal{D})$ and a space of solutions of the system of the noninteracting Maxwell equations and the scalar φ^4-equation:*

$$
\nabla^{AB'} f^+_{AB} = 0, \qquad \nabla^{AB'} f^+_{A'B'} = 0,
\tag{9.18}
$$

$$
\begin{aligned}
f^+_{AB} = f^-_{AB}, \qquad & f^+_{A'B'} = f^-_{A'B'}, \\
\Box\varphi_+ = \lambda\varphi_+^3, \qquad & \varphi_- = \lambda\varphi_+, \quad \lambda \in \mathbb{C}.
\end{aligned}
\tag{9.19}
$$

We illustrate this result with a simple example. We fix a covering of the domain $X = L_+(\mathbb{C}M_0) \times L_-(\mathbb{C}M_0)$ by the domains of the form $X_{AA'} = \{(\zeta, \eta) \in X : \zeta_A \cdot \eta_{A'} - \eta_A\zeta_{A'} \neq 0\}$, and we have on X a holomorphic bundle equivalent to $\mathcal{O}(-1,0) \oplus \mathcal{O}(0,-1)$ on $X \cap L(\mathbb{C}M_0)$, given by the equality

$$
(I + H_1)(I + H_{0'}) = (I + H_{1'})(I + H_0),
$$

where $H_A, H_{A'}$ are holomorphic transition matrices,

$$
H_{A'} = \frac{\langle \zeta \cdot \eta \rangle}{(\zeta_0 \cdot \eta_{A'} - \eta_0\zeta_{A'})(\zeta_1\eta_{A'} - \eta_1\zeta_{A'})}
\begin{pmatrix} \zeta_{A'} \cdot \eta_{A'} & \eta_{A'}^2 \\ \zeta_{A'}^2 & -\zeta_{A'} \cdot \eta_{A'} \end{pmatrix},
$$

where $(\zeta, \eta) \in X_{0A'} \cap X_{1A'}$, and

$$
H_A = \frac{\langle \zeta \cdot \eta \rangle}{(\zeta_A \cdot \eta_{0'} - \eta_A \cdot \zeta_{0'})(\zeta_A \cdot \eta_{1'} - \eta_A\zeta_{1'})}
\begin{pmatrix} \zeta_A \cdot \eta_A & \eta_A^2 \\ \zeta_A^2 & -\zeta_A \cdot \eta_A \end{pmatrix},
$$

where $(\zeta, \eta) \in X_{A0'} \cap X_{A1'}$. The assertion of Theorem 9.3 applied to this bundle gives one of a few known explicit solutions of the system (9.19), where $a_\pm = 0$, $\varphi_+ = \varphi_- = (1 + \det x_{AA'})^{-1}$.

It is possible to obtain a twistor interpretation of solutions of the Yang–Mills equations of the form (1.16), (1.17) as an interesting additional consequence of Propositions (9.2), (9.3), and (9.4).

Proposition 9.5. *Let the fields a^+ and a^- satisfy the Maxwell equations in the domain $\mathcal{D} \subset \mathbb{C}M$; in addition, the field $a^0 = (a^+ - a^-)$ is auto-dual. Let the scalar functions φ_+ and φ_- also satisfy the equations of the form (9.19), where $\square = \square_{+-}$. Finally, let the $(0,1)$-forms $\theta_\pm, \Phi_\pm$, and $\Omega_\pm$ on $L(\mathcal{D})$ be such that their Penrose transforms are equal to $a_\pm$, $\varphi_\pm$, and $(\mp\frac{1}{2}\varphi_+\varphi_-)$, respectively.*

Then:

(a) *the field of the form (1.16), where*

$$\nabla_{BC'} = \frac{\partial}{\partial x^{BC'}} + a^0_{BC'}$$

satisfies the Yang–Mills equation;

(b) *this Yang–Mills field corresponds to such a holomorphic vector bundle on $L^{(3)}(\mathcal{D})$ which is given by the $(0,1)$-form of the $\bar{\partial}$-connection θ on $L_+(\mathcal{D}) \times L_-(\mathcal{D})$ with values in $\mathrm{End}[\mathcal{O}(-1,0) \oplus \mathcal{O}(1,0)]$,*

$$\begin{pmatrix} \theta_+ + \Omega_+\langle \zeta \cdot \eta \rangle & \Phi_+ \\ \Phi_-\langle \zeta \cdot \eta \rangle^2 & \theta_- + \Omega_-\langle \zeta \cdot \eta \rangle \end{pmatrix}.$$

Proposition 9.5 was obtained in 1983 by A. Bell and G. Khenkin (unpublished). For the case when the parameter λ in (9.19) is equal to zero, the result of Proposition 9.5 is contained in Ward [1981]. The assertion (a) is a small generalization of the result of the classical paper Corrigan and Fairlie [1977]. The part of assertion (b) consisting of the fact that the stated deformation given by the form θ of the topologically trivial bundle $\mathcal{O}(-1,0) \oplus \mathcal{O}(1,0)$, in fact, generates some bundle on $L^{(3)}(\mathcal{D})$ results from Propositions 9.2, 9.3, and 9.4. From Theorem 8.3(a), it also follows that this bundle corresponds to some Yang–Mills field. Finally, the coincidence of this field with a field from Proposition 9.5(a) follows from the formulae for the Penrose transform (see also similar computations in Buchdahl [1985]).

References[1]

Ablowitz, M.J., Nachman, A.I.

[1986] Multidimensional nonlinear evolutions and inverse scattering. Physica D*18*, 223–241. Zbl.604.35070

---

[1] For the convenience of the reader, references to reviews in Zentralblatt für Mathematik (Zbl.), compiled using the MATH database, have, as far as possible, been included in this bibliography.

Atiyah, M.F.

[1979] Geometry of Yang–Mills Fields. Pisa: Lezioni Fermiani. Zbl.435.58001

Atiyah, M.F., Hitchin, N.J.

[1988] The Geometry and Dynamics of Magnetic Monopoles. Princeton: Princeton University Press. Zbl.671.53001

Atiyah, M.F., Ward, R.S.

[1977] Instantons and algebraic geometry. Commun. Math. Phys. 55, 117–124. Zbl.362.14004

Baston, R.J., Mason, L.J.

[1987] Conformal gravity, the Einstein equations and spaces of complex null geodesics. Class. Quan. Grav. 4, 815–826.

Beals, R., Coifman, R.R.

[1986] The D-bar approach to inverse scattering and nonlinear evolutions. Physica D18, 242–249. Zbl.619.35090

Belavin, A.A., Polyakov, A.M., Schwarz, A.S., Tyupkin, Yu.S.

[1975] Pseudoparticle solutions of the Yang–Mills equations. Phys. Lett. B 59, No. 1, 85–87.

Belavin, A.A., Zakharov, V.E.

[1978] Yang–Mills equations as inverse scattering problems. Phys. Lett. B 73, 53–57.

Bourguignon, J.P., Lawson, Jr., H.B.

[1981] Stabiblity and isolation phenomena for Yang–Mills fields. Commun. Math. Phys. 79, 189–230. Zbl.475.53060

Buchdahl, N.P.

[1985] Analysis on analytic spaces and non-self-dual Yang–Mills fields. Trans. Am. Math. Soc. 288, No. 2, 431–469. Zbl.533.32014

Choquet-Bruhat, Y., Christodoulou, D.

[1981] Existence of global solutions of the Yang–Mills, Higgs and spinor field equation in $3 + 1$ dimensions. Ann. Sci. Ec. Norm. Super., IV. Ser. 14, No. 4, 481–506. Zbl.499.35076

Corrigan, E., Fairlie, D.B.

[1977] Scalar field theory and exact solution to a classical SU(2)-gauge theory. Phys. Lett. B 67, 69–71.

Deligne, P.

[1970] Equations différentielles à points singuliers reguliers. Lect. Notes in Math. 163. Zbl.244.14004

Eardley, D.M., Moncrief, V.

[1982] The global existence of Yang–Mills–Higgs fields in 4-dimensional Minkowski space I, II. Commun. Math. Phys. 83, 171–191, 193–212. Zbl.496.35061, Zbl.496.35062

Eastwood, M.G., Penrose, R., Wells, Jr., R.O.

[1981] Cohomology and massless fields. Commun. Math. Phys. 78, 305–351. Zbl.465.58031

Eastwood, M.G., Pool, R., Wells, Jr., R.O.

[1985] The inverse Penrose transform of a solution to the Maxwell–Dirac–Weyl field equations. J. Funct. Anal. 60, 16–35. Zbl.552.53046

Faddeev, L.D.
[1974] Inverse problem of quantum scattering, II. Itogi Nauki Tekh., Ser. Sovrem. Probl. Mat. *3*, 93–180. English transl.: J. Sov. Math. *5*, 334–396 (1976). Zbl.299.35027

Faddeev, L.D., Slavnov, A.A.
[1988] Introduction to Quantum Theory of Gauge Fields. 2nd ed. Moscow: Nauka. English transl.: Reading, Mass., London, 1980. Zbl.486.53052,Zbl.663.53060

Flato, M., Simon, I.
[1980] On a linearization program of non-linear field equations. Phys. Lett. B *94*, 518–522.

Forgacs, P., Horvath, Z., Palla, L.
[1981] Towards complete integrability of the self-duality equations. Phys. Rev. D. *23*, No. 8, 1876–1879.

1982] On the linearization of source-free gauge field equations. Phys. Lett. B *115*, No. 6, 463–467.

Gindikin, S.G., Khenkin, G.M. [= Henkin, G.M.]
[1981] Penrose transform and complex integral geometry. Itogi Nauki Tekh., Ser. Sovrem. Probl. Mat. *17*, 57–111. Zbl.482.53053. English transl.: J. Sov. Math. *21*, 508–551 (1983).

Griffiths, P.A.
[1966] The extension problem in complex analysis II; embeddings with positive normal bundle. Am. J. Math. *88*, No. 2, 366–446. Zbl.147,75

Gu, Chaohao
[1981] On classical Yang–Mills fields. Phys. Rep. *80*, No. 4, 251–337.

Harnad, J., Shnider, S.
[1986] Constraints and field equations for ten-dimensional super Yang–Mills theory. Commun. Math. Phys. *106*, 183–199.

Hartshorne, R.
[1977] Algebraic Geometry. New York, Heidelberg, Berlin: Springer-Verlag. Zbl.367.14001

Henkin, G.M. [= Khenkin, G.M.]
[1984] Tangent Cauchy–Riemann equations and the Yang–Mills–Higgs and Dirac fields. Proc. Int. Congr. Math., Warszawa, 1983, 809–827. Zbl.584.58050

Henkin, G.M., Manin, Yu.I.
[1980] Twistor description of classical Yang–Mills–Dirac fields. Phys. Lett. B *95*, No. 3 and 4, 405–408.

[1981] On the cohomology of twistor flag spaces. Compos. Math. *44*, No. 1–3, 103–111. Zbl.495.32004

Henkin, G.M., Novikov, R.G.
[1988] A multidimensional inverse problem in quantum and acoustic scattering. Inverse Problems *4*, 103–121. Zbl.697.35108

Henkin, G.M., Polyakov, P.L.
[1990] Residue integral formulas and the Radon transform for differential forms on q-linearly concave domains. Math. Ann. *286*, 225–254. Zbl.704.32002

Hitchin, N.J.
[1980] Linear field equations on self-dual spaces. Proc. R. Soc. Lond., Ser. A *370*, 173–191. Zbl.436.53058

[1982] Monopoles and geodesics. Commun. Math. Phys. *83*, 579–602. Zb.502.58017

Hughston, L.P., Shaw, W.T.
[1990] Twistors and strings. In: Twistors in Mathematics and Physics (T. Bailey and R. Baston, eds.). London Math. Soc. Lecture Note Series *156*, 218–245. Cambridge: Cambridge University Press.

Isenberg, J., Yasskin, P.
[1979] Twistor description of non-self-dual Yang–Mills fields. Complex manifold techniques in theoretical physics, Res. Notes. Math. *32*, 180–206. Zbl.421.53023

[1986] Ambitwistors and strings. In: Proc. of the Oregon Meeting. Singapore: World Scientific, 787–796.

Isenberg, J., Yasskin, P., Green, P.S.
[1978] Non-self-dual gauge fields. Phys. Lett. B *78*, No. 4, 462–464.

Jaffe, A., Taubes, C.
[1980] Vortices and Monopoles. Boston: Birkhäuser. Zbl.457.53034

Kapranov, M.M., Manin, Yu.I.
[1986] Twistor transform and algebraic-geometric constructions for solutions of field theory equations. Usp. Mat. Nauk *41*, No. 5, 85–107. English transl.: Russ. Math. Surv. *41*, No. 5, 33–61 (1986). Zbl.614.32025

Khenkin, G.M. [= Henkin, G.M.]
[1980] The representation of the Yang–Mills equations in the form of the Cauchy–Riemann equations on the twistor space. Dokl. Akad. SSSR *255*, No. 4, 844–847 (Russian).

[1981] The representation of solutions of the φ^4-equation in the form of holomorphic bundles over the twistor space. Dokl. Akad. Nauk SSSR *260*, No. 5, 1086–1089. English transl.: Sov. Math., Dokl. *24*, 415–419 (1981). Zbl.501.53044

[1982] The Yang–Mills–Higgs fields as holomorphic vector bundles. Dokl. Akad. Nauk SSSR *265*, No. 5, 1081–1085. English transl.: Sov. Math., Dokl. *26*, 224–228 (1982). Zbl.516.32013

[1985] The method of integral representations in complex analysis. Itogi Nauki Tekh., Ser. Sovrem. Probl. Mat., Fundam. Napravleniya 7, 23–124. English transl. in: Several Complex Variables I, Encyc. Math. Sci. 7, 19–117. Berlin: Springer-Verlag, 1990.

Khenkin, G.M., Manin, Yu.I.
[1982] Yang–Mills–Dirac equations as Cauchy–Riemann equations in twistor space. Yad. Fiz. *35*, 1610–1626. English transl.: Sov. J. Nucl. Phys. *35*, 941–950 (1982). Zbl.587.58056

Khenkin, G.M., Novikov, R.G.
[1987] $\bar{\partial}$-equation in the multidimensional inverse scattering problem. Usp. Mat. Nauk *42*, No. 3, 93–152. English transl.: Russ. Math. Surv. *42*, No. 3, 109–180 (1987). Zbl.674.35085

Khudaverdyan, O.M., Roslyj, A.A., Schwarz, A.S.
[1986] Supersymmetry and complex geometry. Itogi Nauki Tekh., Ser. Sovrem. Probl. Mat., Fundam. Napravleniya *9*, 247–284. English transl. in: Several Complex Variables III, Encyc. Math. Sci. 9, 223–261. Berlin: Springer-Verlag, 1989.

LeBrun, C.
[1982] The first formal neighborhood of ambitwistor space for curved space-time. Lett. Math. Phys. *6*, 345–354. Zbl.518.58023

[1983] Spaces of complex null geodesics in complex-Riemannian geometry. Trans. Am. Math. Soc. *278*, 209–231. Zbl.562.53018

[1991] Thickenings and conformal gravity. Commun. Math. Phys. *139*, No. 1, 1–43.

Leiterer, J.
[1986] Holomorphic vector bundles and the Oka–Grauert principle. Itogi Nauki Tekh., Ser. Sovrem. Probl. Mat., Fundam. Napravleniya *10*, 75–121. English transl. in: Several Complex Variables IV, Encyc. Math. Sci. 10, 63–105. Berlin: Springer-Verlag, 1990.

Manakov, S.V., Novikov, S.P., Pitaevskij, L.P., Zakharov, V.E.
[1980] Theory of Solitons. Inverse Problem Method. Moscow: Nauka. English transl.:
Contemp. Sov. Math., New York (1984). Zbl.598.35003
Manakov, S.V., Zakharov, V.E.
[1981] Three-dimensional model of relativistic invariant field theory integrable by
the inverse scattering transform. Lett. Math. Phys. 5, No. 3, 247–253.
Manin, Yu.I.
[1981] Gauge fields and holomorphic geometry. Itogi Nauki Tekh., Ser. Sovrem.
Probl. Mat. 17, 3–55. English transl.: J. Sov. Math. 21, 465–507 (1983).
Zbl.486.53051
[1984] Gauge Fields and Complex Geometry. Moscow: Nauka. English transl.:
Berlin: Springer-Verlag, 1988. Zbl.576.53002
Newman, E.T.
[1986] Gauge theories, the holonomy operator, and the Riemann–Hilbert problem.
J. Math. Phys. 27, 2797–2802.
Newton, R.G.
[1989] Inverse Schrödinger Scattering in Three Dimensions. New York: Springer-
Verlag. Zbl.697.35005
Novikov, R.G.
[1989] Inverse scattering problem for two-dimensional Schrödinger equations under
fixed energy and non linear equations. Thesis. Moscow University (Russian).
English transl.: J. Funct. Anal. 103, 409–463 (1992).
Okonek, C., Van de Ven, A.
[1990] Stable bundles, instantons and differentiable structures on algebraic surfaces.
In: Several Complex Variables VI, Encyc. Math. Sci. 69, Berlin: Springer-
Verlag, 1990.
Onishchik, A.L.
[1986] Methods of the theory of sheaves and Stein spaces. Itogi Nauki Tekh., Ser.
Sovrem. Probl. Mat., Fundam. Napravleniya 10, 5–73. English transl. in:
Several Complex Variables IV, Encyc. Math. Sci. 10, 1–61. Berlin: Springer-
Verlag, 1990.
Penrose, R.
[1969] Solutions of the zero-rest-mass equations. J. Math. Phys. 10, 38–39.
[1976] Non linear gravitons and curved twistor theory. Gen. Relativ. Gravitation 7,
31–52. Zbl.354.53025
Penrose, R., Rindler, W.
[1984,1986] Spinors and Space Time, Vol. I, II. Cambridge, London, New York:
Cambridge University Press. Zbl.538.53024,Zbl.591.53002
Polyakov, P.L., Khenkin, G.M.
[1990] Integral formulae for the solution of the $\bar{\partial}$-equation and interpolation problems in analytical polyhedra. Tr. Mosk. Mat. O.-va 53, 130–170.
Ponomarev, D.A.
[1984] The Germs on $\mathbb{C}P^1 \times \mathbb{C}P^1$ of holomorphic vector bundles on $\mathbb{C}P^3 \times \mathbb{C}P^3$.
Dokl. Akad. Nauk SSSR 276, 292–295. English transl.: Sov. Math., Dokl. 29,
492–495 (1984). Zbl.596.32037
Pool, R.
[1987] Yang–Mills fields and extension theory. Mem. Am. Math. Soc. 358, 67p.
Zbl.604.32020
Segal, I.
[1979] The Cauchy problem for the Yang–Mills equations. J. Funct. Anal. 33, 175–
194. Zbl.416.58027

Sibner, L.M., Sibner, R.J., Uhlenbeck, K.
[1989] Solutions to Yang–Mills equations which are not self-dual. Proc. Natl. Acad. Sci. USA *86*, 8610–8613.
Taubes, C.H.
[1985] Min-max theory for the Yang–Mills–Higgs equations. Commun. Math. Phys. *97*, 473–540. Zbl.585.58016
Ward, R.S.
[1977] On self-dual gauge fields. Phys. Lett. A *61*, 81–82.
[1981] Ansätze for self-dual Yang–Mills fields. Commun. Math. Phys. *80*, 563–574.
[1984] Completely solvable gauge-field equations in dimension greater than four. Nucl. Phys. B *236*, 381–396.
Ward, R.S., Wells, Jr., R.O.
[1990] Twistor Geometry and Field Theory. Cambridge: Cambridge University Press.
Wells, Jr., R.O.
[1981] Hyperfunction solutions of the zero-rest-mass field equations. Commun. Math. Phys. *78*, 567–600. Zbl.465.58032
[1982] Complex Geometry in Mathematical Physics. Montreal: Presses de l'Université de Montréal. Zbl.481.58001
Witten, E.
[1978] An interpretation of classical Yang–Mills theory. Phys. Lett. B *77*, 394–398.
[1986] Twistor-like transform in ten dimensions. Nucl. Phys. B *266*, 245–264. Zbl.608.53068
Woodhouse, N.M.J.
[1985] Real methods in twistor theory. Classical Quantum Gravity *2*, 257–291. Zbl.575.53082
Yang, C.N.
[1977] Condition of self-duality for SU(2)-gauge fields on Euclidean four-dimensional space. Phys. Rev. Lett. *38*, 1377–1379.
Yasskin, P.B.
[1987] An ambitwistor approach to gravity. In: Gravitation and Geometry, Vol. Hon. I. Robinson, Monogr. Textb. Phys. Sci. *4*, 478–495. Zbl.658.53073

III. Complex Geometry and String Theory

A.Yu. Morozov, A.M. Perelomov

Translated from the Russian
by the Authors

Contents

Chapter 1
Introduction

String theory is a rather new field of theoretical physics. It appeared only twenty years ago to describe phenomenology of strong interactions of elementary particles, and until recently, it has been developing rather slowly. This is because string theory has encountered a number of difficulties that were not easy to overcome; in particular, this theory contained the so-called anomalies that hindered construction of self-consistent string theory. Especially, anomalies led to the breakdown of symmetry properties of this theory after its quantization.

A decisive break in string theory came not very long ago. In the papers by Green and Schwarz [1984; 1985], which essentially used the results of the earlier paper by Alvarez-Gaume and Witten [1985], it was shown that there is a unique string theory without anomalies, its gauge symmetry group being the exceptional group $E_8 \times E_8$ or $SO(32)$. Further development turned this new theory into the most probably candidate for a self-consistent quantum theory of all interactions in nature, including gravitation, thus resolving the long-standing puzzle of quantum gravity.

Naturally, during the last years, string theory was impetuously developed and an enormous number of papers appeared dedicated to its various aspects. We mention here only the pioneering papers by Green and Schwarz [1984; 1985], Gross, Harvey, Martinec and Rohm [1985; 1986], Candelas, Horowitz, Strominger and Witten [1985], as well as papers collected in two volumes edited by Schwarz, *Superstrings: The First 15 Years of Superstring Theory* [1985], and the two-volume monograph by Green, Schwarz and Witten [1987], where one can find explicit discussions of a number of subjects of this wide and complicated theory.

The aim of our contribution is to give a review of a particular part of string theory, namely the analyticity properties of this theory. We shall demonstrate that the theory of strings is profoundly connected with rather contemporary fields of complex geometry.

We start with the consideration of a massless classical pointlike particle that moves in the Minkowski space of D-dimensions, i.e., in the pseudo-Euclidean space of the signature $(D-1, 1)$. It is convenient to describe such a particle by using the action functional

$$A = \int \eta_{\mu\nu} \frac{dx^\mu}{d\tau} \frac{dx^\nu}{d\tau} d\tau, \tag{1.1}$$

where $\eta_{\mu\nu}$ is the Minkowski metric; $\mu, \nu = 0, 1, \ldots, (D-1), \tau$ is the real parameter describing the position of the particle on the trajectory; x^μ are coordinates of the particle in the Minkowski space.

The equations of motion of the particle follow from the variational principle $\delta A = 0$ and are of the form $d^2x/d\tau^2 = 0$. The solutions of these equations are straight lines in the Minkowski space. However, since we consider a massless particle, the additional condition

$$\eta_{\mu\nu}\frac{dx^{\mu}}{d\tau}\frac{dx^{\nu}}{d\tau} = 0 \tag{1.2}$$

should also be fulfilled, i.e., we consider only straight lines lying on the light cone in the Minkowski space.

It can be easily seen that condition (1.2) appears automatically if one introduces the independent metrics $g(\tau)$ on the τ-line. Indeed, instead of (1.1), we then have

$$A = \int g(\tau)\eta_{\mu\nu}\frac{dx^{\mu}}{d\tau}\frac{dx^{\nu}}{d\tau}d\tau. \tag{1.3}$$

Equations (1.1) and (1.3) are in fact equivalent.

In the quantum case, it will be convenient to use the functional integral formulation and work in the Euclidean D-dimensional space with the standard metrics instead of the Minkowski space. Equation (1.3) takes the form

$$A = \int g(\tau)\frac{dx^{\mu}}{d\tau}\frac{dx^{\mu}}{d\tau}d\tau, \tag{1.4}$$

In this theory, the simplest observable one is the "partition function" Z, which is the sum over all closed trajectories in the functional integral

$$Z = \int DgDx^{\mu}\exp(-A). \tag{1.5}$$

Let us now proceed with the consideration of strings. A string is a one-dimensional object, i.e., a curve in the Minkowski space or in Euclidean space, and we describe the points of a string by a coordinate σ, which varies within finite limits, say $0 \leq \sigma \leq 1$. In doing so, we should distinguish the cases of closed and open strings.

In order to consider the motion of a string, one needs another parameter τ, which is a time type. During its evolution, the string describes a two-dimensional surface—the so-called world sheet, which is a generalization of the world line in the case of a pointlike particle. The coordinates of a point on a world sheet are denoted by $x_{\mu}(\sigma, \tau)$.

There are various methods to develop string theory. Originally it was proposed by Nambu [1970] and Goto [1971] to describe the string with the help of an action which is proportional to the area of the world sheet, i.e.,

$$A = \alpha \int d\sigma d\tau \sqrt{\dot{x}'^{2}x'^{2} - (\dot{x}x')^{2}}, \tag{1.6}$$

$$\dot{x}^{\mu} = \partial x^{\mu}/\partial\tau, \quad x^{\mu\prime} = \partial x^{\mu}/\partial\sigma, \quad \dot{x}'^{2} = (\partial x^{\mu}/\partial\tau)(\partial x^{\mu}/\partial\tau),$$

where

$$x'^{2} = (\partial x^{\mu}/\partial\sigma)(\partial x^{\mu}/\partial\sigma), \quad (\dot{x}', x') = (\partial x^{\mu}/\partial\tau)(\partial x^{\mu}/\partial\sigma)$$

is a constant.

A drawback of such a formulation is that the equations obtained are strongly nonlinear and are very difficult to deal with. A convenient formulation that we shall use in what follows was given by Polyakov [1981a,b]. Let us introduce a metric g on the world sheet that is an independent variable in the functional space and let the action of the strings have the form

$$A = \alpha \int d^2\xi \sqrt{|g|}\, g^{ab} \partial_a x^\mu \partial_b x^\mu. \tag{1.7}$$

Here $\xi_1 = \sigma$, $\xi_2 = \tau$, $|g|$ is the modulus of the determinant of the matrix g_{ab}, and g^{ab} is the matrix inverse to the matrix g_{ab}.

Since the derivatives of the matrix g_{ab} do not enter the functional of action, the variables may be excluded from (1.7) (or, what is equivalent, g_{ab} may be integrated out); after that, (1.7) takes the form (1.6). The expressions (1.6) and (1.7) are invariant under a large group of transformations. First of all, these expressions are invariant under reparametrization of the world sheet

$$(\sigma, \tau) \to (\sigma', \tau'), \ \sigma' = f_1(\sigma, \tau), \ \tau' = f_2(\sigma, \tau) \tag{1.8}$$

or, in other words, under the diffeomorphism group of the world sheet. In the case of (1.7), one should take into account that the metric g_{ab} also changes according to the well-known transformation of the second-rank tensor.

From the viewpoint of an observer on the world sheet, the functionals (1.6) and (1.7) describe the theory of a field with one spacelike and one timelike dimension, and the D-variables x^μ behave as scalar fields under the reparametrization group of the world sheet.

Now we may use the reparametrization invariance in order to exclude two of the three independent functions in the metric tensor g_{ab} and reduce it to the standard form

$$g_{ab} = \delta_{ab} \exp(\varphi), \tag{1.9}$$

where δ_{ab} are Kronecker symbols.

It can be easily seen now that the functional (1.7) is also independent of φ and takes the form

$$A = \alpha \int d^2\xi \partial_a x^\mu \partial_a x^\mu. \tag{1.10}$$

In other words, the functional (1.7) is invariant also under the transformations of the conformal group

$$g_{ab} \to \exp(\varphi) g_{ab}, \tag{1.11}$$

and as a result is invariant under the transformations of the group

$$H = \mathrm{Diff} \times \mathrm{Conf}, \tag{1.12}$$

which is a semidirect product of the diffeomorphism group of the world sheet and conformal group. This fact allows one to reduce the functional (1.7) to the form of (1.10).

This is the outlook for the classical case. As for the quantum case, the situation is much more complicated because of anomalies that, in general, can break down the invariance under the conformal group. However, it can be shown that the anomaly vanishes if $D = 26$ and the quantum theory remains conformally-invariant.[1]

Let us stress an important property of the string theory under consideration: it naturally describes string interactions.

For instance, the surface of the type shown in Fig. 1 describes the scattering of two closed strings, and it is necessary to calculate the functional integral over all Riemann surfaces of such a type in order to find the quantum scattering amplitude in Polyakov's approach [1981a,b].

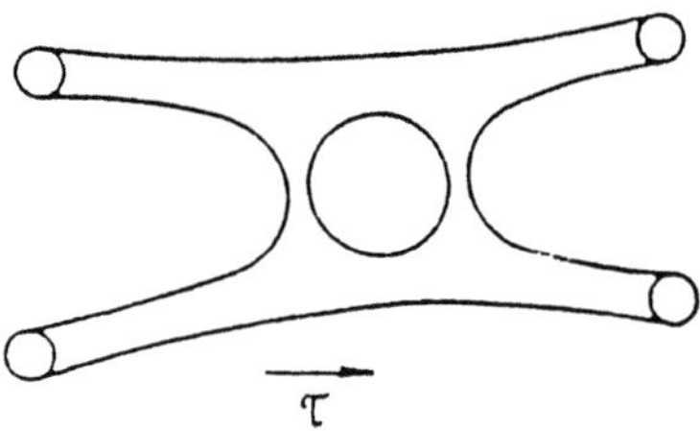

Fig. 1

In this contribution, we restrict ourselves to the consideration of the simplest case of the amplitude of the vacuum–vacuum transitions Z (or, in other words, of the partition function) of a closed string. This corresponds to the calculation of functional integrals over all closed Riemann surfaces S_p of genus p (Riemann surfaces with p handles)

$$Z = \sum_{p=0}^{\infty} Z_p. \tag{1.13}$$

Fig. 2

Here

$$Z_p = \int DgDx \exp\{-A[x,g]\}, \tag{1.14}$$

where the action $A[x, g]$ is given by the formula (1.7).

---

[1] Note that there exists a supersymmetric generalization of a string—a superstring with the ordinary fields $x^\mu(\sigma,\tau)(\mu = 0,\ldots,D-1)$; the theory also contains the anticommuting fields $\psi^\mu(\sigma,\tau)$. Such a theory is conformally invariant in the case $D = 10[10]$.

Thus, we arrive at the problem of calculating the functional integral (1.14). Integration over the coordinates x^μ is a standard problem, since the x^μ-integral is Gaussian. As for integration over metrics on the Riemann surfaces, it is a nontrivial problem.

Let us denote by U_p the space of all metrics on the Riemann surfaces S_p of genus p. The group

$$H_p = \mathrm{Diff}_p \times \mathrm{Conf}_p \tag{1.15}$$

naturally acts on it. Here Diff_p is the diffeomorphism group of the Riemann surface S_p, and Conf_p is the conformal group of this surface S_p. Under the action of the group H_p, the space U_p foliates into orbits (see Fig. 3), and, in fact, the functional integral (1.14) reduces to the integral over the factor space

$$M_p = U_p/H_p, \tag{1.16}$$

known as the moduli space of Riemann surfaces of genus p. Its dimension has been shown by Riemann to be finite and equal to 0 for $p = 0$, equal to 2 for $p = 1$, and equal to $6p - 6$ for $p \geq 2$.

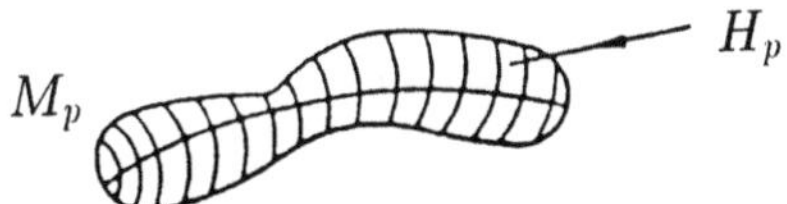

Fig. 3

It is shown in the papers by Teichmüller, Ahlfors, and Bers (see, e.g., Bers [1972] and Schumacher [1990]) that the space M_p has a natural complex structure. Moreover, according to Mumford [1977], the space M_p is algebraic.

Let $p \geq 2$, and $y_1, \dots, y_{3p-3}$ be the local holomorphic coordinates of a point in M_p. Then, after extraction of the volume of the gauge group H_p, the expression for the partition function (1.14) can be rewritten as

$$Z_p = \int d\mu_p(y_1, \dots, y_{3-p3}; \ \bar{y}_1, \dots, \bar{y}_{3p-3}). \tag{1.17}$$

Furthermore, as was shown in the important papers by Belavin and Knizhnik [1986], the measure $d\mu_p(y, \bar{y})$ of interest has the form

$$d\mu_p(y, \bar{y}) = \left(\frac{i}{2}\right)^{3p-3} (\det \mathrm{Im}\, \hat{\tau})^{-13} d\nu_p(y) \wedge \overline{d\nu_p(y)}, \tag{1.17'}$$

where $\hat{\tau}(y, \bar{y})$ is the matrix of the periods of the Riemann surface with the coordinates, $y_1, y_{3p-3}, \bar{y}_1, \dots, \bar{y}_{3p-3}$, in M_p, and $d\nu_p(y)$ is the holomorphic $(3p - 3, 0)$-form

$$d\nu_p(y) = F(y) dy_1 \wedge \cdots \wedge dy_{3p-3}. \tag{1.18}$$

As for the holomorphic form $d\nu_p(y)$, it is nowhere vanishing inside the space M_p. The space M_p is noncompact, and when points of this space approach infinity (which correspond to degeneracy of Riemann surfaces), the $d\nu_p(y)$ form has a pole of second order.

It can be readily seen that the form $d\nu_p(y)$ is uniquely defined by these conditions up to a constant factor. This makes it possible to express $d\nu_p(y)$ in the case of $p = 2$, 3, and 4 by theta-functions (Belavin, Knizhnik, Morozov and Perelomov [1986]). In Manin [1986], the author used the results of the paper by Faltings [1984] and the above-mentioned properties of the measure $d\nu_p(y)$ to express it by theta-functions and Abelian differentials by means of a complicated formula, valid for an arbitrary genus. A somewhat simpler expression was given in the paper by Beilinson and Manin [1986].

Thus, the construction of the partition function Z in the closed string theory is equivalent to the calculation of the functional integral (1.14), where the action A is given by the integral over the Riemann surface (1.7). That is why we need some information about Riemann surfaces. This information is found in the next chapter.

Chapter 2
Riemann Surfaces

2.1. Basic Facts

In this chapter, we recall some facts from the theory of Riemann surfaces necessary for further development. The interested reader can find details and more detailed information in the monographs by Farkas and Kra [1980], Fay [1973], Griffiths and Harris [1978], Mumford [1983; 1984], and Arbarello, Cornalba, Griffiths and Harris [1986].

Let S_p be a compact closed oriented surface of genus p, i.e., the surface with p handles. First of all, let us recall some facts about the topology of the surface S_p.

The homology groups H_j $(j = 0, 1, 2)$ of the surfaces S_p are very simple:

$$H_0(S_p; \mathbb{Z}) = \mathbb{Z}, \qquad H_1(S_p; \mathbb{Z}) = \mathbb{Z}^{2p}, \qquad H_2(S_p, \mathbb{Z}) = \mathbb{Z}, \qquad (2.1)$$

which is obvious from geometrical considerations. Hence, we get the expression for the Euler characteristics of the surface S_p,

$$\chi = b_0 - b_1 + b_2 = 1 - 1\,2p + 1 = 2(1 - p). \qquad (2.2)$$

Here b_j $(j = 0, 1, 2)$ are Betti numbers of S_p.

Furthermore, for any pair of 1-cycles a and b on S_p, the intersection index $a \cdot b$ is defined as

$$a \cdot b = -b \cdot a, \qquad (2.3)$$

and thereby the integer skew-symmetrical bilinear form is defined on $H_1(S_p, \mathbb{Z})$. Though there is no standard basis, on $H_1(S_p, \mathbb{Z})$, there always exists the basis $\{C_j\} = \{a_1, \dots, a_p; b_1, \dots, b_p\}$, which satisfies the condition

$$a_i \cdot a_j = b_i \cdot b_j = 0, \qquad a_i \cdot b_j = \delta_{ij}, \quad i, j = 1, \dots, p, \tag{2.4}$$

where δ_{ij} is the Kronecker symbol.

In a more invariant form,

$$C_i \cdot C_j = J_{ij}, \quad i, j = 1, \dots, 2p, \tag{2.5}$$

where the matrix J is of the block form

$$J = \begin{pmatrix} 0 & I \\ -I & 0 \end{pmatrix}, \tag{2.6}$$

and I is the unit matrix of order p.

We refer to such a basis as a canonical one. Any two canonical bases are related by linear transformations generated by an integer-valued matrix A that leaves the matrix J invariant

$$\tilde{C} = C \cdot A^{-1}, \qquad A'JA = J. \tag{2.7}$$

Thus matrix A belongs to the group $S_p(2p, \mathbb{Z})$ of integer symplectic matrices or, briefly, to the modular group.

In the space of the one-dimensional cohomologies $H^1(S_p)$, we may define the dual basis $\gamma = \{\gamma^j\} = \{\alpha^1, \dots, \alpha^p; \beta^1, \dots, \beta^p\}$, which satisfies the condition

$$\langle \gamma^j, C_k \rangle = \int_{C_k} \gamma^j = \delta_k^j, \quad j, k = 1, \dots, 2p. \tag{2.8}$$

The next important fact is that the surface S_p is a complex manifold. In order to see this, we introduce on S_p any Riemannian metric g_{ab}. Then, in the vicinity U of any point of the surface, this metric can be expressed in the conformal form

$$ds^2 = \exp(\varphi) \left(d\xi_1^2 + d\xi_2^2 \right). \tag{2.9}$$

We may now introduce the local complex coordinates $z = \xi^1 + i\xi^2$, $\bar{z} = \xi^1 - i\xi^2$; and the metric takes the form

$$ds^2 = \exp(\varphi)dzd\bar{z} = g_{z\bar{z}}dzd\bar{z}. \tag{2.10}$$

A more invariant method of introducing the local complex coordinates makes use of the complex structure tensor

$$J_b^a = \sqrt{|g|}\, g^{ac}\varepsilon_{cb}, \tag{2.11}$$

satisfying the condition

$$J^2 = -I. \tag{2.12}$$

The local complex coordinates are then given by the solution of equation

$$J_b^a \partial_a z = i\partial_b z, \quad \partial_a = \partial/\partial\xi^a. \tag{2.13}$$

It can be readily seen that if we introduce on one vicinity U the coordinate z and on the other vicinity V another local complex coordinate w, the transition function from z to w on $U \cap V$ is an analytical function. Thereby the complex coordinate z is globally defined by the whole surface S_p, and the surface S_p is a complex manifold.

Returning to the consideration of the space $H^1(S_p)$, we may now construct linear combinations of 1-form α^j and β^k, which are holomorphic and antiholomorphic. It is convenient to normalize the holomorphic 1-differentials $\omega_j = f_j(z)dz$ (or simply holomorphic differentials), for $j = 1,\ldots,p$, by the condition

$$\int_{a_i} \omega_j = \delta_{ij}. \tag{2.14}$$

Then the integrals over the remaining b-cycles define the matrix of order p,

$$\int_{b_i} \omega_j = \tau_{ij}, \tag{2.15}$$

which is called the period matrix of the Riemann surface S_p. It is known that the matrix τ is symmetric and its imaginary part is a positively defined matrix. If we go from the basis $\{a_j, b_j\}$ to the canonical basis $\{\tilde{a}_j, \tilde{b}_j\}$, the period matrix transforms by the linear fractional transformation

$$\tau \to \tilde{\tau} = (A\tau + B)(C\tau + D)^{-1}. \tag{2.16}$$

Furthermore, the formulae of differential geometry on the Riemann surface written in terms of complex coordinates and the metric agreeing with the complex structure are significantly simplified.

Note first of all that by using the metrics (2.10), one can reduce any tensor on the Riemann surface to the tensor of the type (p, q), i.e., to the tensor of the type

$$\underbrace{t_{z,\ldots,z}}_{p\text{-times}}, \underbrace{\bar{z},\ldots,\bar{z}}_{q\text{-times}}. \tag{2.17}$$

Such a tensor determines the (p, q)-differential

$$t_{z,\ldots z,\, \bar{z},\ldots,\bar{z}}\, dz\ldots dz\, d\bar{z}\ldots d\bar{z}, \tag{2.18}$$

whose transformation law under coordinate changes is determined by

$$\tilde{t}_{z,\ldots,z,\bar{z},\ldots,\bar{z}}\, d\tilde{z}\ldots d\tilde{z}\, d\bar{\tilde{z}}\ldots d\bar{\tilde{z}} = t_{z,\ldots,z,\, \bar{z},\ldots,\bar{z}}\, dz\ldots dz\ldots d\bar{z}\ldots d\bar{z}. \tag{2.19}$$

In other words,

$$\tilde{t}_{z,\ldots,z,\bar{z},\ldots,\bar{z}} = \left(\frac{d\tilde{z}}{dz}\right)^{-p} \left(\frac{d\bar{\tilde{z}}}{d\bar{z}}\right)^{-q} t_{z,\ldots,z,\bar{z},\ldots,\bar{z}}. \tag{2.20}$$

Note that the complexified tangent space to S_p may be decomposed into holomorphic and antiholomorphic subspaces, i.e., the space of holomorphic (respectively, antiholomorphic) vector fields. Analogously, the dual space of 1-form (or 1-differentials) can be expanded into subspaces of holomorphic and antiholomorphic differentials. The holomorphic differentials $f(z)dz$ generate a holomorphic line bundle over S_p called the canonical bundle K.

Thus, any tensor of the type $(p, 0)$ on the surface S_p can be considered as a section of the p-degree of the bundle K, K^p. Holomorphic line bundles will be important for us, and we continue with a more detailed consideration of them.

2.2. Holomorphic Line Bundles on Riemann Surfaces

We give here two different descriptions of holomorphic line bundles: one in terms of transition functions and the other in terms of divisors.

Recall that a holomorphic line bundle L over the complex manifold M may be determined in terms of an M-covering by the vicinities $\{U_\alpha\}$ together with the mappings Φ_α, which define local trivializations:

$$\Phi_\alpha : L|_{U_\alpha} \to U_\alpha \times C. \tag{2.21}$$

The mappings Φ_α determine the transition functions

$$g_{\alpha\beta} = \Phi_\alpha \cdot \Phi_\beta^{-1}, \tag{2.22}$$

which, as can be easily verified, satisfy the co-cycle condition

$$g_{\alpha\beta} g_{\beta\gamma} g_{\gamma\alpha} = 1. \tag{2.23}$$

It is useful to take into consideration that the function $g_{\alpha\beta}$ may be considered as a section of a bundle of holomorphic functions $\mathcal{O}$ over the vicinity $(U_\alpha \cap U_\beta)$.

Let us give the definition of this term that is a generalization of the bundle term for the case when, roughly speaking, the dimension of fibers may vary.

Definition. Let $\{U_\alpha\}$ be a covering of the space M. We say that the sheaf F is given on M if for any U the Abelian group F_U is given, which is called the group of section F over U, so that the following conditions are fulfilled:

1. If $W \subseteq V \subseteq U$, then $z_{U,W} = z_{VW} \cdot z_{UV}$, where z_{UV} is the map $F(U) \to F(V)$, induced by the insertion $V \subseteq U$.
2. If $\delta_1, \delta_2, \ldots$ are sections over $U_1, U_2, \ldots$, respectively, and each pair δ_i, δ_j has the same restriction on $U_i \cap V_j$, then each δ_i is a restriction of some section over $U_1 \cup V_2 \cup \cdots$.

3. If ρ is the section of the bundle F over $U \cup V$ that goes over into identity at a restriction both on U and V, then ρ is the identical section.

It can now be easily seen that the transition functions $g_{\alpha\beta} = \Phi_\alpha \cdot \Phi_\beta^{-1}$ are the sections of a bundle of holomorphic functions $\mathcal{O}$ over the vicinity $U_\alpha \cap U_\beta$.

One may also consider other trivializations by taking the bundle of holomorphic functions $f_\alpha \in \mathcal{O}^*(V_\alpha)$ that do not vanish on V_α and by then defining $\Phi'_\alpha = f_\alpha \Phi_\alpha$ with $h_{\alpha\beta} = f_\alpha/f_\beta$ as transition functions.

Let us introduce the notations: $H^0(M, \mathcal{O}^*)$ is the space of sections of a line bundle L on M. Furthermore, the 1-cocycles $g_{\alpha\beta} \in \mathcal{O}^*(U_\alpha \cap U_\beta)$ define the elements of the group—the first cohomology group $H^1(M, \mathcal{O}^*)$. In fact, $H^1(M, \mathcal{O}^*)$ is the group of classes of isomorphic line bundles over M, i.e., the element of the group $H^1(M, \mathcal{O}^*)$ defines the line bundle up to isomorphism.

In order to obtain more information about the group $H^1(M, \mathcal{O}^*)$, consider the exact sequence of sheaves

$$0 \to Z \xrightarrow{i} \mathcal{O} \xrightarrow{\exp} \mathcal{O}^* \to 0, \tag{2.24}$$

where Z is a constant sheaf, i.e., $Z(V_\alpha) = Z$ for all V_α, i is the mapping of insertion Z into $\mathcal{O}$, and exp is the exponential mapping, i.e., if $f \in \mathcal{O}(V_\alpha$, then $\exp(f) = e^{2\pi i f}$. It is the case to see that $\mathrm{Im}(i) = \mathrm{Ker}(\exp)$ and, consequently, the sequence (2.24) is exact.

Furthermore, any exact sequence of the type (2.24) corresponds to a long exact sequence of cohomology groups (Arbarello, Cornalba, Griffiths and Harris [1986]). In this case, the consequence has the form

$$0 \to H^0(M, Z) \to H^0(M, \mathcal{O}) \to H^0(M, \mathcal{O}^*) \to$$

$$\to H^1(M, Z) \to H^1(M, \mathcal{O}) \to H^1(M, \mathcal{O}^*) \to$$

$$\to H^2(M, Z) \to H^2(M, \mathcal{O}) \to H^2(M, \mathcal{O}^*) \to \cdots \tag{2.25}$$

The first line in (2.25) gives the exact sequence of Abelian groups

$$0 \to Z \to C \to C^* \to 0, \tag{2.26}$$

which is due to the known fact that global holomorphic functions on M are by necessity constants.

And, since for the case of the Riemann surface $(M = S_p)$,

$$H^2(S_p; Z) \simeq Z; \quad H^1(S_p; Z) \simeq Z^{2p},$$

and

$$H^1(S_p; \mathcal{O}) \simeq C^p; \quad H^2(S_p; \mathcal{O}) = 0, \tag{2.27}$$

we have also

$$0 \to H^1(S_p; \mathcal{O})/H^1(S_p; Z) \to H^1(S_p; \mathcal{O}^*) \to H^2(S_p; Z) \to 0, \qquad (2.28)$$

or

$$0 \to C^p/Z^{2p} \to H^1(S_p; \mathcal{O}^*) \to Z \to 0. \qquad (2.29)$$

Thus, we see that the $H^1(S_p; \mathcal{O}^*)$ group (usually denoted by $\mathrm{Pic}(S_p)$) is not connected, and the connected component of unity is the complex torus

$$T^p = C^p/Z^p, \qquad (2.30)$$

which is usually called the Jacobian $J(S_p)$ of the surface S_p.

Thus, we have obtained the complete classification of line bundles on an arbitrary Riemann surface S_p: Such a bundle L is characterized by the integer number $\deg L$—the bundle degree, and by the Jacobian point $J(S_p)$ of the complex torus $T^p = C^p/Z^{2p}$.

If a metric will be given on the bundle L, then the bundle degree can be written as the integral

$$\deg(L) = \int_{S_p} c_1(L), \qquad (2.31)$$

where $c_1(L)$ is called the first Chern class of the bundle L and is presented as the curvature form of the metric on L. Thus, the line bundle on S_p is completely characterized by its first Chern class and by the element of the group

$$H^1(S_p; \mathcal{O})/H^1(S_p; Z) \simeq C^p/Z^{2p} \qquad (2.32)$$

of the group of flat holomorphic line bundles on S_p. This is the complex torus T^p and, in fact, even the Abelian manifold $J(S_p)$ (or simply Jacobian) of the Riemann surface S_p. Note also that the first Chern class may be interpreted as an obstacle for global trivialization of the bundle L.

Note one more useful formula:

$$\deg(L_1 \otimes L_2) = \deg L_1 + \deg L_2. \qquad (2.33)$$

2.3. Divisors

Another (and very useful) description of line bundles is the description in terms of divisors. The divisor D on S_p is the formal finite sum of points on S_p,

$$D = \sum_j n_j P_j, \qquad (2.34)$$

where n_j are integer numbers, and $P_j \in S_p$.

Divisors on S_p form an Abelian group denoted by $\text{Div}(S_p)$. The relationship with line bundles can be established as follows. Suppose that we restrict the divisor D on the vicinity U_α in S_p. Then one can always find a meromorphic function f_α, the zeroes and poles of which coincide with the restriction D on U_α, and we may construct the transition function $g_{\alpha\beta} = f_\alpha/f_\beta \in O^*(U_\alpha \cap U_\beta)$. The line bundle given by such transition functions $\{g_{\alpha\beta}\}$ is called the line bundle associated with the divisor D and is denoted by $[D]$. If we choose another system of functions $\{f'_\alpha\}$, connected with same D, then

$$f_\alpha/f'_\alpha \in \mathcal{O}^*(U_\alpha) \quad \text{and} \quad g'_{\alpha\beta} = f'_\alpha/f'_\beta = g_{\alpha\beta}(f'_\alpha/f_\alpha)(f'_\beta/f_\beta)^{-1},$$

so that these transition functions define the isomorphic line bundle. Hence it also follows that if D and D' are two divisors that correspond to systems of functions $\{f_\alpha\}$ and $\{f'_\alpha\}$, then the divisor $D + D'$ corresponds to the system of functions $\{f_\alpha f'_\alpha\}$. Consequently,

$$[D + D'] = [D] \otimes [D'], \tag{2.35}$$

$$[-D] = [D]^{-1}. \tag{2.36}$$

Now let L be a line holomorphic bundle and s be some meromorphic section of this bundle. Then

$$\text{div}(s) = \sum n_i P_i - \sum m_j Q_j, \tag{2.37}$$

where the $\{P_i\}$ are zeroes of section s of order n_i, and the $\{Q_j\}$ are the poles of a section of order m_j. Any other meromorphic section is of the form $s' = f \cdot s$, where f is a meromorphic function on S_p. Here the divisor of section s' differs from that of section s by the divisor of the meromorphic function (f). Thus, there arises the real equivalence relation between divisors. Namely, two divisors D and D' are called linearly equivalent $D \sim D'$, if $D = D' + (f)$; and $[D] \simeq [D']$. Thus, we see that

$$[\ ] : D \to [D] \tag{2.38}$$

is the map from the divisor group into the group of isomorphic classes of line bundles on the modulus of linear equivalence. As has already been noted, this group is called the Picard group $\text{Pic}(S_p)$ and is isomorphic to the group $H^1(S_p, \mathcal{O}^*)$.

Thus, we have the mapping

$$[\ ] : \text{Div}(S_p) \to H^1(S_p, \mathcal{O}^*). \tag{2.39}$$

It is convenient to use divisors for studying the properties of holomorphic and meromorphic sections of line bundles. To this end, let us associate to any divisor D the space $\mathcal{L}(D)$, i.e., the space of such meromorphic functions on S_p for which

$$D + (f) \geq 0. \tag{2.40}$$

In other words, the space $\mathcal{L}(D)$ is the space of such meromorphic functions that are holomorphic on $S_p/(\cup_j P_j)$ with $\mathrm{ord}_{p_i}(f) \geq -n_i$, where $\mathrm{ord}_{p_i}(f)$ is the order zero or the pole of the function f at the point P_i.

Now let s_0 be the global meromorphic section of the bundle $[D]$ with the divisor $D = (s_0)$. Then for any global holomorphic section s of the bundle $[D]$, the quotient s/s_0 is a meromorphic function on S_p with the divisor

$$(s/s_0) = (s) - (s_0) \geq D, \tag{2.41}$$

i.e.,

$$s/s_0 \in \mathcal{L}(D) \quad \text{and} \quad (s) = D + (s/s_0). \tag{2.42}$$

On the other hand, if $f \in \mathcal{L}(D)$, then the section $s = f \cdot s_0$ of the line bundle $[D]$ is holomorphic, whence it follows that multiplication by s_0 provides an identification with the holomorphic section space of the bundle $[D]$:

$$\mathcal{L}(D) \stackrel{s_0}{\to} H^0(S_p, \mathcal{O}([D])). \tag{2.43}$$

More exactly, let $D = \sum n_j P_j$ be a positive divisor on S_p, and $s_0 \in H^0(S_p, \mathcal{O}[D])$ be the section of the bundle $[D]$ with the divisor D. Then the tensor multiplication by s_0 gives identification with the meromorphic functions on S_p with poles of order $\leq n_j$ at the points P_j and holomorphic sections of the bundle $[D]$.

Let L be a line bundle. Let us denote by $\mathcal{L}(D)$ the sheaf of meromorphic sections of this bundle with poles of order $\leq n_j$ at the point P_j, and by $\mathcal{L}(-D)$ the sheaf of sections with zeroes of order $\geq n_j$ at the point P_j.

We see now that by multiplying by the quantity s_0 or by the inverse quantity s_0^{-1}, we get identification (2.39):

$$\mathcal{L}(D) \simeq \mathcal{O}(L \otimes [D]), \qquad \mathcal{L}(-D) \simeq \mathcal{O}(L \otimes [-D]), \tag{2.44}$$

It can be easily seen that the divisor defines the line bundle and its section. Namely, the function that represents it has zeroes and poles at the points corresponding to this divisor. Note the special role of the so-called unit section s_0 that corresponds to a divisor consisting of a point P. We may define the divisor $D = P$ by using the functions $f(z) = z$ in the neighbourhood U of the point P, where z is the local coordinate on U, such that $f(P) = 0$. Let us introduce $V = M - \{P\}$ and define the transition function $g_{UV} = z$ on $U \cup V$. Then the unit section $1_{\mathcal{O}(P)}$ can be defined similarly to that mentioned above, supposing $s_0 = 1$ and $s = g_{UV} s_0$. Then $1_{\mathcal{O}(P)}$ takes the value 1 on U and has a zero of the first order at $z = P$. It can be meromorphically expanded on the whole surface S_p. Analogously, one can define $1_{\mathcal{O}(-P)}$, which has a first-order pole at $z = P$. Finally, one may define the unit section for an arbitrary divisor D by means of forming tensor products.

Chapter 3
Cauchy–Riemann Operators on Riemann Surfaces

As has been previously noted, the surface S_p is the complex manifold, so that it is convenient to describe it using the complex coordinate z. Thus, let S_p be the Riemann surface of genus p. As is well known (see, e.g., Griffiths and Harris [1978]; Chern [1956]), there are the so-called isothermal coordinates ξ_1 and ξ_2 on S_p, i.e., coordinates such that the metric g is globally conformally flat:

$$ds^2 = \exp(\varphi)(d\xi_1^2 + d\xi_2^2). \tag{3.1}$$

The complex coordinate z is given by the formula

$$z = (\xi_1 + i\xi_2), \qquad \bar{z} = (\xi_1 - i\xi_2), \tag{3.2}$$

and the metric (3.1) takes the form

$$ds^2 = \exp(\varphi)dzd\bar{z} = g_{z\bar{z}}dzd\bar{z}. \tag{3.3}$$

In the case under consideration, it is very convenient to classify various quantities according to irreducible representations of the two-dimensional rotation group—the SO(2) group, which is isomorphic to the group $U(1)$. Under the action of this group, $g : dz \to \exp(i\alpha)dz$ so that the vector quantities of the type V^z and the 1-forms of the type $V_{\bar{z}}d\bar{z}$ can be prescribed by the number $+1$, while the vectors of the type $V^{\bar{z}}$ and the 1-forms of the type $V_z dz$ can be prescribed by the number (-1).

The metric tensor $g_{z\bar{z}}$ allows one to transform the upper $\bar{z}$-indices into lower z-indices. For instance, $V_z = g_{z\bar{z}}V^{\bar{z}}$, and we can establish an isomorphism between the space of all vectors $V^{\bar{z}}$ and all forms of $V_z dz$. An analogous isomorphism exists between the vectors V^z and the forms $V_{\bar{z}}d\bar{z}$. Thus, using the metric tensor $g_{z\bar{z}}$ and the tensor inverse to it, $g^{z\bar{z}}$, we can transform any tensor into a tensor with n upper and m lower z-indices. When transforming $z \to \exp(i\alpha)z$ of the $U(1)$ group, such a tensor is multiplied by the number $\exp[i(n - m)\alpha]$, and we denote the space of all such tensors by T^{n-m}. The T^1 is a complexified tangent space to S_p or a canonical line bundle K. The space T^n is the n-th degree of the canonical line bundle $T^n \simeq K^{\otimes n}$. Here, when transforming the coordinates $z \to w(z)$, the transformation law of such a tensor has the form

$$T(w) = (dw/dz)^n T(z). \tag{3.3'}$$

It is useful to define the two covariant derivatives

$$\nabla_n^z : T^n \to T^{n+1}, \qquad \nabla_n^z T = g^{z\bar{z}}\partial_{\bar{z}}T, \tag{3.4}$$

and

$$\nabla_z^n : T^n \to T^{n-1}, \qquad \nabla_z^n T = (g_{z\bar{z}})^n \partial_z \left((g^{z\bar{z}})^n T\right),$$

$$\nabla_z^n T = \left(\partial_z - n g^{z\bar{z}}(\partial_z g_{z\bar{z}})\right)T = \left(\partial_z - n(\partial_z \varphi)\right)T. \tag{3.5}$$

Let us now give one more useful formula for the scalar curvature

$$R = -\exp(-\varphi)\partial_z\partial_{\bar z}\varphi. \tag{3.6}$$

Taking this formula into account, the Ricci identity for the tensor $T \in \mathcal{T}^n$ takes the form

$$[\nabla_z^{n+1}\nabla_n^z - \nabla_{n-1}^z\nabla_z^n]T = \frac{n}{2}RT. \tag{3.7}$$

Correspondingly, for the Laplace operator acting on the scalar functions, we have

$$\Delta = \nabla_{-1}^z\nabla_z^0 + \nabla_z^1\nabla_0^z = 2\nabla_{-1}^z\nabla_z^0 = 2\nabla_z^1\nabla_0^z. \tag{3.8}$$

In the general case, it is useful to introduce two Laplace operators $\Delta_n^\pm : \mathcal{T}^n \to \mathcal{T}^n$ according to formulae

$$\Delta_n^+ = \nabla_z^{n+1}\nabla_n^z, \qquad \Delta_n^- = \nabla_{n-1}^z\nabla_z^n. \tag{3.9}$$

Then, as can be readily seen, the nonzero eigenvalues in these operators coincide.

For the zero modes, it is convenient to use the duality property of cohomology groups and the Riemann–Roch theorem (Griffiths and Harris [1978]).

According to the so-called Serre duality, we have

$$H^1(S_p, L) \simeq H^0(S_p, \Omega \otimes L^{-1})^*. \tag{3.10}$$

On the other hand, the Riemann–Roch theorem gives

$$\dim H^0(S_p, L) - \dim H^1(S_p, L) = \deg L - p + 1, \tag{3.11}$$

and we obtain with (3.10):

$$\dim H^0(S_p, L) - \dim H^0(S_p, \Omega \otimes L^{-1}) = \deg L - p + 1. \tag{3.12}$$

Hence, in particular, it follows that if we take L as the trivial bundle for which $\dim H^0(S_p, L) = 1$, we have

$$\dim H^0(S_p, \Omega) = p. \tag{3.13}$$

On the other hand, putting $L = \Omega$, we get

$$\deg \Omega = 2p - 2. \tag{3.14}$$

Now we are able to calculate the index of the operator ∇_n^z:

$$\begin{aligned}
\mathrm{Ind}\,\nabla_n^z &= \dim \mathrm{Ker}\,\nabla_n^z - \dim \mathrm{Coker}\,\nabla_n^z \\
&= \dim H^0(S_p, \Omega^{\otimes n}) - \dim H^1(S_p, \Omega^{\otimes n}) \\
&= \dim H^0(S_p, \Omega^{\otimes n}) - \dim H^0(S_p, \Omega^{\otimes(1-n)}) \\
&= n\big(\deg(\Omega)\big) - p + 1 = (2n - 1)(p - 1).
\end{aligned} \tag{3.15}$$

212   A.Yu. Morozov and A.M. Perelomov

Furthermore, owing to the Kodaira theorem (Griffiths and Harris [1978]), if $\deg L < 0$, then $\dim H^0(S_p, L) = 0$. Therefore, if $n > 1$, then $\dim \operatorname{Coker} \nabla_n^z = 0$, and, consequently,

$$\dim \operatorname{Ker} \nabla_n^z = (2n - 1)(p - 1), \quad n > 1. \tag{3.16}$$

Let us now proceed with the description of the local properties of the metric space on the Riemann surface S_p of genus p.

Let the metric have the form $g(\xi) = g_{ab}(\xi)d\xi^a d\xi^b$ in the coordinates ξ^a $(a = 1, 2)$. All the physical models considered in connection with string theory are invariant with respect to simultaneous coordinate changes,

$$\xi \to \tilde{\xi}(\xi), \tag{3.17}$$

and the metrics $g \to \tilde{g}$ connected by the relation

$$\tilde{g}(\tilde{\xi}) = g(\xi), \tag{3.18}$$

i.e.,

$$\tilde{g}_{ab}(\tilde{\xi})d\tilde{\xi}^a d\tilde{\xi}^b = g_{ab}(\xi)d\xi^a d\xi^b, \tag{3.19}$$

or

$$\tilde{g}_{ab}(\xi)\frac{\partial \tilde{\xi}^a}{\partial \xi^{a'}}\frac{\partial \tilde{\xi}^b}{\partial \xi^{b'}} + (\tilde{\xi} - \xi)^c \partial_c \tilde{g}_{a'b'}(\xi) = g_{a'b'}(\xi). \tag{3.20}$$

This invariance allows one to fix the complex coordinates $\xi, \bar{\xi}$ on the surface such that the metric has the conformal form

$$g(\xi, \bar{\xi}) = \rho(\xi, \bar{\xi})d\xi d\bar{\xi}. \tag{3.21}$$

The arbitrary infinitesimal variation of the metric (3.21) can be written as:

$$\delta g(\xi, \bar{\xi}) = \delta g_{\xi, \bar{\xi}} d\bar{\xi} d\xi + \delta g_{\bar{\xi}, \bar{\xi}}(d\bar{\xi})^2 + \delta g_{\xi, \xi}(d\xi)^2. \tag{3.22}$$

Such a transformation defines the vector in the tangent space of all metrics g and its expansion into three components. However, this expansion is nonorthogonal relative to the scalar product

$$||\delta g_{ab}||^2 = \int d^2\xi \rho \delta g_{ab} \delta g^{ab} \tag{3.23}$$

in the metric space.

In order to give an orthogonal expansion, let us go over to the new coordinates $\zeta = \xi + \varepsilon(\xi, \bar{\xi})$, in which the metric takes the conformal form

$$g + \delta g = \tilde{\rho}(\zeta, \bar{\zeta})d\zeta d\bar{\zeta} = \tilde{\rho}(\zeta, \bar{\zeta})\left|\frac{\partial \zeta}{\partial \xi}\right|^2 \left|d\xi + \frac{\partial \zeta/\partial \bar{\xi}}{\partial \zeta/\partial \xi}d\bar{\xi}\right|^2. \tag{3.24}$$

Now rewrite (3.22) as

$$\delta g(\xi, \bar{\xi}) = \delta \rho d\xi d\bar{\xi} + \rho \eta d\bar{\xi}^2 + \rho \bar{\eta} d\xi^2, \tag{3.25}$$

where we denote by η the value of the type $(-1,1)$—the so-called Beltrami differential.

Comparing (3.24) and (3.25), we get

$$\frac{\partial \zeta}{\partial \bar{\xi}} = \eta \frac{\partial \zeta}{\partial \xi}, \quad \text{i.e.,} \quad \bar{\partial}\varepsilon = \eta + O(\varepsilon^2), \tag{3.26}$$

$$\frac{\partial \bar{\zeta}}{\partial \xi} = \bar{\eta} \frac{\partial \bar{\zeta}}{\partial \bar{\xi}}, \quad \text{i.e.,} \quad \partial \bar{\varepsilon} = \bar{\eta} + O(\varepsilon^2). \tag{3.27}$$

It has been shown already by Gauss that equations (3.26) are locally integrable at any $\eta, \bar{\eta}$. This does not, however, mean that they have global solutions as continuous vector fields $\varepsilon(\xi, \bar{\xi})$ for all Beltrami differentials η on non simply-connected surfaces of genus $p > 1$. The subspace of Beltrami differentials orthogonal to differentials of the type $\bar{\partial}\varepsilon$ with continuous ε is given by the equation

$$\langle \eta, \bar{\partial}\varepsilon \rangle = \int \rho \bar{\eta} \, \bar{\partial}\varepsilon \, d^2\xi = 0 \tag{3.28}$$

for all continuous ε.

Such $\eta(\xi, \bar{\xi})$ are arbitrary linear combinations of antiholomorphic quadratic differentials $f_\alpha(\xi)$, $\alpha = 1, \ldots, n_2$, $n_2 = 3p - 3$, for $p \geq 2$. The Beltrami differentials of general type are written as follows:

$$\eta = \sum_{\alpha=1}^{n_2} \delta y_\alpha \left(C_{\alpha\beta} \frac{\overline{f_\beta(\xi)}}{\rho(\xi, \bar{\xi})} + \bar{\partial}\varepsilon_\alpha(\xi, \bar{\xi}) \right) + \bar{\gamma}\tilde{\varepsilon}(\xi, \bar{\xi})$$

$$= \sum_{\alpha=1}^{n_2} \delta y_\alpha \eta_\alpha(\xi, \bar{\xi}) + \bar{\partial}\tilde{\varepsilon}(\xi, \bar{\xi}), \tag{3.29}$$

where $\tilde{\varepsilon}, \varepsilon_\alpha$ are continuous vector fields.

The Beltrami differentials

$$\eta_\alpha = \left(C_{\alpha\beta} f_\beta(\xi)/\rho \right) + \bar{\partial}\varepsilon_\alpha \tag{3.30}$$

are chosen dual to the quadratic differentials

$$\int \eta_\alpha f_\beta d^2\xi = \delta_{\alpha\beta}. \tag{3.31}$$

Then in (3.29)

$$(C^{-1})_{\alpha\beta} = N_2^{\alpha\beta} \equiv \int \frac{f^\alpha \overline{f^\beta}}{\rho} d^2\xi. \tag{3.32}$$

The variation of the metric (3.25) connected with the discontinuous vector fields (with the first term in (3.29)) corresponds to a variation of the complex structure of the surface. The parameters $y_\alpha, \bar{y}_\alpha$ are called holomorphic moduli of the complex structure or simply the Riemann surface moduli. Various choices of moduli $y_\alpha, \bar{y}_\alpha$ distinguished by the choice of vector fields are possible. Sometimes one use moduli connected with Beltrami differentials not satisfying the constraint (3.31).

Chapter 4
Determinants of Cauchy–Riemann Operators

As has been shown above, the index of the operator ∇_n^z is nonzero. As a consequence of this and of the fact that ∇_n^z acts on an infinite-dimensional vector space, the determinant of this operator cannot be determined in the usual way. The correct definition of such a determinant was given in Bismut and Freed [1986a,b], Quillen [1985], Knizhnik [1986], Alvarez-Gaume, Bost, Moore, Nelson and Vafa [1987], Bost [1987], and Smit [1988], which we follow below.

Let V be the finite-dimensional vector space; $\{v_j\}$, $j = 1, \ldots, n$, be a basis in it, and $A : V \to V$ be a linear transformation in V. Then the determinant A can be defined as

$$\det A \cdot (v_1 \wedge \cdots \wedge v_n) = (Av_1 \wedge \cdots \wedge Av_n). \tag{4.1}$$

This definition may be generalized to the case of the linear elliptic operator D acting between two different vector spaces V_1 and V_2 of the same dimension

$$D : V_1 \to V_2. \tag{4.2}$$

This mapping induces the mapping in the space of maximal external degree, which will be called the determinant of the operator D:

$$\det D : \textstyle\bigwedge^{\max} V_1 \to \textstyle\bigwedge^{\max} V_2. \tag{4.3}$$

We emphasize that with such a definition, $\det D$ produces not a number but an element of the one-dimensional linear space

$$\det D \in (\textstyle\bigwedge^{\max} V_1)^* \otimes (\textstyle\bigwedge^{\max} V_2). \tag{4.4}$$

In other words, the value of $\det D$ depends on the choice of the bases in the spaces V_1 and V_2.

This definition of D is also applicable to the infinite-dimensional case, provided D is a Fredholm operator and its index is zero.

In this case, let us consider the exact sequence

$$0 \to \operatorname{Ker} D \to V_1 \to V_2 \to \operatorname{Coker} D \to 0. \tag{4.5}$$

Since both spaces $\operatorname{Ker} D$ and $\operatorname{Coker} D$ are finite-dimensional, it follows that $\det D$ may be defined according to the formula

$$\det D = (\textstyle\bigwedge^{\max} \operatorname{Ker} D)^* \otimes (\textstyle\bigwedge^{\max} \operatorname{Coker} D). \tag{4.6}$$

Let us now consider the case when we have a family of operators D_x, smoothly depending on the complex parameter $x \in X$, where X is the complex manifold. Then for $x \in X$, the formula (4.6) defines a complex one-dimensional linear space. As x varies, so does this space, thereby defining the one-dimensional line bundle over X, which is called the determinant line bundle.

Let us consider an operator $\bar{\partial}$, important for the following example.

Let us determine the Laplace operator acting on the $(j,0)$ differentials (we will often call them simply j-differentials or differentials of degree j) as follows:

$$\Delta_j : \mathcal{A}_{j,0} \to \mathcal{A}_{j,0}; \quad \Delta_j \equiv \bar{\partial}^+\bar{\partial}, \quad \bar{\partial} \equiv \partial_{\bar{z}} = g_{z\bar{z}}\nabla_j^z. \tag{4.7}$$

Just this operator is of interest for the string theory. In the conformal gauge,

$$\Delta_j = \bar{\partial}^+\bar{\partial} = -\rho^{j-1}\partial\rho^{-j}\bar{\partial}. \tag{4.7'}$$

The regularized determinant is equal to

$$\mathrm{det}'_{\mathrm{Reg}}\,\Delta_j = \det N_j \det N_{1-j}|\det\bar{\partial}|^2 \exp\left(\frac{1}{12\pi}c_j S_L\right). \tag{4.8}$$

Here N_j is the matrix of order n_j composed of any possible scalar products of zero modes of the operator Δ_j, i.e., holomorphic j-differentials $\phi_\alpha^{(j)}$ ($\alpha = 1,\dots,n_j = (2j-1)(p-1)$), $\bar{\partial}\phi_\alpha^{(j)} = 0$:

$$\det N_j \equiv \det_{(\alpha\beta)}\langle\phi_\alpha^{(j)},\phi_\beta^{(j)}\rangle = \det_{(\alpha\beta)}\int\rho^{1-j}\bar{\phi}_\alpha^{(j)}\phi_\beta^{(j)}. \tag{4.9}$$

The Liouville action $S_L = \int R\frac{1}{\Delta}R\sqrt{g}$ or, in the conformal gauge,

$$S_L = \int|\partial\log\rho|^2 d^2\xi \tag{4.10}$$

(R is the two-dimensional curvature, in the conformal gauge $\sqrt{g}R = \partial\bar{\partial}\log\rho$), $c_j = 6j^2 - 6j + 1$.

First, let us put the regularization aside and write $\det\Delta_j$ as a functional integral in Grassmanian fields of j-differentials $\phi^{(j)}$. We suppose that $n_j = 0$, i.e., the zero modes Δ_j are absent and $\det N_j = 1$. We use everywhere the conformal gauge $g_{ab} = \rho\delta_{ab}$. Then

$$\det\Delta_j \equiv \int\left|D\phi^{(j)}\right|^2\exp\langle\phi^{(j)},\Delta_j\phi^{(j)}\rangle = \int\left|D\phi^{(j)}\right|^2\exp\int\rho^{-j}\left|\bar{\partial}\phi^{(j)}\right|^2$$

$$= \int\left|D\phi^{(j)}D\chi^{(1-j)}\right|^2\exp\int\chi^{(1-j)}\partial\bar{\phi}^{(j)}$$

$$\times\exp\int\bar{\chi}^{(1-j)}\partial\bar{\phi}^{(j)}\exp\int\rho^j\left|\chi^{(1-j)}\right|^2$$

$$= \int\left|D\phi D\chi e^{\int\chi\bar{\partial}\phi}\right|^2\sum_{n=0}^{\infty}\frac{1}{n!}\left(\int\rho^j|\chi|^2\right)^n$$

$$= \sum_{n=0}^{\infty}\frac{1}{n!}\int_{d^2\xi_1}\rho^j\cdots\int_{d^2\xi_n}\rho^j\left|\int D\phi D\chi e^{\int\chi\bar{\partial}\phi}\chi(\xi_1)\cdots\chi(\xi_n)\right|^2.$$

$$\tag{4.11}$$

We introduce here the auxiliary functional integration over the Grassmanian fields $(1-j)$-differentials $\chi^{(1-j)}$.

The functional integral $\int D\phi D\chi e^{\int \chi\bar\delta\phi}\chi(\xi_1)\ldots\chi(\xi_n)$ is nonzero only in the cases when n coincides with the number of zero modes of χ-holomorphic $(1-j)$-differentials, i.e., at $n = n_{1-j}$. Moreover, the $\xi_1,\ldots,\xi_n$-dependence of the integrals is completely described by a determinant composed from these zero modes $\chi_1(\xi)\ldots\chi_{n_{1-j}}(\xi)$:

$$\int D\phi D\chi e^{\int \chi\bar\delta\phi}\chi(\xi_1)\ldots\chi(\xi_{n_{1-j}}) \equiv \det \bar\partial_j \cdot \det_{(\alpha\beta)}\chi_\alpha(\xi_\beta). \tag{4.12}$$

The $\xi_1\ldots\xi_{n_{1-j}}$-independent proportionality coefficient is denoted by $\det \bar\partial_j = \det \bar\partial_{1-j}$. The n-sum in (4.11) contains just one term with $n = n_{1-j}$ and

$$\det \Delta_j = \int^{d^2\xi_1} \rho^j(\xi_1)\ldots \int^{d^2\xi_{n_{1-j}}} \rho^j(\xi_{n_{1-j}})\left|\det_{(\alpha\beta)}\chi_\alpha(\xi_\beta)\right|^2\left|\det \bar\partial_j\right|^2$$

$$= \det N_{1-j}\left|\det \bar\partial_j\right|^2. \tag{4.13}$$

In the more general case $n_j \neq 0$, it can be analogously shown that

$$\det{}' \Delta_j = \det N_j \det N_{1-j}\left|\det \bar\partial_j\right|^2, \tag{4.14}$$

and

$$\det \bar\partial_j = \det \bar\partial_{1-j} \equiv \frac{\int D\phi^{(j)} D\chi^{(1-j)} e^{\int \chi\bar\delta\phi}\phi(\zeta_1)\ldots\phi(\zeta_{n_j})\chi(\xi_1)\ldots\chi(\xi_{n_{1-j}})}{\det_{(\alpha\beta)}\phi_\alpha(\zeta_\beta)\det_{(\gamma\delta)}\chi_\gamma(\xi_\delta)}$$

$$= \left\langle\prod^{n_j}\phi(\zeta_\alpha)\prod^{n_{j-1}}\chi(\xi_\gamma)\right\rangle\Big/\det_{(\alpha\beta)}\phi_\alpha(\zeta_\beta)\det_{(\gamma\delta)}\chi_\gamma(\xi_\delta). \tag{4.15}$$

This formula for $\det \bar\partial_j$ does not contain the conformal factor ρ and includes only the holomorphic dependence on the moduli.

In the above calculations, we did not explicitly introduce the regularization. When taking the regularization into account, it is convenient to preserve the definition $\det \bar\partial_j$ for the moduli-holomorphic component of the determinant. To this end, it appears necessary to introduce the factor $\exp(\frac{c_j}{2\pi}S_L)$ in the expression for $\det'_{\mathrm{Reg}} \Delta_j$, as it is done in formula (4.8). After extracting this factor, $\det \bar\partial_j$ depends only on the complex structure on the surface, its module-dependence being holomorphic.

Consider now the question of constructing Riemann metrics on the space $\Omega^{\otimes n}$—the space of holomorphic differentials of degree n. Namely, let us describe the construction of a special metric on the Riemann surface surface induces the natural metric on the spaces $\Omega^{\otimes n}$—the so-called Arakelov metric (Arakelov [1974]). Consider first the orthonormalized basis $(\omega_1,\ldots,\omega_p)$

of holomorphic differentials in $H^0(S_p; \Omega)$. We have here the natural metric determined by the equation:

$$(\omega_j, \omega_k) = \frac{i}{2} \int_{S_p} \bar{\omega}_j \wedge w_k, \tag{4.16}$$

and require that

$$(\omega_j, \omega_k) = \delta_{jk}. \tag{4.17}$$

Now define the $(1,1)$-form ω on S_p:

$$\omega = \frac{i}{2p} \sum_{j=1}^{P} \bar{\omega}_j \wedge \omega_j. \tag{4.18}$$

The metric on S_p corresponding to this form is defined by the formula

$$ds^2 = p^{-1} \sum_{j=1}^{P} \bar{\omega}_j \otimes \omega_j \tag{4.19}$$

and is distinguished; it is called the Bergmann metric.

Consider now on S_p some line bundle L together with a hermitean scalar product on the fiber L over the point $z \in S_p$. Denote the induced norm on fibers of the bundle L by $\| \ \|$. Let s be any section of the bundle L. Then the first Chern class of this bundle is given by the formula

$$c_1(L) = \frac{\partial^2}{\partial z \partial \bar{z}} \ln \|s\|^2 dz \wedge d\bar{z}. \tag{4.20}$$

The line bundle L over S_p is called admissible with respect to the metric (4.19) if

$$c_1(L) = 2\pi i \deg(L) \cdot \omega. \tag{4.21}$$

Such a metric on the line bundle L is called the Arakelov metric. It has been introduced and studied in Arakelov [1974], where, in particular, it was proved that any line bundle over S_p possesses an admissible metric that is unique up to a numerical factor.

Fixing this metric, we may then define Green's function connected with this metric. Let P and Q be two points on the Riemann surface S_p and z be the local coordinate on S_p. Green's function is a real nonnegative function $G(P, z)$ of a variable z satisfying the conditions:

1. At $z = P$, it has the first-order zero,
2. $\frac{\partial^2}{\partial z \partial \bar{z}} \ln G(P, z) dz \wedge d\bar{z} = 2\pi i \omega,$ $\qquad\qquad$ (4.22)
   where ω is given by the formula (4.18),
3. $\int_{S_p} \ln G(P, z) \omega = 0,$ $\qquad\qquad$ (4.23)
4. $G(P, Q) = G(Q, P),$ $\qquad\qquad$ (4.24)
   i.e., $G(P, Q)$ is a symmetric function.

Furthermore, it can easily be shown that if Δ is the Laplace operator acting on the scalar function f, then

$$f(P) = \int_{S_p} \ln G(P, Q) \Delta f(Q) \omega. \tag{4.25}$$

The function $\ln G(P, Q)$ is regular everywhere except at the point $Q = P$, where it has the logarithmic singularity

$$\ln G(P, Q) \simeq \ln |z|. \tag{4.26}$$

In accordance with this, we may call the function $g(P, Q) = \ln G(P, Q)$ the Green function connected with the metric (4.19). (Explicit formulae for the Green functions are given in Chapter 7.)

Knowing the function $G(P, Q)$, we may construct the admissible metric on Ω in the following way. Consider $\mathcal{O}_{S_p}(Q)$ to be the bundle of holomorphic functions on $Q \in M$, and $\mathbf{1}_{\mathcal{O}(P)}$ to be the unit section of this bundle at the point P. Define the hermitean metric $\| \ \|$ on $\mathcal{O}_{S_p}(Q)$ by the formula

$$\|\mathbf{1}\|_{\mathcal{O}(P)} = G(P, Q) = \exp\big(g(P, Q)\big), \tag{4.27}$$

where $\|\mathbf{1}\|_{\mathcal{O}(P)}$ is the norm of the unit section $\mathbf{1}_{\mathcal{O}(P)}$, $P \in S_p$. It is clear that the metric is admissible. By creating tensor products, one may construct the admissible metric for any divisor D.

Let us choose on S_p some point P and construct the admissible metrics by using $\mathcal{O}(P)$. Let $f dz$ be 1-differential on S_p with the first-order pole at the point P. Then it is possible to introduce the norm on Ω using the formula

$$\|\Omega\| = |\operatorname{res}_P f dz| \lim_{z \to P} |z - P| / G(z, P)^{-1}. \tag{4.28}$$

By creating tensor degrees, we obtain the admissible metric on $\Omega^{\otimes n}$. These metrics are of interest first of all because they may be "transferred" to the space

$$\det R\Gamma(S_p; L) \equiv \bigwedge{}^{\max} H^0(S_p, L) \otimes (\bigwedge{}^{\max} H^1(S_p, L). \tag{4.29}$$

Such metrics on the space $\det R\Gamma(S_p, \Omega^{\otimes n})$ will be called the Faltings metrics. We present the basic theorem.

Theorem 1 (Faltings [1984]). *For any line bundle L over S_p with an admissible metric on L, there is a unique method of introducing a hermitean metric on the space $\det R\Gamma(S_p, L)$ so that the following properties are fulfilled:*

1. The isometry $L \to \tilde{L}$ induces the isometry

$$\det R\Gamma(S_p, L) \to \det R\Gamma(S_p, \tilde{L}).$$

2. If the metric on L is multiplied by the number $\alpha > 0$, then the metric on $\det R\Gamma(S_p, L)$ is multiplied by $\alpha^{\chi(L)}$,

$$\chi(L) = \dim H^0(S_p, L) - \dim H^1(S_p, L) = \deg L - p + 1. \tag{4.30}$$

3. *The metric on* $\det R\Gamma(S_p, L)$ *is compatible with that induced by the Green function on* $\mathcal{O}(D)[P]$. *More explicitly: suppose that* D_1 *and* D_2 *are divisors such that* $D = D_1 + P$, $P \in S_p$. *Then the isomorphism*

$$\det R\Gamma(S_p, (D)) \simeq \det R\Gamma(S_p(D_1)) \otimes \mathcal{O}(D)[P], \qquad (4.31)$$

which is induced by the exact sequence

$$0 \to \mathcal{O}(D_1) \to \mathcal{O}(D) \to \mathcal{O}(D)[P] \to 0, \qquad (4.32)$$

is in fact an isometry.

4. *The metric on* $\det R\Gamma(S_p, \Omega) = \wedge^p H^0(S_p, \Omega)$ *is the metric defined by the canonical scalar product on* $H^0(S_p, \Omega)$.

Following Alvarez-Gaume, Bost, Moore, Nelson and Vafa [1987], let us present the construction of the Faltings metrics on $\det R\Gamma(S_p, L)$.

But first we explain property 3 of Theorem 1. We restrict ourselves to the consideration of the case when the degree of the bundle $L \geq 2p - 1$ so that according to the Kodaira theorem, $\dim H^1(M, L) = 0$. Let $P_1, \ldots, P_r$ be r different points on S_p such that $r = d + 1 - p = \dim H^0(S_p, L)$, $L(P) = \mathcal{O}(E - \sum_{j=1}^r P_j)$, where E is the fixed divisor.

Let us now form the vector space $\otimes_{j=1}^r L[P_j] = \wedge^r \otimes_{j=1}^r L[P_j]$, which is the product of complex vector spaces of range 1. There is a natural metric $\| \ \|_n$ on this space that is defined by the metric on each component

$$\| \ \|_n = \|\alpha_1 \otimes \cdots \otimes \alpha_r\|_n = \prod_j \|\alpha_j\|. \qquad (4.33)$$

Define the Faltings metric $\| \ \|_F$ on the space $\otimes L(P_j)$ according to the formula

$$\| \ \|_F = \| \ \|_n \prod_{i<j} \frac{1}{G(P_i, P_j)}. \qquad (4.34)$$

Using the exact sequence

$$0 \to L\left(-\sum_{j=1}^r P_j\right) \to L \to \bigoplus_{j=1}^r L[P_j] \to 0, \qquad (4.35)$$

we obtain the isomorphism

$$\det R\Gamma(S_p, L) \simeq \det R\Gamma\left(S_p, L\left(-\sum_{j=1}^r P_j\right)\right) \otimes \left\{\bigoplus_{k=1}^r L[P_k]\right\}. \qquad (4.36)$$

In order to see that this isomorphism is in fact an isometry, one must show that each component in (4.33) possesses the Faltings metric on it. This may

220 A.Yu. Morozov and A.M. Perelomov

be proved by consecutive application of property 3 of Theorem 1. Removing one point P_ℓ, we obtain the isomorphism

$$\det R\Gamma(S_p, L) \simeq \det R\Gamma\left(S_p, L\left(-\sum_{j=1}^{r} P_j\right)\right)$$

$$\otimes\left\{\otimes_{i=1}^{r} L(-P_j)[P_\ell]\right\} \otimes \left\{\otimes_{k=1}^{r} L[P_k]\right\}$$

$$\simeq \det R\Gamma\left(S_p, L\left(-\sum_{j=1}^{r} P_j\right)\right)$$

$$\otimes\left\{\otimes_{k=1}^{r} L[P_k]\right\} \otimes \left\{\otimes_{i=1}^{r} \mathcal{O}(-P_i)[P_\ell]\right\}. \qquad (4.36')$$

On the other hand, the metric on $\mathcal{O}(-P_i)[P_j]$ is given by the formula

$$\left\|\mathcal{O}(-P_i)[P_j]\right\| = \frac{1}{G(P_i, P_j)}. \qquad (4.37)$$

Furthermore, the metric on $\det R\Gamma(S_p, L(-\sum P_j))$ is isometric to the flat metric on $\mathbb{C}$, since the degree of bundle $L(-\sum_{j=1}^{r} P_j)$ is equal to $(p-1)$. Hence it follows that (4.36) is in fact an isometry for the Faltings metric.

Now we can construct the Faltings metric on $\det R\Gamma(S_p, L(-\sum P_j))$. Let U be the open covering of the space S_p so that for $(P_1, \ldots, P_r) \in U$, the exact sequence (4.35) induces the isomorphism

$$\alpha : \det R\Gamma(S_p, L) \to \otimes_{j=1}^{r} L[P_j]. \qquad (4.38)$$

Note that this condition is equivalent to the condition

$$H^0\left(S_p, L\left(-\sum_{j=1}^{r} P_j\right)\right) = H^1\left(S_p, L\left(-\sum_{j=1}^{r} P_j\right)\right) = 0. \qquad (4.39)$$

If $\{\alpha_j\}$ is the basis in $H^0(S_p, L)$, then the isomorphism α is explicitly given by the formula

$$\alpha : \alpha_1 \wedge \cdots \wedge \alpha_r \to \det \alpha_i(P_j). \qquad (4.40)$$

It follows from Theorem 1 that the mapping

$$\beta : S_p \to \mathrm{Pic}_{p-1}(S_p) \qquad (4.41)$$

possesses this property, since $U \subset \beta^{-1}(\mathrm{Pic}_{p-1}(-\theta))$. Besides, the space $\det R\Gamma(S_p, L(P))$ can be identified with the fiber over $\beta(P_1, \ldots, P_r)$ of the bundle $\mathcal{O}(-\theta)$. Denote by $\mathbf{1}_\theta$ the unit section $\mathcal{O}(-\theta)$ in $(\mathrm{Pic}_{p-1}(-\theta))$ and by $\lambda(P_1,\dots,P_r)$ the function on U such that

$$\lambda(P_1,\dots,P_r) = \left\|\mathbf{1}_\theta\beta(P_1,\dots,P_r)\right\|^2. \tag{4.42}$$

Using the exact sequence (4.35), we obtain the Faltings metric on $\det R\Gamma(S_p, L)$,

$$\left\|\det R\Gamma(S_p,L)\right\|_F^2 = \Delta\left\|\mathbf{1}_\theta\beta(P_1,\dots,P_r)\right\|\alpha^*\|\otimes L[P_i]\|_F^2$$

$$= \Delta\left\|\mathbf{1}_\theta\beta(P_1,\dots,P_r)\right\|^2 \frac{1}{\prod_{i\neq j}G(P_i,P_j)}\alpha^*\|\otimes L[P_i]\|_n^2$$

$$= \Delta\left\|\mathbf{1}_\theta\beta(P_1,\dots,P_r)\right\|^2\left[\prod G(P_i,P_j)\right]^{-1}\left\|\det\alpha_i(P_j)\right\|_n^2. \tag{4.43}$$

Since this metric is unique up to a scalar multiplication, one may use property 4 of Theorem 1 in order to normalize it and to describe the proportionality constant in terms of the Faltings invariant $\delta(S_p)$. Following Alvarez-Gaume, Bost, Moore, Nelson and Vafa [1987], let us first choose the divisor D on M such that the degree $\Omega(-D)$ is equal to $(p-1)$. Thus D has the form $D = P_1 + \cdots + P_r - Q$. Making use of the basis $\{w_1,\dots,w_p\}$ for $H^0(S_p,\Omega)$ introduced earlier in analogy with the above, one may get the following identity

$$\left\|\mathbf{1}_\theta\beta(P_1,\dots,P_r,Q)\right\| = \exp\left(-\delta(S_p)/8\right)\times\frac{\|\det w_i(P_j)\|_e}{\prod G(P_i,P_j)}\prod_j G(P_j,Q). \tag{4.44}$$

We have expressed here the proportionality factor via $\delta(S_p)$, which will be useful later.

In conclusion of this chapter, we note that the results obtained may also be generalized to obtain the metrics on determinant line bundles of holomorphic n-differentials.

Using the results of this chapter, one may also construct an analogous determinant line bundle on the moduli space of Riemann surfaces.

Chapter 5
The Moduli Space of Riemann Surfaces

5.1. The Determinant Line Bundle on the Moduli Space

To construct a determinant line bundle connected with a holomorphic n-differential, let us consider an infinitesimal variation of the conformal structure of the Riemann surface. Such a variation corresponds to a tangent vector in the Teichmüller space or, in other words, to a deformation with the zero trace of a metric on the Riemann surface S_p. It can be easily seen that this leads to the following transformation for differentials,

$$\delta\nabla_n^z = \frac{1}{2}\delta g^{zz}\nabla_z^n + \frac{n}{2}\nabla_z^n(\delta g^{zz}), \tag{5.1}$$

this variation being holomorphic with respect to the complex structure of the Teichmüller space $\mathcal{T}_p$. The theory of Teichmüller spaces is represented in the survey of G. Schumacher [1990].

It is also known that there exists a bundle $\mathcal{X}_p$ over $\mathcal{T}_p$ with the fiber S_p. The space $\mathcal{X}_p$ is usually called the Teichmüller curve, and $\mathcal{X}_p$ is the fiber bundle

$$\tilde{\pi} : \mathcal{X}_p \to \mathcal{T}_p. \tag{5.2}$$

The differential of this map

$$d\tilde{\pi} : T\mathcal{X}_p \to T\mathcal{T}_p \tag{5.3}$$

has a one-dimensional kernel that is a one-dimensional space tangent to the fiber. Thus, the map kernel of the $d\tilde{\pi}$ defines the line bundle over $\mathcal{T}_p$, which is usually denoted by $\tilde{\Omega}^{-1}$. Note that it follows from the Serre duality that this bundle is dual to that of holomorphic quadratic differentials. Analogously, one can construct the bundles $\tilde{\Omega}^{\otimes n}$ to which operators ∇_n^z are related.

Furthermore, as follows from (5.1), these bundles holomorphically depend on the holomorphic coordinates at the Teichmüller space $\mathcal{T}_p$, and for the fixed point y, we have

$$\det R\Gamma(S_p, \tilde{\Omega}^{\otimes n})^{-1} = \left(\textstyle\bigwedge^{\max} H^0(S_p, \tilde{\Omega}^{\otimes n})\right)^* \otimes \left(\textstyle\bigwedge^{\max} H^1(S_p, \tilde{\Omega}^{\otimes n})\right). \tag{5.4}$$

Considering the projection $\tilde{\pi}$ of this bundle, we obtain the holomorphic determinant bundle over the Teichmüller space

$$\mathcal{L}_n \to \mathcal{T}_p. \tag{5.5}$$

Note that at $n > 1$, we have due to the Kodaira theorem,

$$\mathcal{L}_n = \left(\textstyle\bigwedge^{\max} \tilde{\pi}_* \tilde{\Omega}^{\otimes n}\right)^* \tag{5.6}$$

(but it should be taken into account that $\mathcal{L}_1$ differs from $(\wedge^p \tilde{\pi}_* \tilde{\Omega})^*$.

Furthermore, since the Teichmüller space is contractible, the line bundles on them are trivial. But we are interested in the corresponding line bundles on the moduli space $\mathcal{M}_p$, which is the factor-space

$$\mathcal{M}_p = \mathcal{T}_p/\Gamma_p. \tag{5.7}$$

It appears that the group Γ_p acts on $\mathcal{L}_n$ holomorphically, but not trivially. That is why the holomorphic bundle $\pi : X_p \to \mathcal{M}_p$ over $\mathcal{M}_p$; $X_p = \mathcal{X}_p/\Gamma_p$ is defined (which we denote as before by $\mathcal{L}_n$), which is nontrivial if it is considered as a holomorphic bundle.

Furthermore, as shown in Belavin and Knizhnik [1986], the partition function of the d-dimensional string is described by the section of the line bundle

$$s = (\mathcal{L}_1)^{-d/2} \otimes \mathcal{L}_2. \tag{5.8}$$

On the other hand, as seen from (5.6), the bundle $\mathcal{L}_2$ is isomorphic to the canonical bundle K over the moduli space $\mathcal{M}_p$:

$$(\mathcal{L}_2)^{-1} = \Lambda^{3p-3} \pi_* \Omega^{\otimes 2} = K_{\mathcal{M}_p}, \tag{5.9}$$

or, in other words, to the determinant bundle for the cotangent bundle over $\mathcal{M}_p$.

Thus, the problem of interest is now reduced to that of constructing a Hermitean metric on S_p with zero curvature. To this end, one may use the Quillen metric on the bundle $\mathcal{L}_n$. To define it, let us choose the basis $\{\chi_j\}$, $j = 1, \ldots, k$; $k = (2n-1)(p-1)$ in the space of holomorphic differentials of degree n. Then the section of the bundle $\mathcal{L}_n$ is given by the formula

$$s = (\chi_1 \wedge \cdots \wedge \chi_k)^{-1}, \qquad k = (2n - 1)(p - 1). \tag{5.10}$$

The Quillen metric $\| \ \|_Q$, according to Alvarez-Gaume, Bost, Moore, Nelson and Vafa [1987] and Smit [1988], is defined by the formula

$$\|s\|_Q^2 = \frac{\det \Delta_n}{\det(\chi_i, \chi_j)}, \quad n > 1, \tag{5.11}$$

where s is the section of $\mathcal{L}_n$, and Δ_n is the Laplace operator acting in the space $\tilde{\Omega}^{\otimes n}$. It is connected with the operator ∇_n^z for the admissible metric described in the preceding chapter. As for the bundle $\mathcal{L}_1$, the Quillen metric is given by the formula

$$\| \ \|_Q^2 = \frac{\det \Delta_1}{\det(\omega_i, \omega_j) \int_{S_p} \sqrt{g}}. \tag{5.12}$$

It is known that the Quillen metric varies holomorphically on the moduli space $\mathcal{M}_p$ and that there exists a single holomorphic connection with the curvature form for it:

$$\frac{\partial^2}{\partial y \partial \bar{y}} \ln \|s\|_Q^2, \quad y \in \mathcal{M}_p, \quad s \in \mathcal{L}_n. \tag{5.13}$$

Let us introduce the following expression, which, as will be shown in Chapter 8, determines the partition function of the bosonic string:

$$Z_p = \int_{\mathcal{M}_p} \prod_{j=1}^{3p-3} dy_j \wedge d\bar{y}_j \det(\omega_i, \omega_j)^{-d/2} \tilde{Z}_p, \tag{5.14}$$

where

$$\tilde{Z}_p = \left(\frac{\det \Delta_1}{\det(\omega_i, \omega_j) \int_{S_p} \sqrt{g}}\right)^{-d/2} \left(\frac{\det \Delta_2}{\det(\chi_i, \chi_j)}\right). \tag{5.15}$$

Using the metric (5.11), we may rewrite it as

$$Z_p = \int_{M_p} \prod_1^{3p-3} dy_j \wedge d\bar{y}_j \det(\omega_i, \omega_j)^{-d/2} \frac{\|s_2\|_Q^2}{\|s_1\|_Q^{-d}}, \tag{5.16}$$

where $s_1 \in \mathcal{L}_1$ and $s_2 \in \mathcal{L}_2$ are the sections of the bundles $\mathcal{L}_1$ and $\mathcal{L}_2$. Note that we use here the fact that quadratic differentials holomorphically depend on the point $y \in \mathcal{M}_p$. We see now that in order to show that the integrand in (5.16) is a function rather than a section of a bundle in the moduli space, it is necessary to prove the existence of a holomorphic isomorphism between $(\mathcal{L}_1)^{-d/2}$ and $\mathcal{L}_2$ for some value of d (or, in other words, the absence of the holomorphic anomaly for some d).

5.2. Measures on the Moduli Space. The Mumford Measure

As was shown in the preceding chapter, the problem of finding a measure on the moduli space $\mathcal{M}_p$ of Riemann surfaces includes as a part the problem of holomorphic isomorphisms of determinant line bundles over $\mathcal{M}_p$. To solve the latter problem, the Riemann–Roch theorem is insufficient and one should use its generalization—the so-called Riemann–Roch–Grothendieck theorem. This theorem is a generalization of the Riemann–Roch theorem to the case of families of Riemann surfaces parametrized by the points of a complex manifold. We shall not present the proof or even an exact formulation of this theorem here, but refer the reader to the monographs by Griffiths and Harris [1978], Mumford [1983; 1984], and Arabello, Cornalba, Griffiths and Harris [1986].

Let $X_p = \mathcal{X}_p/\Gamma$ be a universal curve that describes the family of Riemann surfaces of genus p, parametrized by the points $\mathcal{M}_p$, and

$$\pi : X_p \to \mathcal{M}_p \tag{5.17}$$

is the projection map considered in the preceding chapter. Let $\tilde{\mathcal{L}}$ be a holomorphic line bundle over X_p. Then in the space of vector bundles over X_p and $\mathcal{M}_p$, there exists the map $\pi_!$, which, in particular, transforms the bundle $\tilde{\mathcal{L}}$ into the bundle $\mathcal{L} = \pi_! \tilde{\mathcal{L}}$ over $\mathcal{M}_p$,

$$\pi_! : \tilde{\mathcal{L}} \to \mathcal{L}. \tag{5.18}$$

On the other hand, the bundles $\tilde{\mathcal{L}}$ and $\mathcal{L}$ are characterized by the first Chern class $c_1(\tilde{\mathcal{L}})$ and $c_1(\mathcal{L})$, which can be considered as a cycle of codimension 1 in the space of X_p and $\mathcal{M}_p$, respectively, and the map π induces the map in the space of cycles that will be denoted by π_*.

The Riemann–Roch–Grothendieck theorem applied to the case in which we are interested gives

$$c_1(\mathcal{L}) = \pi_* \left[\frac{c_1(\tilde{\mathcal{L}})^2}{2} - \frac{c_1(\tilde{\mathcal{L}})c_1(\tilde{\Omega})}{2} + \frac{c_1(\tilde{\Omega})^2}{12} \right], \tag{5.19}$$

where $\tilde{\Omega}$ is the bundle over X_p generated by holomorphic differentials.

This formula is just a generalization of the Riemann–Roch theorem for the Riemann surface to the case of a family of Riemann surfaces. Let us apply it to the determinant line bundles $\mathcal{L}_n$, where $\mathcal{L}_n$ is determined in (5.6). Assuming first of all that $\tilde{\mathcal{L}} = \tilde{\Omega}$, we obtain

$$c_1(\mathcal{L}_1)^{-1} = \tfrac{1}{12}\pi_* \left(C_1(\tilde{\Omega})^2 \right). \tag{5.20}$$

The bundle $\mathcal{L}_1^{-1}$ is called the Hodge bundle, and its first Chern class is usually denoted by $\lambda : c_1(\mathcal{L}_1^{-1}) = \lambda$.

Setting $\tilde{\mathcal{L}} = \Omega^{\otimes n}$ in (5.18), we get

$$c_1(\mathcal{L}_n)^{-1} = \left(6(n-1)n + 1 \right)\lambda, \tag{5.21}$$

which implies that the bundle $\mathcal{L}_n$ is holomorphically isomorphic to the bundle $(\mathcal{L}_1)^{6n^2-6n+1}$,

$$\mathcal{L}_n \simeq (\mathcal{L}_1)^{6n^2-6n+1}. \tag{5.22}$$

In particular,

$$\mathcal{L}_2 \simeq (\mathcal{L}_1)^{13}. \tag{5.23}$$

Returning to the consideration of the partition function (5.16), we come to a conclusion of the existence of a critical dimension, namely that at $d = 26$, the integrand expression takes a particularly simple form,

$$\frac{d\mu(y) \wedge \overline{d\mu(y)}}{\|s\|_Q^2}. \tag{5.24}$$

This statement is called the Belavin–Knizhnik theorem (Belavin and Knizhnik [1986]).

Theorem 2.1. *The integration measure in the theory of closed bosonic orientable strings in the dimension $d = 26$ is the square of the modulus of the global holomorphic cross section of the bundle $\mathcal{L}_2 \otimes \mathcal{L}_1^{-13}$ and with no zeroes on $\mathcal{M}_p$ divided by the 13th power of the natural metric on $\mathcal{L}_1$. Here*

$$d\mu = \frac{\chi_1 \wedge \cdots \wedge \chi_{3p-3}}{(w_1 \wedge \cdots \wedge w_p)^{13}} \tag{5.25}$$

is the global holomorphic section of the bundle $\mathcal{L}_2 \otimes \mathcal{L}_1^{-13}$. This value is the modular form with respect to the action of the Teichmüller group Γ_p. The form

$d\mu$ is called the Mumford form. Note that the formal use of the formula (5.21) gives

$$(\mathcal{L}_{1/2})^{-2} \simeq \mathcal{L}_1, \tag{5.26}$$

$$(\mathcal{L}_{1/2})^8 \simeq (\mathcal{L}_2)^{-1} \otimes (\mathcal{L}_1)^9, \tag{5.27}$$

which can be proven strictly as well.

The bundle $(\mathcal{L}_{1/2})$ is the spinor bundle. Its degree is equal to $(p-1)$, and it is defined by the choice of the odd theta-characteristic.

It can be further shown that the norm of this section in the Quillen metric is given by the formula

$$||s||_Q^2 = \left[\frac{\det' \Delta_1}{\det(\omega_i, \omega_j) \int \sqrt{g}} \right]^{-1/2} \frac{\prod_{i \neq j} G(P_i, P_j)}{\prod G(P_i, Q) || \det \omega_i(P_j) ||^2}$$
$$\times \left\| \mathbf{1}_\theta \beta(P_1, \dots, P_p, Q) \right\|^2. \tag{5.28}$$

Analogously, for the norm of a section over $\mathcal{L}_2$, we have

$$||s||_Q^2 = \frac{\det' \Delta_2}{\det(\chi_i, \chi_j)} = \exp\left(\frac{\delta}{12} \right) \frac{\prod G(P_i, P_j)}{|| \det \chi_i(P_j) ||^2}$$
$$\times \left\| \mathbf{1}_\theta \beta(P_1, \dots, P_n) \right\|^2, \tag{5.29}$$

where

$$s = (\chi_1 \wedge \cdots \wedge \chi_m)^{-1}, \quad m = 3p - 3. \tag{5.30}$$

Therefore, the expression for the partition function can be rewritten as

$$|\chi_1 \wedge \cdots \wedge \chi_{3p-3}|^2 \exp\left(\frac{\rho\delta}{4} \right) \frac{\prod G(P_i, P_j)}{|| \det X_i(P_j) ||^2} \left\| \mathbf{1}_\theta \beta(P_1, \dots, P_m) \right\|^2. \tag{5.31}$$

Now, using the basis $(\omega_1, \dots, \omega_p)$ of holomorphic differentials, we can define the map τ of the Teichmüller space $\mathcal{T}_p$ into the "Siegel upper half-plane" $\mathcal{H}_p$:

$$\tau : \mathcal{T}_p \to \mathcal{H}_p, \quad \tau_{ij} = \int_{b_j} \omega_i. \tag{5.32}$$

Here the matrix τ is the complex symmetric matrix of order p, and therefore, it depends on $p(p+1)/2$ parameters. In terms of these parameters, the partition function may be rewritten as

$$|\chi_1 \wedge \cdots \wedge \chi_{3p-3}|^2 \left| F(\tau) \right|^2 (\det \operatorname{Im} \tau)^{-13}. \tag{5.33}$$

Here $F(\tau)$ is the modular form relative to the group Γ_p. It should be taken into account that as $p > 4$, these parameters cannot be independent ones, and hence, they cannot be considered as coordinates in the moduli space $\mathcal{M}_p$.

Let us proceed to the derivation of the Belavin–Knizhnik theorem. See also Bost [1987].

5.3. Regularization and Anomalies

Let us regularize the determinant of the Laplace operator in the following way:

$$\log \mathrm{Det}'_\varepsilon \Delta_j \equiv -\int_\varepsilon^\infty \frac{dt}{t}\,\mathrm{Tr}\,e^{-t\Delta_j}. \tag{5.34}$$

The regularized determinant in (4.8) is defined as

$$\log \mathrm{Det}_{\mathrm{Reg}} \Delta_j \equiv \lim_{\varepsilon \to +0} \left[\log \mathrm{Det}_\varepsilon \Delta_j - \frac{a_j}{\varepsilon} - b_j \log \varepsilon \right], \tag{5.35}$$

with counterterms adjusted in such a way that the limit exists.

Now examine the ultraviolet behavior of the determinant: the asymptotic of $\mathrm{Det}_\varepsilon \Delta_j$ at small ε. The trace in (5.36) may be evaluated in the following way:

$$\mathrm{Tr}\,e^{-t\Delta_j} = \int^{d^2\xi_0} \rho(\xi_0)\langle \xi_0 | e^{-t\Delta_j} | \xi_0 \rangle. \tag{5.36}$$

Expand the operator $\Delta_j(\xi) = -\rho(\xi)^{j-1}\partial\rho(\xi)^{-j}\bar\partial$ in the vicinity of ξ_0, using the representation

$$\rho(\xi) = \rho(\xi_0)\bigl(1 - |\xi - \xi_0|^2 \rho\mathcal{R}(\xi_0) + \cdots\bigr), \tag{5.37}$$

where $\mathcal{R} = -\rho^{-1}\partial\bar\partial \log \rho$ is the two-dimensional curvature, and the dots stand for the terms $\sim o\bigl((\xi - \xi_0)^3\bigr)$, which are proportional to the derivatives and powers of $\mathcal{R}$. Then

$$\Delta_j = -\frac{1}{\rho(\xi_0)} \left[\bigl(1 + |\xi - \xi_0|^2 \rho\mathcal{R}(\xi_0)\bigr)\partial_\xi\partial_{\bar\xi} + j\overline{(\xi - \xi_0)}\rho\mathcal{R}(\xi_0)\partial_{\bar\xi} + \cdots \right]. \tag{5.38}$$

After Fourier transformation with respect to the difference $\xi - \xi_0$,

$$\langle \xi_1 | e^{-t\Delta_j} | \xi_0 \rangle = \int \frac{d^2 p}{\rho_0 (2\pi)^2} e^{\frac{i}{2}p\overline{(\xi_1-\xi_0)}} e^{\frac{i}{2}\bar p(\xi_1 - \xi_0)}$$

$$\times \exp\left[-\frac{t}{\rho_0}\left[\left(\frac{1}{4} + \rho_0\mathcal{R}_0 \frac{\partial}{\partial p}\cdot\frac{\partial}{\partial \bar p}\right)p\bar p - j\rho_0\mathcal{R}_0 \frac{\partial}{\partial p}p \right] \right]$$

$$= \frac{e^{-|\xi_1-\xi_0|^2 \rho_0/t}}{\pi t}\left(1 + \frac{t\mathcal{R}_0}{6}(1 - 3j) + O(t^2)\right). \tag{5.39}$$

Thus,

$$\log \mathrm{Det}_\varepsilon \, \Delta_j \equiv - \int_\varepsilon^\infty \frac{dt}{t} \int^{d^2\xi_0} \rho(\xi_0) \langle \xi_0 | e^{-t\Delta_j} | \xi_0 \rangle$$

$$= -\frac{1}{\pi\varepsilon} \int \rho(\xi_0) d^2\xi_0 - \frac{3j-1}{6\pi} \log\varepsilon \int \rho(\xi_0)\mathcal{R}(\xi_0) d^2\xi_0 + \cdots,$$

$$(5.40)$$

and the coefficients a_j and b_j in (5.35) are equal to

$$a_j = -\frac{1}{\pi} \int \rho(\xi) d^2\xi,$$

$$b_j = \frac{3j-1}{6\pi} \int \rho\mathcal{R}(\xi) d^2\xi = \frac{(3j-1)(p-1)}{3}$$

$$(5.41)$$

(the Gauss–Bonnet formula for the Euler characteristic of a genus-p surface was applied, $\int \rho\mathcal{R}(\xi) d^2\xi = -2\pi(p-1)$).

Appendix: The Proof of the Riemann–Roch Formula (3.11). As the simplest application of the asymptotic formulae (5.40), let us give a proof of the Riemann–Roch formula (3.11), which relates the numbers of zero modes of the operators $\Delta_j = \bar\partial_j^\dagger \bar\partial_j$ and $\Delta_{1-j} = \bar\partial_{1-j}^\dagger \bar\partial_{1-j}$. First, it is necessary to recall that

$$\bar\partial_j : \mathcal{A}_{j,0} \to \mathcal{A}_{j,1}; \qquad \bar\partial_j^\dagger : \mathcal{A}_{j,1} \to \mathcal{A}_{j,0}. \tag{5.42}$$

The nonzero modes of the self-conjugate operators $\Delta_j = -\rho^{j-1}\partial\rho^{-j}\bar\partial$ and $\Delta_{1-j} = -\rho^{-j}\partial\rho^{j-1}\bar\partial$ are in one-to-one correspondence and have the same eigenvalues: If

$$\Delta_j \phi_\lambda^{(j)} = -\rho^{j-1}\partial\rho^{-j}\bar\partial\phi_\lambda^{(j)} = \lambda\phi_\lambda^{(j)}, \tag{5.43}$$

then

$$\overline{\chi_\lambda^{(1-j)}} \equiv \rho^{-j}\bar\partial\phi_\lambda^{(j)} \tag{5.44}$$

satisfies the equation

$$-\rho^{-j}\bar\partial\left(\rho^{j-1}\partial\overline{\chi_\lambda^{(1-j)}}\right) = \lambda\overline{\chi_\lambda^{(1-j)}}, \quad \text{i.e.,} \quad \Delta_{1-j}\overline{\chi_\lambda^{(1-j)}} = \lambda\overline{\chi_\lambda^{(1-j)}}. \tag{5.45}$$

Because of this, only zero modes of Δ_j and Δ_{1-j} contribute to the difference

$$n_j - n_{1-j} = \mathrm{Tr}_{\mathcal{A}_{j,0}} e^{-t\Delta_j} - \mathrm{Tr}_{\mathcal{A}_{1-j,0}} e^{-t\Delta_{1-j}} \tag{5.46}$$

at any positive value of t. Equations (5.38) and (5.40) imply that

$$n_j - n_{1-j} = \int \rho(\xi) d^2\xi \left[\frac{1}{\pi t}\left(1 + \frac{t\mathcal{R}(\xi)}{6}(1-3j) + O(t^2)\right) \right.$$

$$\left. - \frac{1}{\pi t}\left(1 + \frac{t\mathcal{R}(\xi)}{6}(3j-2) + O(t^2)\right) \right]$$

$$= b_j - b_{1-j} = \frac{1-2j}{2\pi} \int \rho\mathcal{R} d^2\xi = (2j-1)(p-1). \tag{5.47}$$

5.4. Anomalies

We proceed with the study of coordinate and metric dependencies of the regularized determinant. For the sake of brevity, we introduce the following notation:

$$\mathcal{Z}_j\big[g(\xi)\,|\,\xi\big] \equiv \mathrm{Det}'_{\mathrm{Reg}}\,\Delta_j/\det N_j \cdot \det N_{1-j}, \tag{5.48}$$

where $\mathcal{Z}_j$ is the gauge invariant with respect to the general coordinate transformation (3.16)–(3.18):

$$\mathcal{Z}_j\big[\tilde{g}(\zeta)\,|\,\zeta\big] = \mathcal{Z}_j\big[g(\xi)\,|\,\xi\big] \quad \text{for} \quad \tilde{g}(\zeta) = g(\xi). \tag{5.49}$$

This property allows one to fix the conformal gauge (3.20) and to consider only the variations of $\log \mathcal{Z}_j$ under three independent variations of the metric (3.1). If $\mathcal{Z}_j$ is defined as the functional integral (4.11), these variations are related to the regularized stress tensor:

$$\delta \log \mathcal{Z}_j = \int \Big[\delta g^{\xi\bar{\xi}}\langle T_{\xi\bar{\xi}}^{(\mathrm{Reg})}\rangle + \delta g^{\xi\xi}\langle T_{\xi\bar{\xi}}^{(\mathrm{Reg})}\rangle \delta g^{\bar{\xi}\bar{\xi}}\langle T_{\xi\bar{\xi}}^{(\mathrm{Reg})}\rangle \Big] \sqrt{\det g}\, d^2\xi; \tag{5.50}$$

moreover, according to (4.11),

$$T_{\xi\xi} \equiv T = j\phi^{(j)}\partial\chi^{(1-j)} - (1-j)\partial\phi^{(j)}\chi^{(1-j)}. \tag{5.51}$$

Substituting the Beltrami differential $\eta = \delta g_{\xi\xi}/g_{\xi\bar{\xi}}$ in the form of (3.28), we make use of the conservation law

$$\bar{\partial}T_{\xi\xi} = -\partial T_{\xi\bar{\xi}} \tag{5.52}$$

for the following transformation:

$$\int \delta g^{\xi\xi}\langle T_{\xi\xi}^{(\mathrm{Reg})}\rangle \sqrt{\det g}\, d^2\xi = \int \eta\langle T_{\xi\xi}^{(\mathrm{Reg})}\rangle d^2\xi$$

$$= \sum_{\alpha} \delta y_\alpha \int \eta_\alpha\langle T_{\xi\xi}^{(\mathrm{Reg})}\rangle d^2\xi + \int \bar{\partial}\tilde{\varepsilon}\langle T_{\xi\xi}^{(\mathrm{Reg})}\rangle d^2\xi$$

$$= \sum_{\alpha} \delta y_\alpha \int \eta_\alpha\langle T_{\xi\xi}^{(\mathrm{Reg})}\rangle d^2\xi - \int \partial\tilde{\varepsilon}\langle T_{\xi\bar{\xi}}^{(\mathrm{Reg})}\rangle d^2\xi. \tag{5.53}$$

Thus,

$$\delta \log \mathcal{Z}_j = \sum_{\alpha=1}^{n_2} \delta y_\alpha \int \eta_\alpha\langle T_{\xi\xi}^{(\mathrm{Reg})}\rangle d^2\xi + \text{c.c.} + \int \Big[\frac{\delta\rho}{\rho} - \partial\tilde{\varepsilon} - \bar{\partial}\bar{\tilde{\varepsilon}} \Big]\langle T_{\xi\bar{\xi}}^{(\mathrm{Reg})}\rangle d^2\xi. \tag{5.54}$$

Two anomalous relations are valid for the variations of $\log \mathcal{Z}_j$:

$$\delta_\rho \log \mathcal{Z}_j = \int \frac{\delta\rho}{\rho}\langle T^{(\mathrm{Reg})}_{\xi\bar{\xi}}\rangle d^2\xi = -\frac{c_j}{6\pi}\int \delta\rho R d^2\xi = \frac{c_j}{6\pi}\int \frac{\delta\rho}{\rho}\partial\bar{\partial}\log\rho\, d^2\xi$$

$$(5.55)$$

$$= -\frac{c_j}{12\pi}\delta\rho S_L, \qquad (5.56)$$

and

$$\delta_\eta d_{\bar{\eta}}\log \mathcal{Z}_j = -\frac{c_j}{6\pi}\int\left[\left|\frac{\partial(\rho\eta)}{\rho}\right|^2 + |\eta|^2\partial\bar{\partial}\log\rho\right]d^2\xi \qquad (5.57)$$

$$= -\frac{c_j}{12\pi}\delta_\eta\delta_{\bar{\eta}}S_L \qquad (5.58)$$

(in (5.54) $\frac{\delta\rho}{\rho} - \partial\tilde{\varepsilon} - \bar{\partial}\bar{\tilde{\varepsilon}} = 0$). Here $c_j = 6j^2 - 6j + 1$, and the Liouville action

$$S_L\left[g(\xi)\,|\,\xi\right] = \int \mathcal{R}\frac{1}{\Delta}\mathcal{R}\sqrt{\det g}\, d^2\xi \qquad (5.59)$$

contains the two-dimensional curvature $\mathcal{R}$. In the conformal gauge, $\mathcal{R} = -\rho^{-1}\partial\bar{\partial}\log\rho$, and

$$S_L\left[\rho(\xi)d\xi d\bar{\xi}\,|\,\xi\right] = \int \left|\partial_\xi\log\rho\right|^2 d^2\xi. \qquad (5.60)$$

The proof of (5.55)–(5.58) is given below in the Appendix following this section.

Let us note from the very beginning that

$$S_L\left[g(\xi)e^{\phi(\xi)}\,|\,\xi\right] - S_L\left[g(\xi)\,|\,\xi\right] = \int \sqrt{\det g}(g^{ab}\partial_a\phi\partial_b\phi + 2\mathcal{R}_g\phi)d^2\xi. \quad (5.61)$$

Equations (5.55)–(5.58) imply that $\mathrm{Det}\,\bar{\partial}_j$, related to $\mathcal{Z}_j$ in the following way,

$$\mathcal{Z}_j \equiv \frac{\mathrm{Det}'_{\mathrm{Reg}}\,\Delta_j}{\det N_j \det N_{1-j}} = \exp\left(-\frac{c_j}{12\pi}S_L\right)\left|\mathrm{Det}\,\bar{\partial}_j\right|^2, \qquad (5.62)$$

satisfies:

$$\delta_\rho \log \mathrm{Det}\,\bar{\partial}_j = 0, \qquad (5.63)$$

$$\delta_{\bar{\eta}}\log \mathrm{Det}\,\bar{\partial}_j = 0, \qquad (5.64)$$

$$\delta_\eta \log \mathrm{Det}\,\bar{\partial}_j = \sum_{\alpha=1}^{n_2}\delta y_\alpha \int \eta_\alpha\langle \hat{T}_{\xi\xi}\rangle d^2\xi,$$

$$\hat{T}_{\xi\xi} \equiv T^{(\mathrm{Reg})}_{\xi\xi} + \frac{c_j}{12\pi}T^{(L)}_{\xi\xi}. \qquad (5.65)$$

5.5. Gravitational Anomaly (Alvarez-Gaume and Witten [1985]; Knizhnik [1986])

The analytic coordinate change $\xi \to \zeta = \xi + \varepsilon(\xi)$, $\bar{\partial}\varepsilon = 0$, preserves the conformal gauge of the metric

$$\tilde{\rho}(\zeta)d\zeta d\bar{\zeta} = \rho(\xi)d\xi d\bar{\xi},$$

$$\tilde{\rho}(\zeta) = \frac{\rho(\xi)}{|\partial\zeta/\partial\xi|^2} = \rho(\zeta)\left(1 - \frac{\partial(\varepsilon\rho)}{\rho} - \frac{\bar{\partial}(\bar{\varepsilon}\rho)}{\rho} + o(\varepsilon^2)\right). \tag{5.66}$$

Since

$$\mathcal{Z}_j\big[\tilde{\rho}(\zeta)d\zeta d\bar{\zeta}\,|\,\zeta\big] = \mathcal{Z}_j\big[\rho(\xi)d\xi d\bar{\xi}\,|\,\xi\big], \tag{5.67}$$

and

$$\mathcal{Z}_j\big[\tilde{\rho}(\zeta)d\zeta d\bar{\zeta}|\zeta\big] = \mathcal{Z}_j\big[\rho(\zeta)d\zeta d\bar{\zeta}\,|\,\zeta\big]\exp\frac{c_j}{6\pi}\int\left[\frac{\partial(\varepsilon\rho)}{\rho} + \frac{\bar{\partial}(\bar{\varepsilon}\rho)}{\rho}\right]\partial\bar{\partial}\log\rho\,d^2\xi, \tag{5.68}$$

the coordinate dependence of $\operatorname{Det}\bar{\partial}_j$ is given by the following formula:

$$\operatorname{Det}\bar{\partial}_j[\zeta] = \operatorname{Det}\bar{\partial}_j[\xi]\exp\frac{c_j}{6\pi}\int\frac{\partial(\varepsilon\rho)}{\rho}\partial\bar{\partial}\log\rho\,d^2\xi. \tag{5.69}$$

The fact that the determinant does depend on coordinates is known as gravitational anomaly. Under a change of metric, $\rho \to e^{\phi}\rho$ with nonsingular function ϕ, the right-hand side of (5.69) acquires an extra factor of

$$\exp\frac{c_j}{6\pi}\int\big(\partial\varepsilon\partial\bar{\partial}\phi + \varepsilon\bar{\partial}(\partial\phi\partial\log\rho)\big)d^2\xi. \tag{5.70}$$

Integration by parts, together with analyticity of ε, $\bar{\partial}\varepsilon = 0$, ensures that (5.70) vanishes and the right-hand side is in fact invariant under the change $\rho \to e^{\phi}\rho$ in accordance with (5.63). However, if the conformal factor ρ is multiplied by a function with zeros or poles, i.e., under the transformation to a singular metric, the gravitational anomaly and $\operatorname{Det}\bar{\partial}_j$ itself can vary.

Appendix: The Derivation of Anomalous Relations (5.55)–(5.58).
Let us start with formulas (5.55) and (5.56); see Polyakov [1981a,b]:
Under the variation of the conformal factor,

$$\delta_p\Delta_j = -\frac{\delta\rho}{\rho}(j - 1)\rho^{j-1}\partial\rho^{-j}\bar{\partial} + j\rho^{j-1}\partial\frac{\delta\rho}{\rho}\rho^{-j}\bar{\partial}, \tag{5.71}$$

232   A.Yu. Morozov and A.M. Perelomov

and

$$
\delta_\rho \log \mathrm{Det}'_{\mathrm{Reg}} \Delta_j \equiv \delta_\rho \lim_{\varepsilon \to +0} \left[-\int_\varepsilon^\infty \frac{dt}{t} \mathrm{Tr}' \, e^{-t\Delta_j} + \int \frac{\sqrt{\det g}\, d^2\xi}{\pi \varepsilon} \right.
$$

$$
\left. -\frac{3j-1}{6\pi} \log \varepsilon \int \rho \mathcal{R} d^2\xi \right] = \lim_{\varepsilon \to +0} \left\{ \int_\varepsilon^\infty dt \left[(j-1)\, \mathrm{Tr}_{A_{j,0}}\, \frac{\delta\rho}{\rho} \Delta_j e^{-t\Delta_j} \right. \right.
$$

$$
\left. \left. - j\, \mathrm{Tr}_{A_{1-j,0}}\, \frac{\delta\rho}{\rho} \overline{\Delta_{1-j}}\, e^{-t\overline{\Delta_{1-j}}} \right] + \frac{1}{\pi\varepsilon} \int \delta\rho\, d^2\xi \right\}
$$

$$
= \lim_{\varepsilon \to +0} \int \delta\rho\, d^2\xi \left[\frac{1}{\pi\varepsilon} - (j-1)\int_\varepsilon^\infty dt \frac{\partial}{\partial t} \langle\xi| e^{-t\Delta_j} |\xi\rangle \right.
$$

$$
\left. + j \int_\varepsilon^\infty dt \frac{\partial}{\partial t} \langle\xi| e^{-t\overline{\Delta_{1-j}}} |\xi\rangle \right] = \lim_{\varepsilon \to +0} \int \delta\rho\, d^2\xi \left[\frac{1}{\pi\varepsilon} + (j-1)\langle\xi| e^{-t\Delta_j} |\xi\rangle \right.
$$

$$
\left. - j \langle\xi| e^{-t\overline{\Delta_{1-j}}} |\xi\rangle \right] - \lim_{t \to +\infty} \int \delta\rho\, d^2\xi \left[(j-1)\langle\xi| e^{-t\Delta_j} |\xi\rangle - j \langle\xi| e^{-t\overline{\Delta_{1-j}}} |\xi\rangle \right].
$$

$$
\tag{5.72}
$$

The last two items in this sum get contributions only from the zero modes $\phi_\alpha^{(j)}$, $\alpha = 1 \ldots n_j$, $\bar\partial \phi_\alpha^{(j)} = 0$; $\phi_\beta^{(1-j)}$, $\beta = 1 \ldots n_{1-j}$, $\bar\partial \phi_\beta^{(1-j)} = 0$:

$$
\lim_{t \to +\infty} \langle\xi| e^{-t\Delta_j} |\xi\rangle = \rho^{-j}(\xi, \bar\xi) \sum_{\alpha,\gamma=1}^{n_j} \overline{\phi_\alpha^{(j)}(\xi)} (N_j^{-1})_{\alpha\gamma} \phi_\gamma^{(j)}(\xi), \tag{5.73}
$$

with

$$
(N_j)_{\alpha\gamma} \equiv \int \rho^{1-j}(\xi, \bar\xi) \overline{\phi_\alpha^{(j)}(\xi)} \phi_\gamma^{(j)}(\xi) d^2\xi \tag{5.74}
$$

being the matrix of scalar products of holomorphic j-differentials. Equation (5.73) implies that

$$
- \lim_{t \to +\infty} (j-1) \int \delta\rho d^2\xi \langle\xi| e^{-t\Delta_j} |\xi\rangle = -(j-1) \int \frac{\delta\rho}{\rho} \rho^{1-j} \overline{\phi_\alpha^{(j)}} (N_j^{-1})_{\alpha\gamma} \phi_\gamma^{(j)} d^2\xi
$$

$$
= \delta_\rho \log \det N_j. \tag{5.75}
$$

The second and third items in (5.72) may be substituted by their asymptotic approximations (5.40). Finally, we obtain for (5.72):

$$
\delta_\rho \log \mathrm{Det}'_{\mathrm{Reg}} \Delta_j = \delta_\rho \log \det N_j + \delta_\rho \log \det N_{1-j}
$$

$$
+ \lim_{\varepsilon \to +0} \int \delta\rho d^2\xi \left\{ \frac{1}{\pi\varepsilon} + (j-1)\left[\frac{1}{\pi\varepsilon} - \frac{3j-1}{6\pi}\mathcal{R} + o(\varepsilon) \right] \right.
$$

$$
\left. - j\left[\frac{1}{\pi\varepsilon} + \frac{3j-2}{6\pi}\mathcal{R} + o(\varepsilon) \right] \right\}, \tag{5.76}
$$

i.e., for $\mathcal{Z}_j \equiv \mathrm{Det}'_{\mathrm{Reg}} \Delta_j / \det N_j \cdot \det N_{1-j}$,

$$\delta_\rho \log \mathcal{Z}_j = -\frac{6j^2 - 6j + 1}{6\pi} \int \delta\rho \mathcal{R} d^2\xi = -\frac{c_j}{12\pi} \delta_\rho \int \left|\partial \log \rho\right|^2 d^2\xi. \tag{5.77}$$

For generic two-dimensional Laplace operator $\Delta = -f\partial h\bar\partial$ (Kallosh and Morozov [1988]),

$$\delta \log \frac{\mathrm{Det}'_{\mathrm{Reg}} \Delta}{\det N_f \det N_h} = -\frac{1}{12\pi} \delta \int^{d^2\xi} \left[|\partial \log f|^2 - 4\partial \log f \cdot \bar\partial \log h + |\partial \log h|^2 \right]. \tag{5.78}$$

We now proceed to the proof of (5.58).

Under an infinitesimal variation of complex structure metric transforms according to (3.24):

$$\rho(\xi, \bar\xi) d\xi d\bar\xi \to \tilde\rho(\zeta, \bar\zeta) d\zeta d\bar\zeta = \rho(\xi, \bar\xi) d\xi d\bar\xi + \rho\eta d\bar\xi^2 + \rho\bar\eta d\xi^2, \tag{5.79}$$

implying that

$$\partial\zeta/\partial\bar\xi = \eta \cdot \partial\zeta/\partial\xi, \tag{5.80}$$

$$\rho(\xi, \bar\xi) = \tilde\rho(\zeta, \bar\zeta) \left|\frac{\partial\zeta}{\partial\xi}\right|^2 \left(1 + \left|\frac{\partial\zeta/\partial\bar\xi}{\partial\zeta/\partial\xi}\right|^2\right) = \tilde\rho(\zeta, \bar\zeta) \left|\frac{\partial\zeta}{\partial\xi}\right|^2 (1 + \eta\bar\eta). \tag{5.81}$$

Also

$$\frac{\partial}{\partial\zeta} = (1 - \eta\bar\eta)^{-1} \left(\frac{\partial\zeta}{\partial\xi}\right)^{-1} \left(\frac{\partial}{\partial\xi} - \bar\eta \cdot \frac{\partial}{\partial\bar\xi}\right), \tag{5.82}$$

$$d^2\zeta = \left|\frac{\partial\zeta}{\partial\xi}\right|^2 (1 - \eta\bar\eta) d^2\xi. \tag{5.83}$$

After the variation (5.79), the Liouville action $S_L = \int |\frac{\partial}{\partial\xi} \log \rho(\xi, \bar\xi)|^2 d^2\xi$ turns into

$$\tilde S_L = \int \left|\frac{\partial}{\partial\zeta} \log \tilde\rho(\zeta, \bar\zeta)\right|^2 d^2\zeta$$

$$= \int \frac{d^2\xi}{1 - \eta\bar\eta} \left|\left(\frac{\partial}{\partial\xi} - \bar\eta \cdot \frac{\partial}{\partial\bar\xi}\right) \log \frac{\rho(\xi, \bar\xi)}{|\partial\zeta/\partial\xi|^2 (1 + \eta\bar\eta)}\right|^2. \tag{5.84}$$

For the sake of brevity, we write $\partial/\partial\xi = \partial$, $\partial/\partial\bar{\xi} = \bar{\partial}$. The difference

$$
\begin{aligned}
\tilde{S}_L - S_L = \int d^2\xi(1 + \eta\bar{\eta})\Big\{ &|\partial\log\rho - \bar{\eta}\bar{\partial}\log\rho|^2 \\
&- (\partial\log\rho - \bar{\eta}\bar{\partial}\log\rho)(\bar{\partial} - \eta\partial)(\eta\bar{\eta} + \log\partial\zeta + \log\bar{\partial}\bar{\zeta}) \\
&- (\bar{\partial}\log\rho - \eta\partial\log\rho)(\partial - \bar{\eta}\bar{\partial})(\eta\bar{\eta} + \log\partial\zeta + \log\bar{\partial}\bar{\zeta}) \\
&+ \big|(\partial - \bar{\eta}\bar{\partial})(\eta\bar{\eta} + \log\partial\zeta + \log\bar{\partial}\bar{\zeta})\big|^2 \Big\} - \int d^2\xi|\partial\log\rho|^2 + o(\eta^3).
\end{aligned}
$$

$$(5.85)$$

For the proof of (5.58), we need only contributions proportional to $\eta\bar{\eta}$. Moreover, according to (5.72),

$$
\bar{\partial}\log\partial\zeta = \bar{\partial}\partial\zeta/\zeta = \partial\eta + \eta\partial\log\partial\zeta,
\tag{5.86}
$$

and

$$
(\partial - \bar{\eta}\bar{\partial})(\eta\bar{\eta} + \log\partial\zeta + \log\bar{\partial}\bar{\zeta}) = (1 - \eta\bar{\eta})\partial\log\partial\zeta + \bar{\partial}\eta + \eta\partial\bar{\eta} + o(\eta^3).
\tag{5.87}
$$

Neglecting the boundary terms in the integration by parts, on also obtains

$$
\int |\partial\log\partial\zeta|^2 \to \int |\bar{\partial}\log\partial\zeta|^2 = \int |\partial\eta + \eta\partial\log\partial\zeta|^2.
\tag{5.88}
$$

Similarly,

$$
\int (\partial\log\partial\zeta)\bar{\partial}(\eta\bar{\eta}) \to o(\eta^3).
\tag{5.89}
$$

Applying (5.86)–(5.89), we derive from (5.85):

$$
\begin{aligned}
\delta_\eta\delta_{\bar{\eta}}S_L \\
= 2\int d^2\xi\big\{ &\eta\bar{\eta}|\partial\log\rho|^2 + |\partial\eta|^2 + \eta\bar{\eta}\partial\bar{\partial}\log\rho + \eta\bar{\partial}\bar{\eta}\partial\log\rho + \bar{\eta}\partial\eta\bar{\partial}\log\rho\big\} \\
= 2\int d^2\xi\Big\{ &\left|\frac{\partial(\rho\eta)}{\rho}\right|^2 + |\eta|^2\partial\bar{\partial}\log\rho\Big\},
\end{aligned}
\tag{5.90}
$$

i.e., exactly what was stated in (5.58).

For a detailed proof of (5.57), see Belavin and Knizhnik [1986]. We give here a brief version of it.

Let us work out the contribution to a determinant variation under (5.79), proportional to $\eta\bar{\eta}$. After the transformation(5.79), the Laplace operator Δ_j acquires the form of

$$
\begin{aligned}
\tilde{\Delta}_j[\zeta] = &-\rho^{j-1}\big(1 + (2 - j)\eta\bar{\eta}\big)\big[\partial - \bar{\eta}\bar{\partial} - (1 - j)(\bar{\partial}\bar{\eta})\big]\rho^{-j} \\
&\times \big(1 + (j + 1)\eta\bar{\eta}\big)\big[\bar{\partial} - \eta\partial - j(\partial\eta)\big].
\end{aligned}
\tag{5.91}
$$

This expression is easily derived from formulae for the transformed action

$$\tilde{S} \equiv \int \tilde{\rho}^{-j} \left| \frac{\partial \tilde{\phi}}{\partial \bar{\zeta}} \right|^2 d^2\zeta = \int \rho^{-j} \big(1 + (j+1)\eta\bar{\eta}\big) \big| [\bar{\partial} - \eta\partial - j(\partial\eta)]\phi \big|^2 d^2\xi, \tag{5.92}$$

and the norm

$$\tilde{\|}\delta\tilde{\phi}\tilde{\|}^2 \equiv \int \tilde{\rho}^{1-j} |\delta\tilde{\phi}|^2 d^2\zeta = \int \rho^{j-1}\big(1 + (2-j)\eta\bar{\eta}\big) |\delta\tilde{\phi}|^2 d^2\xi,$$

$$\tilde{\phi} = \phi \left| \frac{\partial\zeta}{\partial\xi} \right|^{-j}. \tag{5.93}$$

We shall also need the projector $\mathbb{P}_j;\ \mathcal{A}_{j,0} \to \mathcal{H}_j$ on the space of zero modes (holomorphic j-differentials). $\mathbb{P}_j$ is defined by its kernel

$$\mathbb{P}_j(\xi, \xi') = \sum_{\alpha,\gamma=1}^{n_j} \phi_\alpha^{(j)}(\xi)\, \big(N_j^{-1}\big)_{\alpha\gamma}\, \overline{\phi_\gamma^{(j)}(\xi')}. \tag{5.94}$$

The variation of the holomorphic j-differential $\delta\phi_\alpha^{(j)}$ satisfies the equation

$$(\Delta_j + \delta\Delta_j)(\phi_\alpha^{(j)} + \delta\phi_\alpha^{(j)}) = 0, \tag{5.95}$$

i.e.,

$$\delta\phi_\alpha^{(j)} = -\frac{1}{\Delta_j}(\mathbb{I} - \mathbb{P}_j)\delta\Delta_j\phi_\alpha^{(j)}, \tag{5.96}$$

and is orthogonal to the space of zero modes $\mathcal{H}_j$. This last requirement specifies the meaning of $1/\Delta_j$. The ultraviolet specification is

$$\left[\frac{1}{\Delta_j} \right]_\varepsilon = \int_\varepsilon^\infty dt\, e^{-t\Delta_j} = \frac{e^{-\varepsilon\Delta_j}}{\Delta_j}. \tag{5.97}$$

Then the variation of $\det N_j = \det_{(\alpha\gamma)} \int \rho^{1-j} \overline{\phi_\alpha^{(j)}} \phi_\gamma^{(j)}$ is equal to

$$\delta_\eta \delta_{\bar{\eta}} \log\det N_j = \mathrm{Tr}_{\mathcal{H}_j}(j-2)\eta\bar{\eta} + \int \rho^{1-j} \overline{\delta\phi_\alpha^{(j)}}\, \big(N_j^{-1}\big)_{\alpha\gamma}\, \delta\phi_\gamma^{(j)}, \tag{5.98}$$

or

$$\delta_\eta \delta_{\bar{\eta}} \log\det N_j = \mathrm{Tr}_{\mathcal{H}_j}(j-2)\eta\bar{\eta} + \lim_{\varepsilon\to+0} \int \delta_{\bar{\eta}}\Delta_j \left[-\frac{1}{\Delta_j} \right]_\varepsilon$$

$$\times \sum_{\alpha,\gamma} \overline{\phi_\alpha^{(j)}}\, \big(N_j^{-1}\big)_{\alpha\gamma}\, \phi_\gamma^{(j)}. \tag{5.99}$$

Note now, that the variation of the kernel

$$\delta_\eta \mathbb{P}_j(\xi, \xi') = \sum_{\alpha,\gamma=1}^{n_j} \overline{\phi_\alpha^{(j)}(\xi')}\, \big(N_j^{-1}\big)_{\alpha\gamma}\, \delta_\eta\phi_\gamma^{(j)}(\xi'). \tag{5.100}$$

This is because $\delta_\eta \overline{\phi_\alpha^{(j)}} = 0$ and $\phi N^{-1}\delta N \cdot N^{-1}\bar\phi = \phi N^{-1}(\delta\bar\phi \cdot \phi + \bar\phi\delta\phi)N^{-1}\bar\phi = \phi N^{-1}\delta\bar\phi \cdot \mathbb{P} + \mathbb{P} \cdot \delta\phi)N^{-1}\bar\phi = 0$, since $\delta\bar\phi \cdot \mathbb{P} = 0 = \mathbb{P}\delta\phi$. Thus,

$$\delta_\eta\delta_{\bar\eta} \log \det N_j = \mathrm{Tr}_{\mathcal{H}_j}(j-2)\eta\bar\eta + \lim_{\varepsilon\to+0} \mathrm{Tr}_{\mathcal{H}_j} \delta_{\bar\eta}\Delta_j = j\left[-\frac{1}{\Delta_j}\right]_\varepsilon \delta_\eta\mathbb{P}_j. \quad (5.101)$$

On the other hand,

$$\delta_\eta\delta_{\bar\eta} \log {\det}'_\varepsilon \Delta_j$$

$$= \mathrm{Tr}_{\mathcal{A}_{j,0}}(\mathbb{I} - \mathbb{P}_j)(2 - j)\eta\bar\eta\Delta_j\left[\frac{1}{\Delta_j}\right]_\varepsilon$$

$$+ \mathrm{Tr}_{\mathcal{A}_{1-j,0}}(\mathbb{I} - \mathbb{P}_{1-j})(1 + j)\eta\bar\eta\Delta_{1-j}\left[\frac{1}{\Delta_{1-j}}\right]_\varepsilon$$

$$+ \delta_\eta\, \mathrm{Tr}_{\mathcal{A}_{j,0}}\, \rho^{j-1}(\delta_{\bar\eta}\partial_{1-j})(\mathbb{I} - \mathbb{P}_{1-j})(\rho^{-j}\bar\partial_j)\left[\frac{1}{\Delta_j}\right]_\varepsilon(\mathbb{I} - \mathbb{P}_j)$$

$$= \mathrm{Tr}_{\mathcal{A}_{j,0}}(2 - j)\eta\bar\eta\Delta_j\left[\frac{1}{\Delta_j}\right]_\varepsilon + \mathrm{Tr}_{\mathcal{A}_{1-j,0}}(1 + j)\eta\bar\eta\Delta_{1-j}\left[\frac{1}{\Delta_{1-j}}\right]_\varepsilon$$

$$+ (j - 2)\, \mathrm{Tr}_{\mathcal{H}_j}\, \eta\bar\eta - (j + 1)\, \mathrm{Tr}_{\mathcal{H}_{1-j}}\, \eta\bar\eta + \mathrm{Tr}_{\mathcal{H}_j}\, \delta_{\bar\eta}\Delta_j\left[-\frac{1}{\Delta_j}\right]_\varepsilon \delta_\eta\mathbb{P}_j$$

$$+ \mathrm{Tr}_{\mathcal{H}_{1-j}}\, \delta_{\bar\eta}\Delta_{1-j}\left[-\frac{1}{\Delta_{1-j}}\right]_\varepsilon \delta_\eta\mathbb{P}_{1-j}$$

$$+ \mathrm{Tr}_{\mathcal{A}_{j,0}}\, \rho^{j-1}(\delta_{\bar\eta}\partial_{1-j})\rho^{-j}\delta_\eta\left(\bar\partial_j\left[\frac{1}{\Delta_j}\right]_\varepsilon\right)$$

$$= \delta_\eta\delta_{\bar\eta} \log \det N_j + \delta_\eta\delta_{\bar\eta} \log \det N_{1-j}$$

$$+ (2 - j)\, \mathrm{Tr}_{\mathcal{A}_{j,0}}\, \eta\bar\eta\Delta_j\left[\frac{1}{\Delta_j}\right]_\varepsilon + (1 + j)\, \mathrm{Tr}_{\mathcal{A}_{1-j,0}}\, \eta\bar\eta\Delta_{1-j}\left[\frac{1}{\Delta_{1-j}}\right]_\varepsilon$$

$$+ \varepsilon \cdot \mathrm{Tr}_{\mathcal{A}_{j,0}}\, \rho^{j-1}(\delta_{\bar\eta}\partial_{1-j})e^{-\varepsilon\overline{\Delta_{1-j}}}\rho^{-j}\delta_\eta\bar\partial_j e^{-\varepsilon\Delta_j}. \quad (5.102)$$

Functional traces of the form $\mathrm{Tr}_{\mathcal{A}_{j,0}} = \int \rho_0 d^2\xi_0\langle\xi_0|\ \ |\xi_0\rangle$, which remains in (5.102), are evaluated with the help of the Fourier transformation with respect to the difference $\xi - \xi_0$, as it was done in (5.38)–(5.41). Then,

$$\mathrm{Tr}_{\mathcal{A}_{j,0}}\, \eta\bar\eta\Delta_j\left[\frac{1}{\Delta_j}\right]_\varepsilon = \mathrm{Tr}_{\mathcal{A}_{j,0}}\, \eta\bar\eta e^{-\varepsilon\Delta_j}$$

$$= \frac{1}{\pi\varepsilon}\int \rho\eta\bar\eta + \frac{3j - 1}{6\pi}\int \eta\bar\eta\partial\bar\partial\log\rho + o(\varepsilon),$$

$$\quad (5.103)$$

$$\mathrm{Tr}_{\mathcal{A}_{1-j,0}}\, \eta\bar{\eta}\Delta_{1-j}\left[\frac{1}{\Delta_{1-j}}\right]_{\varepsilon} = \mathrm{Tr}_{\mathcal{A}_{1-j,0}}\, \eta\bar{\eta}e^{-\varepsilon\Delta_{1-j}}$$

$$= \frac{1}{\pi\varepsilon}\int \rho\eta\bar{\eta} - \frac{3j-2}{6\pi}\int \eta\bar{\eta}\partial\bar{\partial}\log\rho + o(\varepsilon),$$

$$(5.104)$$

and the last trace in (5.102) is equal to

$$\varepsilon\cdot\mathrm{Tr}_{\mathcal{A}_{j,0}}\, \rho^{j-1}(\delta_{\bar{\eta}}\partial_{1-j})e^{-\varepsilon\overline{\Delta_{1-j}}}\rho^{-j}\delta_\eta\bar{\partial}_J e^{-\varepsilon\Delta_j} = -\frac{1}{6\pi}\int \eta\bar{\eta}\partial\bar{\partial}\log\rho$$

$$- \frac{c_j}{6\pi}\int \left[|\partial\eta|^2 + (\bar{\partial}\bar{\eta})\eta\partial\log\rho + (\partial\eta)\bar{\eta}\bar{\partial}\log\rho + \eta\bar{\eta}|\partial\log\rho|^2\right].$$

$$(5.105)$$

A combination of these formulas gives (5.57).

5.6. Compactified Moduli Space

The moduli space is a noncompact one, and in order to consider the asymptotic behaviour of the partition function near infinity, it is useful to compactify this space.

There are different ways to compactify the moduli space.

A compactification convenient for our purposes is the Deligne–Mumford compactification (Deligne and Mumford [1969]), which arises by adding to the class of smooth surfaces nonsmooth Riemann surfaces obtained by pinching the surfaces in such a way that finitely many noncontractible loops shrink to points, called nodes. This new moduli space $\bar{\mathcal{M}}_p$ is called the moduli space of stable curves. The boundary of this space

$$[\Delta] = \bar{\mathcal{M}}_p - \mathcal{M}_p$$

is the union

$$[\Delta] = \Delta_0 \cup \Delta_1 \cup \cdots \cup \Delta_{[p/2]};$$

the Δ_q, $q \neq 0$ correspond to the degenerate surface, which separate on the surfaces of genus q and genus $p-q$, the Δ_0 correspond to the surface with one degenerate bundle.

It is not difficult to show that the holomorphic section on the compactified moduli space will necessarily have a pole of certain order.

There exists an essential difference between the space $\mathcal{M}_p$ and the space $\bar{\mathcal{M}}_p$. For example, the bundles $\mathcal{L}_n$ over $\mathcal{M}_p$ are not defined over the space $\bar{\mathcal{M}}_p$. In particular, the bundle $\mathcal{L}_1$ is not a bundle over $\bar{\mathcal{M}}_p$. Let us now consider the space $\bar{X}_p$ and let us denote by sing $[\bar{X}_p]$ the subset in $\bar{X}_p$, which becomes

$[\Delta]$ after the application of map π. It can be shown that the object Ω can be extended from X_p to $\bar{X}_p$. And in this case, $\operatorname{sing}[\bar{X}_p]$ has the codimension 2 in $\bar{X}_p$, so that the second Chern class $c_2(\Omega) \neq 0$, and hence, we cannot consider Ω as a line bundle over $\bar{X}_p$.

Taking into account this fact and using the Riemann–Roch–Grothendieck theorem, we have

$$\lambda = \tfrac{1}{12}\pi_*\big(c_1(\Omega)^2 + [\operatorname{sing}\bar{X}_p]\big), \tag{5.106}$$

where

$$\lambda = c_1(\mathcal{L}_1^{-1}). \tag{5.107}$$

Let us denote the quantity $\pi_*[\operatorname{sing}[X]]$ as δ. Then after some calculations, we have

$$\delta = \Delta_0 + \tfrac{1}{2}\Delta_1 + \Delta_2 + \cdots + \Delta_{(p-q)}. \tag{5.108}$$

And we obtain the Mumford formula for the bundles $\mathcal{L}_n$,

$$c_1(\mathcal{L}_n) = -\left(\lambda + \frac{n}{2}(12\lambda - \delta)\right), \tag{5.109}$$

from which it follows that we have the holomorphic isomorphism

$$\mathcal{L}_{n+1} \simeq (\mathcal{L}_1)^{6n^2+6n+1} \otimes \mathcal{O}(\Delta)^{-n(n+1)/2}. \tag{5.110}$$

Here we denote by $\mathcal{O}(\Delta)$ the line bundle over $\bar{M}_p - \mathcal{M}_p$, which corresponds to the divisor, related to compactification.

It should also be taken into account that the line bundle $(\mathcal{L}_2)^{-1}$ now does not coincide with the canonical bundle over $\bar{\mathcal{M}}_p$. Indeed, the calculations give the formula (Mumford [1977]; Faltings [1984])

$$c_1(K_{\bar{\mathcal{M}}_p}) = 13\lambda - 2\delta, \tag{5.111}$$

but at the same time

$$c_1\big((\mathcal{L}_2)^{-1}\big) = 13\lambda - \delta. \tag{5.112}$$

Hence, the global section of the bundle $\mathcal{L}_1^{-13} \otimes \mathcal{L}_2 \otimes \mathcal{O}(-\Delta)$, which describes the partition function, has a second-order pole at infinity, and the asymptotic behaviour of the partition function at infinity has the form

$$\chi_1 \wedge \ldots \wedge \bar{\chi}_{3p-3}(\det \operatorname{Im}\tau)^{-13}|F(\tau)|^{-2} \sim \frac{1}{|\tau|^{14}}. \tag{5.113}$$

In the simplest cases of genus $p = 1$ and $p = 2$, it is possible to obtain the explicit formulae for the modular form in (5.26).

So, for $p = 1$, from (5.48), we have $\lambda = \tfrac{1}{12}\Delta_0$, so that

$$\mathcal{O}(\Delta) \simeq (\pi_*\Omega)^{\otimes 12}. \tag{5.114}$$

The section of this bundle is a modular form of weight 12 on $\mathcal{H}_1$ and is given by the formula

$$\left(\eta(\tau)\right)^{24} = \exp(2\pi i\tau) \prod_{j=1}^{\infty} \left(1 - \exp(2\pi i\tau n)\right)^{24}. \tag{5.115}$$

It is easy to prove now that because (5.57) is the discriminant of the elliptic curve, the quantity (5.57) cannot be zero, with the exception of subset Δ_0 where this quantity has a zero of first-order. And now using (5.52), we obtain

$$(\pi_!\omega^{\otimes 2}) \simeq (\pi_*\omega)^{\otimes 14}. \tag{5.116}$$

Therefore, we obtain the well-known expression for the integrand in (5.25), namely,

$$\frac{d\tau \wedge d\bar{\tau}}{|\eta(\tau)^{24}|^2}(\operatorname{Im}\tau)^{-14}. \tag{5.117}$$

(Note, that $\exp\delta(S) = (\eta(\tau))^{24}$, in correspondence with Faltings [1984].)

For genus $p = 2$, we have, except for (5.50), once more the relation $10\lambda = \Delta_0 + \Delta_1$, from which it follows that the modular form (5.26) may, in fact, be considered as the section of the bundle

$$\mathcal{O}(\Delta) \simeq (\Lambda^2\pi_*\Omega)^{\otimes 10}. \tag{5.118}$$

Or, using analytical terminology, we have

$$\chi_{10}(\tau) = \left[\prod_{m',m''} \theta\!\left[\begin{matrix} m' \\ m'' \end{matrix}\right](0,\tau) \right]^2, \tag{5.119}$$

and this function has the first-order zero on $\Delta_0 + \Delta_1$.

From the Riemann–Roch–Grothendieck theorem, it also follows that

$$\Lambda^3(\pi_*\Omega^{\otimes 2}) \simeq (\Lambda^2\pi_*\Omega)^3, \tag{5.120}$$

so that for the quantity (5.35), we have

$$\frac{|d\tau_{11} \wedge d\tau_{12} \wedge d\tau_{22}|^2}{|\chi_{10}|^2}(\det\operatorname{Im}\tau)^{-13}. \tag{5.121}$$

Thus, we now obtain some results from the papers by Belavin, Knizhnik, Morozov and Perelomov [1986], Shapiro [1970], Alessandrini and Amati [1971], and Moore [1986], using the pure algebraic method. In principle, it is also possible to consider the case $p = 3$. One should, however, take into account the fact that in this case the divisor of the Hodge class λ will contribute also on the inside of the domain, that is, in the points where the algebraic curves under consideration become hyperelliptic ones. And because of this fact, the calculations become more complicated.

Chapter 6
Examples of the Calculation of the Mumford Measure

6.1. The One-loop Case

Here the Riemann surface S_1 is the torus T, which can be considered as the factor space $T = \mathbb{C}/L$, where L is the regular lattice in a complex plane. The lattice L is generated by two periods ω_1 and ω_2, which by means of similarity and rotation transformations may be reduced to the form

$$\omega_1 = 1, \qquad \omega_2 = \tau, \tag{6.1}$$

where

$$\operatorname{Im}\tau > 0. \tag{6.2}$$

We also choose the standard metric on the torus T,

$$ds^2 = dz \cdot d\bar{z}. \tag{6.3}$$

Let us now consider the moduli space for this case. Note first of all that a choice of a new pair of basis vectors instead of the original ones ω_1 and ω_2 generate the same lattice, and the set of such combinations is generated by the group $\mathrm{SL}(2, Z)$, i.e., the automorphism group of the lattice and this group transform the lattice vectors in the upper half-plane to the vectors in the same half-plane. In the case under consideration, this group acts on the upper half-plane H by means of fractional linear transformations,

$$\tau \to (a\tau + b)(c\tau + d)^{-1}, \tag{6.4}$$

where a, b, c, and d are integers and $ab - cd = 1$.

The group $\mathrm{SL}(2, Z)$ is generated by the transformations

$$\tau \to \tau + 1, \qquad \tau \to -1/\tau \tag{6.5}$$

Here the transformation $\tau \to \tau + 1$ corresponds to the Dehn twist around an a-cycle, and the transformation $\tau \to -1/\tau$ corresponds to the interchange of the cycles a and b.

As for the $\mathrm{SL}(2, Z)$ group, this group is important because it describes unconnected diffeomorphisms of the torus and is usually called the modular group. Therefore, the moduli space is the factor space $H/\mathrm{SL}(2, Z)$.

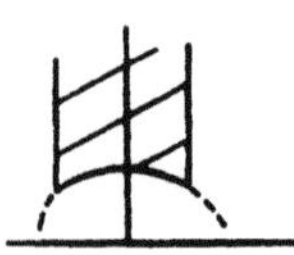

Due to invariance of the partition function for the torus T relative to the $SL(2, Z)$ group, we have

$$Z_1 = \int_{\mathcal{F}} d\tau \frac{\det' \nabla_z^{-1} \det' \nabla_{(1)}^z}{V} \frac{\langle \mu | \varphi \rangle \langle \varphi | \mu \rangle}{\langle \varphi | \varphi \rangle} \left(\frac{\det' \Delta}{\int d^2 z \sqrt{g}} \right)^{-13}, \qquad (6.6)$$

where we use the fact that the quantities μ and φ in this case have only one component and the metric has the form $ds^2 = |dz|^2$. Here V is the volume of the conformal Killing group.

In the case under consideration, the conformal Killing group is $U(1) \times U(1)$ and is generated by two vectors a_1 and a_2, where

$$a_1 = \frac{1}{2} \left(\frac{\partial}{\partial z} + \frac{\partial}{\partial \bar{z}} \right), \qquad a_2 = \frac{1}{2}. \qquad (6.7)$$

This quantity up to the normalization factor is given by the formula $V = \det(a_i, a_j)$, where the scalar product in $(\ ,\ )$ is given by the formula

$$(a, b) = \int d^2 z a_\alpha b^\alpha. \qquad (6.8)$$

Therefore,

$$(a_i, a_j) = \begin{pmatrix} \tau_2, & \tau_1 \tau_2 \\ \tau_1 \tau_2, & \tau_2 |\tau|^2 \end{pmatrix}, \qquad (6.9)$$

and $V = \tau_2^2$.

Furthermore, in the torus case, the quadratic differential has the form $\varphi_{zz} = a$, where a is a complex number. And the Beltrami differential is determined by the metric change under the change τ,

$$ds^2 = |dz|^2 \rightarrow \left| dz + \delta\tau \frac{i}{2\tau_2} d\bar{z} \right|^2, \qquad (6.10)$$

Hence, the Beltrami differential is given by the formula

$$\mu_{zz} = i/2\tau_2. \qquad (6.11)$$

Therefore, up to the normalizing factor, we have

$$\langle \mu | \varphi \rangle = a, \qquad \langle \varphi | \varphi \rangle = a^2 \tau_2, \qquad (6.12)$$

and

$$\frac{\langle \mu | \varphi \rangle \langle \varphi | \mu \rangle}{\langle \varphi | \varphi \rangle} = \frac{1}{\tau_2}. \qquad (6.13)$$

The rest of the problem now is to calculate the determinants in (6.6). For the flat metric on the torus, we have

$$\det' \nabla_z^{(-1)} \det' \nabla_{(1)}^z = \det' \partial_z \cdot \det' \partial_{\bar{z}} = \det' \Delta, \qquad (6.14)$$

so that the entire one-loop partition function has the form

$$Z_1 = \int \frac{d\tau}{\tau_2^2} \tau_2^{12} (\det' \Delta)^{-12}. \qquad (6.15)$$

And to calculate $(\det' \Delta)$ let us find the eigenvalues of the operator Δ for the torus.

As is easy to see, the eigenfunction of the operator Δ for the torus has the form

$$\psi(\sigma_1, \sigma_2) = \exp\left[2\pi i\left(n\sigma_1 - \frac{\tau_1}{\tau_2}n\sigma_2 - \frac{1}{\tau_2}m\sigma_2\right)\right], \qquad (6.16)$$

and the corresponding eigenvalue is given by the formula

$$\frac{4\pi^2}{\tau_2^2}\big[(m + \tau n)(m + \bar{\tau}n)\big]. \qquad (6.17)$$

Therefore,

$$\det' \Delta = \prod_{m,n}{}' \frac{4\pi^2}{\tau_2^2}\big[(m + \tau n)(m + \bar{\tau}n)\big]. \qquad (6.18)$$

However, this product is divergent and should be regularized. And usually the ζ-function regularization is used, which is defined by the formula

$$\zeta(s) = \sum_{m,n}{}' \lambda_{m,n}^{-s}, \quad \lambda_{m,n} = \big[(m + n\tau)(m + \bar{\tau}n)\big]. \qquad (6.19)$$

This function is analytic for a sufficiently large value $\mathrm{Re}\, s$. And the quantity $\zeta(-1)$, which is interesting for us, is defined by means of analytic continuation in s.

In the case under consideration, we have

$$\prod_{m,n}{}' \frac{1}{\tau_2^2} = \tau_2^2, \quad \prod_{m,n}(m + \tau n) = \eta(\tau)^2, \qquad (6.20)$$

where $\eta(\tau)$ is the so-called Dedekind η-function

$$\eta(\tau) = \exp(i\pi\tau/12) \prod_{n>1}\big(1 - \exp(2\pi in\tau)\big). \qquad (6.21)$$

Collecting all the separate factors, we obtain

$$Z_1 = \int_{\mathcal{F}} \frac{d\tau}{\tau_2^2}\tau_2^{-12}\big|\eta(\tau)\big|^{-48}. \qquad (6.22)$$

It is note too difficult to check that the quantity Z_1 is invariant relative to the transformations $\tau \to \tau + 1$ and $\tau \to -\frac{1}{\tau}$ and, therefore, relative to all modular transformations.

6.2. The Measure for the Cases $p = 2$, 3, and 4

In Chapt. 5, it was shown that this measure is determined unambiguously up to a constant multiplier. The main difficulty in the derivation of explicit formulas for general p is the absence of a good parametrization of the space $\mathcal{M}_p$. However, for $p = 2, 3$ (and, in fact, for $p = 1$), the complex structures may be parametrized by period matrices. In conjunction with the analytical properties of the measure established in Chapt. 5, this allows us to express it through θ-constants.

From the results of Chapt. 5, it follows that the measure in question has the form

$$d\mu_p = d\nu_p \left| \chi_{12-p}(\tau) \right|^{-2} (\det \operatorname{Im} \tau)^{p-12}, \quad p = 2, 3. \tag{6.23}$$

Here

$$d\nu_p = \prod_{k \leq j} \frac{i}{2} d\tau_{kj} \wedge d\bar{\tau}_{kj} (\det \operatorname{Im} \tau)^{-(p+1)} \tag{6.24}$$

is the modular invariant measure on $\mathcal{H}_p$. The requirement for (6.23) to be invariant under modular transformations implies that

$$\chi_{12-p}\bigl(M(\tau)\bigr) = \bigl(\det(C\tau + D)\bigr)^{12-p} \chi_{12-p}(\tau). \tag{6.25}$$

Thus, for $p = 2$, χ_{10} is the modular form of weight 10 with no zeroes inside $\mathcal{H}_2$. Besides, at infinity Δ_0 (where τ_{11} of τ_{12} tends to $i\infty$), the measure $\prod_{i \leq j} d\tau_{ij}$ has a first-order pole. Hence the analytical properties of the entire measure $\left\{ \prod_{i \leq j} d\tau_{ij} \cdot (\chi_{12-p}(\tau))^{-1} \right\}$, cited in Chapt. 5, imply that the form χ_{10} has a first-order zero at Δ_0 and a double zero at Δ_1 (where $\tau_{12} \to 0$), i.e., χ_{10} is parabolic. The form χ_{10} is unambiguously defined by the weight and location of the zeroes (up to a constant multiplier) and is equal to a product of all the even θ-constants (Belavin, Knizhnik, Morozov and Perelomov [1986]; Moore [1986]):

$$\chi_{10} = \prod_m \theta_m^2(\tau). \tag{6.26}$$

Theta-constants are defined as

$$\theta_m(\tau) = \sum_{n \in Z^p} \exp\left\{ i\pi \left(n + \frac{m'}{2} \right)' \tau \left(n + \frac{m'}{2} \right) + 2\pi i \left(n + \frac{m'}{2} \right) \frac{m''}{2} \right\}, \tag{6.27}$$

for $p = 2$, and components of the vectors m', m'', which form the "theta-characteristic" $m = (m', m'')$, take the values $0, 1$. The number $e(m) = (m')^t m'' \pmod 2$ is referred to as the parity of the characteristic m, and the product in (6.26) is over all even characteristics. For genus p, there are as many as $2^{p-1}(2^p + 1)$ even and $2^{p-1}(2^p - 1)$ odd characteristics. For odd $e(m) = 1$, $\theta_m(\tau) \equiv 0$. Using (6.27), one can easily check that χ_{10} has the first- and second-order zeroes at Δ_0 and Δ_1, respectively. Also, the nonvanishing of χ_{10} inside $\mathcal{H}_2$ may be easily demonstrated. For this purpose, we make use of the following formula for the fermionic determinant (Knizhnik [1986]):

$$\left(\det_m \bar{\partial}_{1/2} \right) \cdot \left(\det \bar{\partial}_0 \right)^{1/2} = \theta_m(\tau). \tag{6.28}$$

Here $\det_m \bar{\partial}_{1/2}$ is the determinant of the Dirac operator that acts on the space of the left-handed Weyl fermions "living" on S_p and satisfying the periodic

244 A.Yu. Morozov and A.M. Perelomov

(antiperiodic) boundary conditions under transitions along the cycles a_i for $m' = 1(0)$ and b_j for $m'' = 1(0)$,

$$|\det \bar{\partial}_0|^2 \stackrel{\text{def}}{=} \frac{\det'\left(-\frac{1}{\sqrt{g}}\partial_a\sqrt{g}g^{ab}\partial_b\right)}{(\det \operatorname{Im}\tau)\cdot\left(\int\sqrt{g}d^2\xi\right)}. \tag{6.29}$$

From (6.28), it follows that $\theta_m(\tau)$ vanishes at the surfaces S_2 where $\bar{\partial}_{1/2}$ acquires fermionic zero-modes with boundary conditions, specified by m, i.e., when there are holomorphic $\frac{1}{2}$-differentials with characteristic m on S_2. According to the Riemann singularity theorem, the parity of the number of such $\frac{1}{2}$-differentials coincides with the parity $e(m)$ of the characteristic m. Moreover, their number is equal to $e(m)$ for a generic surface. The vanishing of $\theta_m(\tau)$ for even m, $e(m) = 0$, on the surface S_2 thus suggests that there are at least two holomorphic $\frac{1}{2}$-differentials $\psi_1(z)(dz)^{1/2}$ and $\psi_2(z)(dz)^{1/2}$ with the characteristic m on S_2. Besides, any holomorphic k-differential on a surface of genus p is known to have $2k(p-1)$ zeroes. There is, therefore, a meromorphic function $f(z) = \psi_1(z)/\psi_2(z)$ on S_2 with one pole and one zero, which is possible only for surfaces of genus zero. This discrepancy indicates that $\theta_m(\tau)$ has no zeroes inside $\mathcal{H}_2$ for $e(m) = 0$. The absence of holomorphic $\frac{1}{2}$-differentials with an even characteristic seems to also ensue from the fact that any Riemann surface of genus 2 may be represented as a hyperelliptic curve

$$y^2 = (z - a_1)\ldots(z - a_6) \tag{6.30}$$

in $\mathbb{C}^2 = \{(y, z)\}$. It can be easily ascertained that such a surface has exactly six holomorphic $\frac{1}{2}$-differentials:

$$\psi_i = \sqrt{(z - a_i)/y}\,(dz)^{1/2}, \tag{6.31}$$

i.e., one for each for the six odd characteristics, with none left for the even ones.

Let us now turn to $p = 3$. In this case, the measure $\prod_{i\leq j} d\tau_{ij}$ has a first-order pole at Δ_0 and a first-order zero at Δ_1. In addition, the χ_9 form of weight 9 may only be defined as taking values in the character of Γ_3, since in case of $p = 3$, the right-hand side of (6.25) appears to change the sign with substitution of $-M$ for M, while the left-hand side does not. The "real" modular form is

$$\chi_9^2 = \chi_{18}. \tag{6.32}$$

It should have the second-order zero at Δ_0 and the sixth-order one at Δ_1. Such a form exists and is equal to (Belavin, Knizhnik, Morozov and Perelomov [1986])

$$\chi_{18} = \prod_m \theta_m(\tau), \tag{6.33}$$

with the product over all the 36 even characteristics. However, χ_{18} vanishes not only at Δ_0 and Δ_1, but also at a manifold Δ_* of hyperelliptic curves inside $\mathcal{H}_3$ (Igusa [1972]). Indeed, on a hyperelliptic curve of genus 3,

$$y^2 = (z - a_1)\ldots(z - a_8), \tag{6.34}$$

besides 28 fermionic zero-modes

$$\psi^{ij} = \sqrt{(z - a_1)(z - a_j)/y}(dz)^{1/2} \tag{6.35}$$

(one for each of the 28 odd characteristics), there are two additional zero-modes

$$\psi^{(0)} = \frac{1}{\sqrt{y}}(dz)^{1/2}, \qquad \psi^{(1)} = \frac{z}{\sqrt{y}}(dz)^{1/2}, \tag{6.36}$$

with the common even characteristic, its specific value being dependent on the choice of the basis of cycles on the curve (6.34). Thus, χ_{18} actually vanishes at Δ_*; this zero in the coordinates y is of second order. Vice versa, if the form χ_{18} vanishes for some surface Δ_*, there are at least two holomorphic $\frac{1}{2}$-differentials $\psi^{(1)}(z)$ and $\psi^{(2)}(z)$ with the same even characteristic on S_3, their ratio $f(z) = \psi^{(1)}(z)/\psi^{(2)}(z)$ being a meromorphic function on S_3 with two zeroes and two poles, i.e., S_3 is in fact a two-sheet covering of CP^1 and, consequently, a hyperelliptic surface.

To sum up, χ_{18} vanishes inside $\mathcal{H}_3 = M_3$ exactly on hyperelliptic surfaces. This zero is of second order in the coordinates y_j, and the square root $\chi_9 = \sqrt{\chi_{18}}$ may be taken. It remains to be shown that the measure $\prod_{i \leq j} d\tau_{ij}$ has the first-order zero on hyperelliptic curves. For this purpose, we choose the basis of holomorphic quadratic differentials on surface (6.34) as follows:

$$f_k = \frac{z^{k-1}}{y^2}(dz)^2, \quad k = 1, \dots, 5; \quad f_6 = \frac{(dz)^2}{y}. \tag{6.37}$$

Also take the metric $\rho\, dz \cdot d\bar{z}$ symmetric under the transform $P: (y, z) \to (-y, z)$. Then η^6 may be chosen odd under P:

$$\eta^6 = \text{const} \cdot f_6/\rho\, dz \cdot d\bar{z}. \tag{6.38}$$

Furthermore, the holomorphic abelian differentials $\omega_j(z) = \varphi_j(z)dz$ are linear combinations of

$$dz/y, \quad zdz/y, \quad z^2dz/y. \tag{6.39}$$

Hence, (6.37) and the formula for the variation of τ,

$$\partial\tau_{ab}/\partial y_j = \int_{S_p} \eta^j \omega_a \wedge \omega_b, \tag{6.40}$$

which is valid for any genus, suggests that

$$\partial\tau_{ab}/\partial y_6 = 0. \tag{6.41}$$

Taking $y_6(\Delta_*) = 0$, we find

$$\prod_{i \leq j} d\tau_{ij} \underset{y_6 \to 0}{\sim} y_6 dy_1 \wedge \cdots \wedge dy_6.$$

Consequently, the measure

$$\prod_{i \leq j} d\tau_{ij} \chi_{18}^{-1/2} \sim dy_1 \wedge \cdots \wedge dy_6 \tag{6.42}$$

is holomorphic in the coordinates y_j in the vicinity of Δ_* and does not vanish, in accordance with Belavin and Knizhnik [1986]. The integrable rootlike singularity intrinsic in $d\mu_3$, when expressed in terms of the period matrix τ, does not contradict the holomorphic properties of the measure, because in the vicinity of hyperelliptic curves (those possessing the symmetry P), $\mathcal{M}_3$ embeds into $\mathcal{H}_3$ in a nonsmooth way. To summarize, we have thus proved the validity of formulae (6.23), (6.29), and (6.30) for the measure for genus $p = 3$. It is worth noting that analytical properties of the forms χ_{10}, χ_{18} have been studied by Igusa [1972] with the use of other methods.

As far as genus 2 is concerned, the moduli space M_2 may also be parametrized by the coordinates of the ramification points $\lambda_1, \lambda_2, \lambda_3$ of the curve

$$y^2 = z(z - 1)(z - \lambda_1)(z - \lambda_2)(z - \lambda_3) \tag{6.43}$$

in $\mathbb{C}^2 = \{(y, z)\}$. In these coordinates, the measure looks like

$$d\Omega = \prod_{i \leq j} d\tau_{ij} [\chi_{10}(\tau)]^{-1} = [P(\lambda)]^{2/5} [\chi_{10}(\tau)]^{-13/10} \times d\lambda_1 d\lambda_2 d\lambda_3, \tag{6.44}$$

$$P(\lambda) = \lambda_1 \lambda_2 \lambda_3 (1 - \lambda_1)(1 - \lambda_2)(1 - \lambda_3)(\lambda_1 - \lambda_2)(\lambda_2 - \lambda_3)(\lambda_3 - \lambda_1),$$

and the period matrix τ may be expressed by $\lambda_1, \lambda_2, \lambda_3$ in terms of hyperelliptic integrals. To define the partition function

$$Z_3 \sim \int d\Omega \wedge d\bar{\Omega} (\det \operatorname{Im} \tau)^{-13},$$

one may integrate over each $d^2\lambda_i$ throughout the entire complex plane, since this accounts for the contribution of almost each Riemann surface (for the general case) as many as 720 times, thus giving rise to a simple numerical factor of 720.

Chapter 7
Multiloop Calculations in the
Theory of Free Fields on Riemann Surfaces

In order to derive general formulae for any number of loops, it will be necessary to develop the formalism of free fields on Riemann surfaces. This chapter is taken from the paper by Gerasimov, Marshakov, Morozov, Olshanetskij and Shatashvili [1989].

7.1. Differential Geometry of Riemann Surfaces
(Fay [1973]; Mumford [1983; 1984]; Clemens [1980])

Here we collect some facts from the theory of Riemann surfaces that appear useful for multiloop calculations.

The *Jacobian map*,

$$\xi \to \vec{\xi} \equiv \int_{\xi_0}^{\xi} \vec{\omega}; \quad \vec{\xi} = \{\xi_1, \ldots, \xi_p\}, \tag{7.1}$$

may be considered as a map of a genus-p Riemann surface S_p into a p-dimensional torus (Jacobian), which is a factor of $\mathbb{C}^p$ over a group of translations $\xi_i \to \xi_i + \delta_{ij}$, $\xi_i \to \xi_i + T_{ij}$. The exact choice of the point ξ_0 in (7.1) is usually unessential.

The image of a Riemann surface after the map (7.1) is described by Riemann's vanishing theorem in terms of theta-functions. Namely, on S_p there are $(p-1)$ points $R_1^*, \ldots, R_{p-1}^*$, such that for *arbitrary* $p-1$ points $\xi_1, \ldots \xi_{p-1}$ on S_p,

$$\theta_*(\vec{\xi_1} + \cdots + \vec{\xi}_{p-1} - \vec{R}_1^* - \cdots - \vec{R}_{p-1}^*) = 0. \tag{7.2}$$

(The parameter * is an arbitrary nonsingular half-integer theta-characteristic; the theta-function with the characteristic $\begin{bmatrix} \vec{\delta} \\ \vec{\varepsilon} \end{bmatrix}$ is defined as

$$\theta \begin{bmatrix} \vec{\delta} \\ \vec{\varepsilon} \end{bmatrix} (\vec{z}|T) \equiv \sum_{\vec{n} \in \mathbb{Z}^p} \exp\left[i\pi(n+\delta)_i T_{ij}(n+\delta)_j + 2\pi i(n+\delta)_i(z+\varepsilon)_i\right]. \tag{7.3}$$

It is easy to derive from this theorem that the holomorphic 1-differential

$$\nu_*^2(\xi) \equiv \sum_{i=1}^{p} \theta_*(\vec{0}), \; {}_i\omega^i(\xi) \tag{7.4}$$

has double zeros at the points $R_1^*, \ldots, R_{p-1}^*$ and is, in fact, a square of the holomorphic $\frac{1}{2}$-differential $\nu_*(\xi)$. Another corollary is that the *Prym bidifferential*,

$$E(\xi, \xi') \equiv \frac{\theta_*(\vec{\xi} - \vec{\xi}')}{\nu_*(\xi)\nu_*(\xi')}, \tag{7.5}$$

possesses a simple zero when $\xi = \xi'$ and has no poles at all. $E(\xi, \xi')$ is invariant under the shifts of ξ along any A-periods, and changes under a shift of ξ along the B_j-period as

$$E(\xi, \xi') \longrightarrow E(\xi, \xi') \exp\left[2\pi i(\xi - \xi')_j + \pi i T_{jj}\right]. \tag{7.6}$$

There is another useful object on a Riemann surface: a holomorphic $p/2$-differential without poles and zeros:

$$\sigma_*(\xi) \equiv \nu_*(\xi)/\prod_{\alpha=1}^{p-1} E(\xi, R_\alpha'^*). \tag{7.7}$$

248 A.Yu. Morozov and A.M. Perelomov

For an even nonsingular theta-characteristic e, the *Szegő kernel* is defined as

$$G_e^{(1/2)}(\xi, \xi') \equiv \frac{\theta_*(\vec{\xi} - \vec{\xi'})}{\theta_*(\vec{0})E(\xi, \xi')}. \tag{7.8}$$

It may be interpreted as the Green function of $\frac{1}{2}$-differentials (spinors) on a Riemann surface with appropriate boundary conditions:

$$\frac{\langle \tilde{\psi}(\xi)\psi(\xi')\rangle_e}{\det_e \bar{\partial}_{1/2}} \equiv \langle\langle \tilde{\psi}(\xi)\psi(\xi')\rangle\rangle_e = G_e^{(1/2)}(\xi, \xi'). \tag{7.9}$$

For these Green functions, the following analog of Wick's theorem holds:

$$\langle\langle \tilde{\psi}(\xi_1)\ldots\tilde{\psi}(\xi_n)\psi(\xi_1')\ldots\psi(\xi_n')\rangle\rangle_e$$

$$= \frac{\theta_e(\vec{\xi}_1 + \cdots + \vec{\xi}_n - \vec{\xi}_1' - \cdots - \vec{\xi}_n')}{\theta_e(\vec{0})\prod_{i,j}E(\xi_i, \xi_j')}\prod_{i<j}E(\xi_i, \xi_j)E(\xi_i', \xi_j')$$

$$= \det_{(ij)}\frac{\theta_*(\vec{\xi}_i - \vec{\xi}_j')}{\theta_*(\vec{0})E(\xi_i, \xi_j')} = \det_{(ij)}\langle\langle \tilde{\psi}(\xi_i)\psi(\xi_j')\rangle\rangle_e. \tag{7.10}$$

We shall also need Green functions of the Laplace operator Δ_0. Usually, the Green functions $\langle \xi|\Delta_0^{-1}|\xi'\rangle$ on a surface with metric $\gamma(\xi)$ is defined as a solution of the following equation:

$$\Delta_0^{\{\gamma\}}(\xi)\log G^{\{\gamma\}}(\xi, \xi') = \pi\left(\delta^{\{\gamma\}}(\xi, \xi') - \left[\int\sqrt{\gamma(\xi)}\,d^2\xi\right]^{-1}\right). \tag{7.11}$$

Here the δ-function is normalized by the condition

$$\int \delta^{\{\gamma\}}(\xi, \xi')\sqrt{\gamma(\xi)}\,d^2\xi = 1, \quad \text{,i.e., } \delta^{\{\gamma\}}(\xi, \xi') = \delta(\xi, \xi')/\sqrt{\gamma(\xi)}. \tag{7.12}$$

The second term on the right-hand side of (7.11) is due to zero modes: The Green function $\log G(\xi, \xi') = \pi\sum_{\lambda_n \neq 0}\phi_n(\xi)\frac{1}{\lambda_n}\overline{\phi_n(\xi')}$ with normalized eigenfunctions $\phi_n(\xi)$, $\Delta_0\phi_n = \lambda_n\phi_n$, satisfies the equation

$$\Delta_0(\xi)\log G(\xi, \xi') = \pi\sum_{\lambda_n \neq 0}[\Delta_0\phi_n](\xi)\frac{1}{\lambda_n}\overline{\phi_n(\xi')} = \pi\sum_{\lambda_n \neq 0}\phi_n(\xi)\overline{\phi_n(\xi')}$$

$$= \pi\sum_{\lambda_n}\phi_n(\xi)\overline{\phi_n(\xi')} - \pi\sum_{\lambda_n = 0}\phi_n(\xi)\overline{\phi_n(\xi')}$$

$$= \pi\left[\delta(\xi, \xi') - \sum_{\lambda_n = 0}\phi_n(\xi)\overline{\phi_n(\xi')}\right]. \tag{7.13}$$

It is hard to write down an explicit formula for $G^{\{\gamma\}}$ for an arbitrary metric γ. In string theory, however, we need slightly different Green functions, which are solutions of two other equations, (7.14) and (7.19), below. The first of these equations looks like

$$\Delta_0^{\{\gamma\}}(\xi)\log G^{(0)}\big(\xi|\{A_I,\xi_I\}\big) = \sum_{I=1}^{N}\pi\cdot A_I\delta^{\{\gamma\}}(\xi,\xi_I),\tag{7.14}$$

or, in conformal gauge, $\Delta_0 = -\gamma^{-1/2}\partial\bar{\partial}$,

$$\partial\bar{\partial}\log G^{\{0\}}\big(\xi|\{A_I,\xi_I\}\big) = \pi\sum_{I=1}^{N}A_I\delta^{\{\gamma\}}(\xi,\xi_I),\tag{7.15}$$

with the additional constraint

$$\sum_{I=1}^{N}A_I = 0.\tag{7.16}$$

The explicit solution of (7.15) is (note that $\partial\bar{\partial}\log|z|^2 = -\pi\delta(z)$):

$$G^{(0)}\big(\xi|\{A_I,\xi_I\}\big) = f\{A_I,\xi_I\}\prod_{I=1}^{N}\big|E(\xi,\xi_I)^{A_I}\big|^2\cdot e^{2\pi A_I\,\mathrm{Im}(\xi-\xi_I)_i\left(\frac{1}{\mathrm{Im}\,T}\right)_{ij}\mathrm{Im}(\xi-\xi_I)_j}.\tag{7.17}$$

Equation (7.17) defines a single-valued function on S_p. The factor $f\{A_I,\xi_I\}$ is equal to

$$\begin{aligned}
f\{A_I,\xi_I\} &= \prod_{I=1}^{N}\big|\sigma_*(\xi_I)\big|^{2A_I/p}\cdot e^{2\pi\cdot\frac{A_I}{p(p-1)}\,\mathrm{Im}(\tilde{\xi}_I)_i\left(\frac{1}{\mathrm{Im}\,T}\right)_{ij}\mathrm{Im}(\tilde{\xi}_I)_j}\\
&= \prod_{I=1}^{N}g_{\mathrm{Ar}}(\xi_I)^{A_I/2}.
\end{aligned}\tag{7.18}$$

It accounts for the proper dependence of $G^{(0)}(\xi|\{A_I,\xi_I\})$ on ξ_I and is unessential in applications. (The vector $\vec{\tilde{\xi}}_I$ and metric $g_{\mathrm{Ar}}(\xi)$, entered in (7.18), are defined in (7.22) and (7.30).)

The second sort of relevant Green functions is defined by the equation

$$\Delta_0^{\{\gamma\}}(\xi)\log G^{\{\mathcal{R}\}}(\xi,\xi') = \pi\cdot\left(\delta^{\{\gamma\}}(\xi,\xi') + \kappa\mathcal{R}\sqrt{\gamma}(\xi)\right),\tag{7.19}$$

with $\kappa\int\mathcal{R}\sqrt{\gamma}d^2\xi = -1$, i.e.,

$$\kappa = \frac{1}{2\pi(p-1)}.\tag{7.20}$$

The solution of (7.19) looks like

$$\begin{aligned}
G^{\{\mathcal{R}\}}(\xi,\xi') = F\{R_\alpha^*\}&\left|E(\xi,\xi')\sigma_*^{\frac{1}{p-1}}(\xi)\sigma_*^{\frac{1}{p-1}}(\xi')\right|^2\\
&\times e^{\pi\cdot\frac{1}{(p-1)^2}\,\mathrm{Im}\,\tilde{\xi}_i\left(\frac{1}{\mathrm{Im}\,T}\right)_{ij}\mathrm{Im}\,\tilde{\xi}_j'}\big(\gamma(\xi)\gamma(\xi')\big)^{-\frac{1}{2(p-1)}},
\end{aligned}\tag{7.21}$$

with

$$\vec{\tilde{\xi}} \equiv (p-1)\vec{\xi} - \vec{\Delta}_*; \qquad \vec{\tilde{\xi}}' \equiv (p-1)\vec{\xi}' - \vec{\Delta}_*; \qquad \vec{\Delta}_* \equiv \sum_{\alpha=1}^{p-1} \vec{R}_\alpha^*, \qquad (7.22)$$

and is a single-valued function of ξ and ξ'. The factor

$$F\{R_\alpha^*\} = \prod_{\alpha=1}^{p-1} g_{\mathrm{Ar}}(R_\alpha^*)^{\frac{1}{p-1}} \qquad (7.23)$$

turns $G^{\{\mathcal{R}\}}$ into a 0-differential with respect to all R_α^*. Let us note that (7.21) provides a principal way to find a Green function of the type (7.11). For this purpose, one should find a new metric $\tilde{\gamma}$, related to the original γ through

$$\mathcal{R}_{\tilde{\gamma}}\sqrt{\det \tilde{\gamma}} = \partial\bar{\partial}\log\tilde{\gamma} = -2\pi(p-1)\gamma / \int \sqrt{\det\gamma}\, d^2\xi. \qquad (7.24)$$

Then the substitution of this $\tilde{\gamma}$ into (7.21) gives $G^{\{\gamma\}}$.

Define the Green function at coincident points as

$$\log G_{\mathrm{reg}}^{\{\gamma\}}(\xi,\xi') \equiv \lim_{\varepsilon\to 0}\left(\int_\varepsilon^\infty dt\langle\xi\,|e^{-t\Delta_0}|\,\xi'\rangle - \log\frac{|\xi-\xi'|^2\sqrt{\det\gamma(\xi)}}{\varepsilon}\right). \qquad (7.25)$$

The counterterm is chosen to maintain two-dimensional covariance. In fact, we have

$$G_{\mathrm{reg}}^{\{\gamma\}}(\xi,\xi) = \frac{1}{\sqrt{\gamma(\xi)}}\lim_{\xi'\to\xi}\frac{G_{\mathrm{reg}}^{\{\gamma\}}(\xi,\xi')}{|\xi-\xi'|^2}. \qquad (7.26)$$

The following metrics on the Riemann surface are of special interest in string theory.

Bergmann's metric:

$$g_{\mathrm{Berg}}(\xi) = \frac{1}{2ip}\sum_{k,l=1}^{p}\omega_k(\xi)\left(\frac{1}{\mathrm{Im}\,T}\right)_{kl}\overline{\omega_l(\xi)}, \qquad (7.27)$$

normalized so that

$$\int \sqrt{g_{\mathrm{Berg}}(\xi)}\, d^2\xi = 1. \qquad (7.28)$$

Arakelov's metric, related to Bergmann's according to (7.24):

$$\tilde{g}_{\mathrm{Berg}}(\xi) = g_{\mathrm{Ar}}(\xi), \qquad \mathcal{R}_{\mathrm{Ar}}(\xi)\sqrt{\det g_{\mathrm{Ar}}(\xi)} = -2\pi(p-1)g_{\mathrm{Berg}}(\xi); \qquad (7.29)$$

$$g_{\mathrm{Ar}}(\xi) = |\sigma_*(\xi_I)|^{4/p}\cdot e^{\frac{4\pi}{p(p-1)}\,\mathrm{Im}\,\tilde{\xi}_i\left(\frac{1}{\mathrm{Im}\,T}\right)_{ij}\mathrm{Im}\,\tilde{\xi}_j}; \qquad (7.30)$$

$$\vec{\tilde{\xi}} \equiv (p-1)\vec{\xi} - \vec{\Delta}_*. \qquad (7.31)$$

Singular metrics:

$$g_W(\xi) = \left|W(\xi)\right|^2, \tag{7.32}$$

which are squares of moduli of holomorphic or meromorphic 1-differentials $W(\xi)$. These metrics have zeros at some points Q_a and poles at P_b. The curvature is concentrated at these points,

$$\mathcal{R}_W \sqrt{\det G_W} = \partial\bar{\partial}\log g_W = -\pi\left(\sum_{a=1}^{n_Q}\delta(\xi, Q_a) - \sum_{b=1}^{n_P}\delta(\xi, P_b)\right). \tag{7.33}$$

There are constraints on Q_a and P_b:

$$n_Q - n_P = 2(p-1); \qquad \sum_{a=1}^{n_Q}\vec{Q}_a - \sum_{b=1}^{n_P}\vec{P}_b = 2\vec{\Delta}. \tag{7.34}$$

From (7.34), it follows that

$$\int \mathcal{R}_W \sqrt{\det g_W}\, d^2\xi = -2\pi(p-1). \tag{7.35}$$

7.2. The Scalar Field on a Riemann Surface

Let us consider the functional integral

$$A\{k_I\} = \int \mathcal{D}\phi \exp\left(-\frac{1}{2\pi}\int\sqrt{\det g}\, g^{ab}\partial_a\phi\partial_b\phi + \sum_{I=1}^{N}k_I\phi(\xi_I)\right), \tag{7.36}$$

where $\phi(\xi)$ is a scalar field on the surface S_p. Integration over zero mode $\phi = \text{const}$ gives rise to the condition $\sum_I k_I = 0$. Thus, we may use Green functions of the type (7.14). One easily verifies that this Gaussian functional integral is equal to

$$A\{k_I\} = \left(\frac{\det N_0}{\det' \Delta_0}\right)^{1/2}\prod_{I<J}^{N}G(\xi_I, \xi_J)^{k_I k_J}\prod_{I=1}^{N}\sqrt{\det\gamma(\xi_I)}^{-k_I^2/2}\delta\left(\sum_I k_I\right). \tag{7.37}$$

Here,

$$G(\xi, \xi') \equiv \left|E(\xi, \xi')\right|^2 \cdot e^{2\pi\cdot\text{Im}(\xi-\xi')_i\left(\frac{1}{\text{Im}\,T}\right)_{ij}\text{Im}(\xi-\xi')_j}. \tag{7.38}$$

Now consider a somewhat more complicated functional integral:

$$A_\lambda\{k_I\} = \int \mathcal{D}\phi \exp\left(-\frac{1}{2\pi}\int\sqrt{\det g}(g^{ab}\partial_a\phi\partial_b\phi + 2\lambda\mathcal{R}_g\phi) + \sum_{I=1}^{N}k_I\phi(\xi_I)\right). \tag{7.39}$$

252 A.Yu. Morozov and A.M. Perelomov

Integration over zero mode leads to the following constraint:

$$\sum_{I=1}^{N} k_I - \frac{1}{\pi}\lambda \int \sqrt{\det g}\cdot R_g = \sum_{I=1}^{N} k_I + 2\lambda(p-1) = 0. \qquad (7.40)$$

This time, we need the Green functions of the type (7.19). It is useful to shift the integration variable $\phi \to \phi+\phi_0$, with ϕ_0 being the solution of the equation

$$\partial\bar\partial\phi_0 = \lambda\partial\bar\partial \log g(\xi) - \pi\sum_{I=1}^{N} k_I\delta(\xi,\xi_I). \qquad (7.41)$$

Let us introduce an auxiliary singular metric $g_*(\xi,\bar\xi) = |\nu_*(\xi)|^4$ with double zeros at the points R_α^*. We may rewrite (7.41) in the following form:

$$\partial\bar\partial\phi_0 = \lambda\partial\bar\partial \log[g/g_*] - \pi\cdot\left[\sum_{I=1}^{N} k_I\delta(\xi,\xi_I) + 2\lambda\sum_{\alpha=1}^{p-1}\delta(\xi,R_\alpha^*)\right]. \qquad (7.42)$$

The solution of this equation looks like

$$\phi_0(\xi,\bar\xi) = \lambda\cdot\log\frac{g}{g_*} + \sum_{I=1}^{N} k_I\log\left(\frac{G(\xi,\xi_I)F\{R_\alpha^*;\xi_I\}}{\prod_{\alpha=1}^{p-1} G(\xi,R_\alpha^*)^{1/(p-1)}}\right). \qquad (7.43)$$

Insertion of $F\{R_\alpha^*;\xi_I\}$ makes the whole expression scalar at the points R_α^*,ξ_I and is unessential in what follows.

Now it is easy to evaluate the functional integral (7.39):

$$A_\lambda\{k_I\} = \left(\frac{\det N_0}{\det'\Delta_0}\right)^{1/2}\left[\exp\frac{1}{2\pi}\int\sqrt{\det g}\,\phi_0\Delta_0\phi_0\right]\cdot\delta\left(\sum_{I=1}^{N} k_I + 2\lambda(p-1)\right).$$

$$(7.44)$$

Let us begin with the calculation of

$$-\frac{1}{2\pi}\int\sqrt{\det g}\phi_0\Delta_0\phi_0 = (1/2)$$

$$\times\int\left(\lambda\cdot\log\frac{g}{g_*} + \sum_{I=1}^{N} k_I\log\left[\frac{G(\xi,\xi_I)}{\prod_\alpha G(\xi,R_\alpha^*)^{1/(p-1)}}F\{R_\alpha^*;\xi_I\}\right]\right)$$

$$\times\left(\frac{\lambda}{\pi}\cdot\partial\bar\partial\log\frac{g}{g_*} - \sum_{I=1}^{N} k_I\delta(\xi,\xi_I) - 2\lambda\sum_{\alpha=1}^{p-1}\delta(\xi,R_\alpha^*)\right)$$

$$= -\frac{\lambda^2}{2\pi}S_L\left(\frac{g}{g_*}\right) - 2\lambda^2\sum_\alpha\log\frac{g}{g_*}(R_\alpha^*) - \lambda\sum_I k_I\log\frac{g}{g_*}(\xi_I)$$

$$
- \frac{1}{2} \sum_{I,J} k_I k_J \log \frac{G(\xi_I, \xi_J)}{\prod_\alpha G(\xi_I, R_\alpha^*)^{1/(p-1)}}
$$

$$
- \sum_{I,\alpha} \lambda k_I \log \frac{G(\xi_I, R_\alpha^*)}{\prod_\beta G(R_\alpha^*, R_\beta^*)^{1/(p-1)}}. \tag{7.45}
$$

Exponentiation of this expression gives

$$
\exp \frac{1}{2\pi} \int \sqrt{\det g}\, \phi_0 \Delta_0 \phi_0
$$

$$
= \left\{ \exp \frac{\lambda^2}{2\pi} S_L\left(\frac{g}{g_*} \right) \right\} \cdot \prod_\alpha \left(\frac{g(R_\alpha^*)}{g_*(R_\alpha^*)} \right)^{2\lambda^2} \prod_I \left(\frac{g(\xi_I)}{g_*(\xi_I)} \right)^{\lambda k_I}
$$

$$
\times \prod_{I<J} G(\xi_I, \xi_J)^{k_I k_J} \prod_I G(\xi_I, \xi_I)^{k_I^2/2} \prod_{\alpha,I} G(\xi_I, R_\alpha^*)^{2\lambda k_I}
$$

$$
\times \prod_{\alpha<\beta} G(R_\alpha^*, R_\beta^*)^{4\lambda^2} \prod_\alpha G(R_\alpha^*, R_\alpha^*)^{2\lambda^2}, \tag{7.46}
$$

where $S_L(\frac{g}{g_*}) = \int |\partial \log \frac{g}{g_*}|^2 d^2\xi$ is the Liouville action. Taking the regularization rule (7.26) into account, we obtain the final answer:

$$
A_\lambda\{k_1\} = \left\{ \exp \frac{\lambda^2}{2\pi} S_L\left(\frac{g}{g_*} \right) \right\} \cdot \left(\frac{\det N_0}{\det' \Delta_0} \right)^{1/2}
$$

$$
\times \prod_{I<J} G(\xi_I, \xi_J)^{k_I k_J} \prod_{I=1}^N \left(\frac{g_*(\xi_I)}{\prod_\alpha G(\xi_I, R_\alpha^*)^2} \right)^{-\lambda k_I}
$$

$$
\times \prod_{I=1}^N \sqrt{\det g(\xi_I)}^{-(k_I^2 - 2\lambda k_I)/2} \left(\frac{\prod_{\alpha<\beta} G(R_\alpha^*, R_\beta^*)}{|\prod_\alpha \nu'_*(R_\alpha^*)|^2} \right)^{4\lambda^2}.
$$

$$
\tag{7.47}
$$

Now consider the scalar field ϕ, which takes values in a circle of radius r, $\phi \sim \phi + 2\pi r$. On a non-simply-connected surface, this field is not necessarily single-valued. Instead, we have

$$
\phi(\xi, \bar{\xi}) = \bigcup_{\vec{m},\vec{n}} \{\phi_{\vec{m},\vec{n}}(\xi, \bar{\xi})\},
$$

$$
\phi_{\vec{m},\vec{n}}(\xi, \bar{\xi}) = \phi_S(\xi, \bar{\xi}) + i\pi r \left[(\vec{m} + \vec{n}\bar{T}) \frac{1}{\operatorname{Im} T} \vec{\xi} - (\vec{m} + \vec{n}T) \frac{1}{\operatorname{Im} T} \vec{\bar{\xi}} \right], \tag{7.48}
$$

where $\phi_S(\xi, \bar{\xi})$ is single-valued on S_p. The values k_I of the momenta are no longer arbitrary, instead:

$$
k_I \cdot 2\pi r = 2\pi i k_I, \quad k_I \in \mathbb{Z}. \tag{7.49}
$$

The functional integral under consideration is, in fact, an infinite sum, with each item related to a definite homotopic class of mapping of S_p into a circle. Mapping classes are labeled by two integer-valued p-vectors m_i, n_i, and

$$A(r) \equiv \int \mathcal{D}\phi \exp\left(-\frac{1}{2\pi}\int \sqrt{\det g}\, g^{ab}\partial_a\phi\partial_b\phi + \frac{i}{r}\cdot\sum_{I=1}^{N} k_I\phi(\xi_I)\right)$$

$$= \int \mathcal{D}\phi_S \exp\left(-\frac{1}{2\pi}\int \sqrt{\det g}\, g^{ab}\partial_a\phi_S\partial_b\phi_S + \frac{i}{r}\cdot\sum_{I=1}^{N} k_I\phi_S(\xi_I)\right)$$

$$\times \sum_{\{m_i,n_i\}\in \mathbb{Z}^{2p}} \exp\left(-\frac{\pi}{2}r^2(\vec{m}+\vec{n}\bar{T})\frac{1}{\operatorname{Im}T}(\vec{m}+\vec{n}T)\right.$$

$$\left. - \pi\left[(\vec{m}+\vec{n}\bar{T})\frac{1}{\operatorname{Im}T}\vec{z} - (\vec{m}+\vec{n}T)\frac{1}{\operatorname{Im}T}\vec{\bar{z}}\right]\right), \tag{7.50}$$

where

$$\vec{z} \equiv \sum_{I=1}^{N} k_I\vec{\xi}_I. \tag{7.51}$$

We obtain the result of the integration over ϕ_S, making use of (7.40) and the constraint $\sum_I k_I = 0$:

$$\left(\frac{\det N_0}{\det' \Delta_0}\right)^{1/2} \prod_{I<J} |E(\xi_I,\xi_J)|^{-k_I k_J/r^2} e^{\frac{2\pi}{r^2}\operatorname{Im}\vec{z}\frac{1}{\operatorname{Im}T}\operatorname{Im}\vec{z}}$$

$$\times \delta\left(\sum_I k_I\right) \prod_I^N \sqrt{\det g(\xi_I)}^{-k_I^2/2}. \tag{7.52}$$

The sum $I[\vec{z},\vec{\bar{z}}]$ in (7.50) is usually referred to as instantonic contribution (Alvarez-Gaume, Bost, Moore, Nelson and Vafa [1987]), because the nontrivial solutions of equations of motion $\partial\bar{\partial}\phi = 0$ are known as instantons. Instantonic contribution is calculated in the Appendix of this section. According to (7.58) and (7.66) of this Appendix, we have

$$I[\vec{z},\vec{\bar{z}}] = r^{-p}(\det \operatorname{Im}T)^{1/2} e^{-\frac{2\pi}{r^2}\operatorname{Im}\vec{z}\frac{1}{\operatorname{Im}T}\operatorname{Im}\vec{z}} \cdot \tilde{I}[\vec{z},\vec{\bar{z}}]. \tag{7.53}$$

Taking into account that $\det \operatorname{Im}T = \det N_1^{(\mathrm{can})}$, we obtain:

$$A(r) = \left(\frac{\det N_0 \det N_1^{(\mathrm{can})}}{\det' \Delta}\right)^{1/2}$$

$$\times \prod_{I<J}^{N} |E(\xi_I,\xi_J)|^{-k_I k_J/r^2} \tilde{I}[\vec{z},\vec{\bar{z}}] \cdot \delta(\sum_I k_I) \cdot \prod_I^N \sqrt{\det g(\xi_I)}^{-k_I^2/2}.$$

$$\tag{7.54}$$

It is also demonstrated in the Appendix that whenever $\beta^2 \equiv r^2/2$ is a rational number, the sum $\tilde{I}[\vec{z}, \vec{\bar{z}}]$ is a *finite* bilinear combination of theta-functions.

The most important result of the consideration of circle-valued scalars instead of the ordinary scalar fields is the absence of nonholomorphic contributions like $\exp(\mathrm{Im}\, z)\frac{1}{\mathrm{Im}\, T}(\mathrm{Im}\, z)$ in the final answer.

All these considerations can be generalized in a straightforward manner to the case of a multiplet of scalar fields, taking values in a torus (see, for example, Morozov and Olshanetskij [1988]). The main new ingredient is the occurrence of lattice theta-functions, associated with the torus $\mathbb{C}^n/\Gamma$ (Γ being a translation group),

$$\Theta_\Gamma\big(\vec{z}\,|T\big) = \sum_{\vec{\Lambda}_i \in \Gamma \times \mathbb{Z}^p} \exp\{i\pi(\vec{\Lambda}_i T_{ij} \vec{\Lambda}_j) + 2\pi i(\vec{\Lambda}_i \vec{z}_i). \tag{7.55}$$

Appendix: Evaluation of an Instantonic Sum. Let us consider the instantonic sum depending on two real p-vectors μ_i, ν_i, one complex p-vector z_i, and two parameters β, γ:

$$
\begin{aligned}
I_{\mu,\nu}[z, z] \equiv \sum_{\substack{n_i \in \mathbb{Z}+\nu_i \\ m_i \in \mathbb{Z}+\mu_i}} \exp\bigg(&-\pi\beta^2(\vec{m} + \vec{n}\,\bar{T})\frac{1}{\mathrm{Im}\, T}(\vec{m} + \vec{n}\,T) \\
&- 2\pi\gamma\bigg[(\vec{m} + \vec{n}\,\bar{T})\frac{1}{\mathrm{Im}\, T}\vec{z} - (\vec{m} + \vec{n}\,T)\frac{1}{\mathrm{Im}\, T}\vec{\bar{z}}\bigg]\bigg),
\end{aligned}
\tag{7.56}
$$

and express it in terms of theta-functions whenever β^2 is rational. Let us apply the Poisson transformation with respect to the variables m_i,

$$\sum_{m_i \in \mathbb{Z}+\mu_i} f\{m_i\} = \sum_{\{m_i\} \in \mathbb{Z}^p} f\{m_i + \mu_i\} = \sum_{\{M_i\} \in \mathbb{Z}^p} e^{2\pi i M_j \mu_j} \int d^p t\, e^{2\pi i M_j t_j} f\{t_j\},$$
$$\tag{7.57}$$

in order to obtain:

$$I_{\mu,\nu}[\vec{z}, \vec{\bar{z}}] = \beta^{-p}(\det \mathrm{Im}\, T)^{1/2} e^{-\frac{4\pi\gamma^2}{\beta^2}\mathrm{Im}\, \vec{z}\frac{1}{\mathrm{Im}\, T}\mathrm{Im}\, \vec{z}} \cdot \tilde{I}_{\mu,\nu}[\vec{z}, \vec{\bar{z}}],$$

$$\tilde{I}_{\mu,\nu}[\vec{z}, \vec{\bar{z}}] = \sum_{\substack{n_i \in \mathbb{Z}+\nu_i \\ M_i \in \mathbb{Z}}} \exp[2\pi i \vec{M}\vec{\mu}]$$

$$\times \exp\frac{i\pi}{2}\bigg[\bigg(\frac{\vec{M}}{\beta} + \beta\vec{n}\bigg)T\bigg(\frac{\vec{M}}{\beta} + \beta\vec{n}\bigg) - \bigg(\frac{\vec{M}}{\beta} - \beta\vec{n}\bigg)\bar{T}\bigg(\frac{\vec{M}}{\beta} - \beta\vec{n}\bigg)\bigg]$$

$$\times \exp\frac{2\pi i\gamma}{\beta}\bigg[\bigg(\frac{\vec{M}}{\beta} + \beta\vec{n}\bigg)\vec{z} - \bigg(\frac{\vec{M}}{\beta} - \beta\vec{n}\bigg)\vec{\bar{z}}\bigg]. \tag{7.58}$$

If β^2 is rational,

$$\beta^2 = P/Q, \tag{7.59}$$

and further simplifications arise:

$$\tilde{I}_{\mu,\nu}[\vec{z},\vec{\bar{z}}] = \sum_{\substack{n_i \in \mathbb{Z}+\nu_i \\ M_i \in \mathbb{Z}}} \exp[2\pi i \vec{M}\vec{\mu}]$$

$$\times \exp\left[\frac{i\pi PQ}{2}\left(\frac{\vec{M}}{P}+\frac{\vec{n}}{Q}\right)T\left(\frac{\vec{M}}{P}+\frac{\vec{n}}{Q}\right)+2\pi i\gamma Q\left(\frac{\vec{M}}{P}+\frac{\vec{n}}{Q}\right)\vec{z}\right]$$

$$\overline{\times \exp\left[\frac{i\pi PQ}{2}\left(\frac{\vec{M}}{P}-\frac{\vec{n}}{Q}\right)T\left(\frac{\vec{M}}{P}-\frac{\vec{n}}{Q}\right)+2\pi i\gamma Q\left(\frac{\vec{M}}{P}-\frac{\vec{n}}{Q}\right)\vec{z}\right]}.$$

$$(7.60)$$

Let us use the following substitution:

$$\frac{\vec{M}}{P} = \frac{\vec{a}-\vec{b}}{2}+\vec{\varepsilon}_P, \qquad \frac{\vec{n}}{Q} = \frac{\vec{a}+\vec{b}}{2}+\vec{\varepsilon}_Q+\frac{\vec{\nu}}{Q}, \tag{7.61}$$

where a_i and b_i are simultaneously even or odd, and the components of the p-vectors $\vec{\varepsilon}_P$ and $\vec{\varepsilon}_Q$ take values in the sets $\{0,1/P,\ldots,(P-1)/P\}$ and $\{0,1/Q,\ldots,(Q-1)/Q\}$, respectively:

$$\vec{\varepsilon}_P \in \mathbb{Z}_P^p, \qquad \vec{\varepsilon}_Q \in \mathbb{Z}_Q^p$$
$$\left(\mathbb{Z}_n \equiv \frac{1}{n}\cdot\mathbb{Z}[\mathrm{mod}\ n]\right). \tag{7.62}$$

Restrictions on $\vec{a}$ and $\vec{b}$ may be encoded by insertions of δ-functions:

$$\sum_{\mathrm{even}\ m_i}\delta(\vec{a}-\vec{b}-\vec{m}) = 2^{-p}\sum_{\vec{\zeta}\in\mathbb{Z}_2^p}\exp 2\pi i(\vec{a}-\vec{b})\vec{\zeta}. \tag{7.63}$$

The sum (7.60) now turns into

$$\tilde{I}_{\mu,\nu}[\vec{z},\vec{\bar{z}}] = \sum_{\substack{\vec{\zeta}\in\mathbb{Z}_2^p \\ \vec{\varepsilon}_Q\in\mathbb{Z}_Q^p \\ \vec{\varepsilon}_P\in\mathbb{Z}_P^p}}\sum_{\substack{\vec{b}\in\mathbb{Z}_p \\ \vec{a}\in\mathbb{Z}_p}}\exp 2\pi i(\vec{a}-\vec{b})\left(\vec{\zeta}+\frac{\vec{\mu}P}{2}\right)\cdot\exp(-4\pi i\vec{\zeta}\vec{\varepsilon}_P)$$

$$\times \exp\left(\frac{i\pi PQ}{2}\vec{\tilde{a}}T\vec{\tilde{a}}+2\pi i\gamma Q\vec{\tilde{a}}\vec{z}\right)\cdot\overline{\exp\left(\frac{i\pi PQ}{2}\vec{\tilde{b}}T\vec{\tilde{b}}+2\pi\gamma Q\vec{\tilde{b}}\vec{z}\right)},$$

$$(7.64)$$

where

$$\vec{\tilde{a}} \equiv \vec{a}+\vec{\varepsilon}_P+\vec{\varepsilon}_Q+\vec{\nu}/Q, \qquad \vec{\tilde{b}} \equiv \vec{b}-\vec{\varepsilon}_P+\vec{\varepsilon}_Q+\vec{\nu}/Q.$$

Making use of the definition of the theta-function,

$$\theta\begin{bmatrix}\vec{\alpha}\\\vec{\beta}\end{bmatrix}(z|T) \equiv \sum_{\vec{a}\in\mathbb{Z}^P}\exp[i\pi(\vec{a}+\vec{\alpha})T(\vec{a}+\vec{\alpha})+2\pi i(\vec{a}+\vec{\alpha})(\vec{z}+\vec{\beta})], \tag{7.65}$$

we arrive at the final expression:

$$\tilde{I}_{\mu,\nu}[\vec{z},\vec{\tilde{z}}] = \sum_{\vec{\varepsilon}_P,\vec{\varepsilon}_Q,\vec{\zeta}} e^{-4\pi i \vec{\varepsilon}_P \vec{\zeta}} \theta \begin{bmatrix} \vec{\varepsilon}_P + \vec{\varepsilon}_Q + \vec{\nu}/Q \\ \vec{\zeta} + \vec{\mu}P/2 \end{bmatrix} \left(\gamma Q \vec{z} \Big| \frac{PQ}{2} T \right)$$

$$\times \overline{\theta \begin{bmatrix} \vec{\varepsilon}_P + \vec{\varepsilon}_Q + \vec{\nu}/Q \\ \vec{\zeta} + \vec{\mu}P/2 \end{bmatrix} \left(\gamma Q \vec{z} \Big| \frac{PQ}{2} T \right)}. \tag{7.66}$$

7.3. A Circle-valued Field with Modified Lagrangian

Let us consider the fourth type of the functional integral:

$$A_\lambda\{k_I\} \equiv \int \mathcal{D}\phi \exp\left(-\frac{1}{2\pi} \int \sqrt{\det g}\,(g^{ab}\partial_a\phi\partial_b\phi + 2\lambda R_g\phi) + \sum_{I=1}^{N} k_I \phi(\xi_I) \right), \tag{7.67}$$

with the scalar field ϕ taking values in a circle. With a given value of the coefficient λ in (7.67), the possible values of the radius r and momenta k_I are restricted by the single-valuedness condition for $\exp S(\phi)$:

$$2\lambda r = -in, \qquad k_I = ik_I/r; \qquad n, k_I \in \mathbb{Z}. \tag{7.68}$$

Taking these restrictions into account, we can rewrite (7.67) in the form of

$$A_\lambda\{k_I\} = \int \mathcal{D}\phi \exp\left(-\frac{1}{2\pi} \int \sqrt{\det g}\,g^{ab}\partial_a\phi\partial_b\phi \right.$$

$$\left. + \frac{i}{r} \cdot \left[\sum_{I=1}^{N} k_I \phi(\xi_I) + \frac{n}{2\pi} \int \left(\sqrt{\det g}\,R_g\phi \right) \right] \right). \tag{7.69}$$

Divide the field ϕ into homotopically trivial ϕ_S and nontrivial $\phi_{\vec{m},\vec{n}}$ parts, as has been done above. Then

$$A_\lambda\{k_I\} = A_\lambda^q\{k_I\} \cdot A_\lambda^{\text{inst}}\{k_I\}, \tag{7.70}$$

$$A_\lambda^q\{k_I\} = \int \mathcal{D}\phi_S \exp\left(-\frac{1}{2\pi} \int \sqrt{\det g}\,g^{ab}\partial_a\phi_S\partial_b\phi_S \right.$$

$$\left. + \frac{i}{r} \cdot \left[\sum_{I=1}^{N} k_I \phi_S(\xi_I) + \frac{n}{2\pi} \int \left(\sqrt{\det g}\,R_g\phi_S \right) \right] \right), \tag{7.71}$$

$$A_\lambda^{\text{inst}}\{k_I\} = \sum_{\{\vec{m},\vec{n}\}} \exp\left(-\frac{1}{2\pi} \int \sqrt{\det g}\,g^{ab}\partial_a\phi_{\vec{m},\vec{n}}\partial_b\phi_{\vec{m},\vec{n}} \right.$$

$$\left. + \frac{i}{r} \cdot \left[\sum_{I=1}^{N} k_I \phi_{\vec{m},\vec{n}}(\xi_I) + \frac{n}{2\pi} \int \left(\sqrt{\det g}\,R_g\phi_{\vec{m},\vec{n}} \right) \right] \right). \tag{7.72}$$

258   A.Yu. Morozov and A.M. Perelomov

The quantity A_λ^q has already been calculated in (7.47). Now we turn to the instantonic contribution. To begin with, let us note that (7.67) is not generically a properly defined formula. Indeed, the field ϕ itself and not only its derivatives does enter (7.67). However, the field ϕ in not single-valued on S_p and does not take definite value at any given point. To make the field ϕ single-valued, we cut the surface S_p an define the single-valued ϕ on this simply-connected surface S_p^c. Now A_λ^{inst} is well defined, but instead it depends on the choice of cuts. A small deformation of the cut changes the contribution of instantons with the parameters $\vec{m}, \vec{n}$ by an integral of the jump $\Delta\phi_{\vec{m},\vec{n}}$ of the field at the cut over the domain δU, covered by the moving line of the cut,

$$\delta S\left[\phi_{\vec{m},\vec{n}}\right] = \frac{in}{2\pi r} \int_{\delta U} \sqrt{\det g}\, R_g \Delta\phi_{\vec{m},\vec{n}}. \tag{7.73}$$

It deserves note that there are no problems of such kind with the terms $\frac{i}{r} \cdot \sum_{I=1}^{N} k_I \phi(\xi_I)$ because of conditions (7.68). In order to make (7.72) well-defined, one should add some boundary terms at the position of the cut to it. It is easy to verify that the proper expression for the action of the field ϕ is

$$\tilde{S}\left[\phi_{\vec{m},\vec{n}}\right] = \frac{1}{2\pi} \int_{S_p^{\text{cut}}} |\partial\phi_{\vec{m},\vec{n}}|^2$$

$$+ \frac{i}{r} \cdot \left[\sum_{I=1}^{N} k_I \phi_{\vec{m},\vec{n}}(\xi_I) + \frac{n}{2\pi} \int_{S_p^{\text{cut}}} (\partial\bar{\partial}\log g)\phi_{\vec{m},\vec{n}} - \oint_{S_p^{\text{cut}}} \Omega\phi_{\vec{m},\vec{n}}\right]\Bigg), \tag{7.74}$$

where Ω is defined from the equation

$$d\Omega = \frac{n}{2\pi}\partial\bar{\partial}\log g + \sum_a l_a \delta(\xi, P_a), \tag{7.75}$$

and $D = \sum_a l_a P_a$ is an arbitrary divisor of appropriate degree. It is also easy to demonstrate that (7.74) does not depend on the choice of the divisor D and the metric g. Given a section $\omega(\xi)$ of line bundle, associated with D, we may write down the explicit expression for Ω:

$$\Omega = \frac{1}{\pi}{}^*d\log\frac{g(\xi)^{n/2}}{|\omega(\xi)|^2}. \tag{7.76}$$

If we change D and $\omega(\xi)$ for another divisor D' and section $\omega'(\xi)$, the change of (7.74) is

$$\delta\tilde{S} = \frac{i}{\pi r} \oint_{\partial S_p^{\text{cut}}} {}^*d\log\left|\frac{\omega(\xi)}{\omega'(\xi)}\right|^2 \phi(\xi)$$

$$= \frac{i}{\pi r} \cdot \left(\sum_{i=1}^{p} \oint_{A_i} {}^*d\log\left|\frac{\omega(\xi)}{\omega'(\xi)}\right|^2 \Delta_{B_i}\phi - \sum_{i=1}^{p} \oint_{B_i} {}^*d\log\left|\frac{\omega(\xi)}{\omega'(\xi)}\right|^2 \Delta_{A_i}\phi\right)$$

$$= 0[\text{mod } 2\pi i], \tag{7.77}$$

where $\Delta_{A_i}\phi, \Delta_{B_i}\phi \in 2\pi\mathbb{Z}$ are the jumps of the field ϕ at the cuts.

Let us choose a special divisor $D = K^{n/2}$ (K being the canonical divisor of the bundle of holomorphic 1-differentials) and an associated section $(\nu_*(\xi))^n$. Let us show that (7.74) is indeed independent of the metric g. The change of metric, $g \to g'$, gives rise to the following change of (7.74):

$$\tilde{S}_{g'} - \tilde{S}_g = \frac{in}{2\pi r} \left(\int_{S_p^{\text{cut}}} \left(\partial\bar{\partial} \log\left[\frac{g'}{g}\right] \right)\phi \right) - \frac{in}{2\pi r} \left(\oint_{\partial S_p^{\text{cut}}} {}^*d\log\left[\frac{g'}{g}\right]\phi \right)$$

$$= \frac{in}{2\pi r} \left(\oint_{\partial S_p^{\text{cut}}} {}^*d\log\left[\frac{g'}{g}\right]\phi \right) - \frac{in}{2\pi r} \left(\oint_{\partial S_p^{\text{cut}}} {}^*d\log\left[\frac{g'}{g}\right]\phi \right) = 0.$$

$$(7.78)$$

For the sake of convenience, take $g(\xi) = |\nu_*(\xi)|^4$. Then we have the following expression:

$$A_\lambda^{\text{inst}}\{k_I\} = \sum_{\{\vec{m},\vec{n}\}} \exp\left(-\frac{1}{2\pi} \int |\partial\phi_{\vec{m},\vec{n}}|^2 \right.$$

$$\left. + \frac{i}{r} \left[\sum_{I=1}^{N} k_I \phi_{\vec{m},\vec{n}}(\xi_I) - n \sum_{\alpha=1}^{p-1} \phi_{\vec{m},\vec{n}}(R_\alpha^*) \right] \right). \qquad (7.79)$$

This sum has already been calculated in the previous section. The answer is a bilinear combination of theta-functions (if $r^2/2 = P/Q$, $P, Q \in \mathbb{Z}$). In order to obtain a single theta-function with given characteristic $\left[\begin{smallmatrix}\vec{\delta}\\\vec{\varepsilon}\end{smallmatrix}\right]$, one should consider a boundary condition of a more general type:

$$\phi(\xi + A_k) = \phi(\xi) + 2\pi r\left(m_k + \frac{\alpha_k}{PQ} \right),$$

$$\phi(\xi + B_k) = \phi(\xi) + 2\pi r\left(m_k + \frac{\beta_k}{PQ} \right).$$

$$(7.80)$$

Some linear combinations of the quantities $A(\alpha_k, \beta_k)$, which arise in this case instead of (7.79), are equal to the squared moduli of theta-functions. However, in these situations, $\exp(\frac{ik_I}{r}\phi)$ is not well-defined. Also our discussion above, concerning the term $\int \sqrt{\det g}\, \mathcal{R}_g \phi$ appears inapplicable. The proper prescription for (7.70) in this situation is

$$\tilde{S} = -\frac{1}{2\pi} \int |\partial\phi|^2 + \frac{i}{r} \int \Omega \wedge d\phi, \qquad (7.81)$$

where

$$\delta\Omega = \sum_{I=1}^{N} k_I \delta(\xi - \xi_I) + \frac{n}{2\pi} \partial\bar{\partial} \log g. \qquad (7.82)$$

This expression is obviously invariant with respect to the shifts (7.80) and does not depend on any cuts.

7.4. b,c-Systems with Arbitrary (Half)-integer Spin (Knizhnik [1986]; Alvarez-Gaume, Bost, Moore, Nelson and Vafa [1987])

In this section, we reproduce the formulas for conformal blocks of Grassmannian b, c-systems of fields with the spins $(j, 1 - j)$ for arbitrary $j \in \mathbf{Z}/2$. We shall use the following strategy. First, we obtain correlation functions for the simplest case of $j = \frac{1}{2}$, making use of local bosonization. Then, by means of a change of variables in the functional integral, we shall treat the case of the arbitrary half-integer j.

So, let us consider a special case of a b, c-system: fermions $\tilde{\psi}(\xi), \psi(\xi)$ with the spin $\frac{1}{2}$ and the following OPE,

$$\tilde{\psi}(\xi)\psi(\xi') = \frac{1}{\xi - \xi'} + \text{regular terms}. \tag{7.83}$$

The stress tensor has the form

$$T_\psi = \tfrac{1}{2}\left[\tilde{\psi}(\xi)\partial\psi(\xi) - \partial\tilde{\psi}(\xi)\psi(\xi)\right]. \tag{7.84}$$

Locally this theory can be easily bosonized in terms of one scalar field, which takes values in a circle of unit radius:

$$\tilde{\psi} = e^{i\phi}, \qquad \psi = e^{-i\phi};$$
$$T_\psi = T_\phi = \tfrac{1}{2}(\partial\phi)^2. \tag{7.85}$$

Indeed, let us compare the correlation functions on a sphere in the theory of $\tilde{\psi}, \psi$-fields and their bosonic analogies:

$$\left\langle \prod_{i=1}^{n} \tilde{\psi}(\xi_i) \prod_{j=1}^{n} \psi(\xi_j') \right\rangle = \det_{(ij)} \frac{1}{\xi_i - \xi_j'}; \tag{7.86}$$

$$\left\langle \prod_{i=1}^{n} e^{i\phi(\xi_i)} \prod_{j=1}^{n} e^{-i\phi(\xi_j')} \right\rangle = \left| \frac{\prod_{i<j}(\xi_i - \xi_j)(\xi_i' - \xi_j')}{\prod_{i,j}(\xi_i - \xi_j')} \right|^2. \tag{7.87}$$

It is easy to realize that the rational function (7.86) and the one under the module sign in (7.87) possess zeros and poles at the same points on the Riemann sphere and therefore coincide. The holomorphic square root on the right-hand side of (7.87) is referred to as a correlator in the theory of chiral bosonic fields. Note also that the central charges in the fermionic and bosonic theories are the same:

$$-2(6j^2 - 6j + 1)_{j=1/2} = -2c_{1/2} = c_0 = \tfrac{1}{2} \cdot 2(6j^2 - 6j + 1)_{j=0} \tag{7.88}$$

(it is $\frac{1}{2}$ on the right-hand side because ϕ is a chiral boson, i.e., "half" of a normal complex field).

Let us now consider a couple of fermions on an arbitrary Riemann surface. In order to define the theory on an arbitrary surface, one has to choose phases, which fermions acquire when they move along noncontractible cycles. This freedom is fixed by the choice of some "theta-characteristic," i.e., of two p-vectors $\vec{\varepsilon}, \vec{\delta}$. When the fermion is shifted along the $A_k(B_k)$-cycle, it is multiplied by a factor of $e^{i\pi(\varepsilon_k+1)}$ $\left(e^{i\pi(\delta_k+1)}\right)$.

Before we discuss the concrete bosonization formulas, one comment is in order. When a functional integral in bosonic theory is calculated, one should integrate over the momenta p of intermediate states, $F(\partial\xi)e^{ip\xi}$. However, from (7.86), we see that only integer-valued momenta are allowed if we want to make contact with fermionic theory. This is exactly the reason why ϕ should be considered as a field that takes values in a circle of unit radius, $\phi \sim \phi + 2\pi$.

We have already discussed how the correlators of circle-valued scalar fields are calculated. It is easy to derive

$$
A = \left\langle \prod_{i=1}^{N} e^{i\phi(\xi_i)} \prod_{j=1}^{n} e^{-i\phi(\xi_j')} \right\rangle
$$

$$
= \left| \frac{\prod_{i<j} E(\xi_i, \xi_j) E(\xi_i', \xi_j')}{\prod_{i,j} E(\xi_i, \xi_j')} \right|^2 \cdot \frac{\sum\limits_{\substack{\vec{\delta}, \vec{\varepsilon} \\ (\delta\varepsilon)\in 2\mathbb{Z}}} \left| \theta\!\begin{bmatrix} \vec{\delta} \\ \vec{\varepsilon} \end{bmatrix}\!\left(\sum_{i=1}^{n} \vec{\xi}_i - \sum_{i=1}^{n} \vec{\xi}_i'\right) \right|}{|\det \bar{\partial}_0|}.
$$

$$(7.89)$$

In terms of fermions, this formula may be interpreted as follows:

$$
A = \sum_{\substack{\vec{\delta}, \vec{\varepsilon} \\ (\delta\vec{\varepsilon})=\text{even}}} \left| \left\langle \prod_{i=1}^{n} \tilde{\psi}(\xi_i) \prod_{j=1}^{n} \psi(\xi_j') \right\rangle \right|^2. \tag{7.90}
$$

Thus, the correlators in fermionic theory are

$$
\left\langle \prod_{i=1}^{n} \tilde{\psi}(\xi_i) \prod_{j=1}^{n} \psi(\xi_j') \right\rangle_e = \frac{\prod_{i<j} E(\xi_i, \xi_j) E(\xi_I', \xi_j')}{\prod_{i,j} E(\xi_i, \xi_j')} \cdot \frac{\theta_e(\sum_i \vec{\xi}_i - \sum_i \vec{\xi}_i')}{(\det \partial_0)^{1/2}}. \tag{7.91}
$$

It is easy to verify that (7.91) transforms properly under the shifts of ξ_i and ξ_i' along the A or B-cycles. Fay's identity

$$
\frac{\prod_{i<j} E(\xi_i, \xi_j) E(\xi_i', \xi_j')}{\prod_{i,j} E(\xi_i, \xi_j')} \cdot \frac{\theta_e(\sum_i \vec{\xi}_i - \sum_i \vec{\xi}_i')}{(\det \partial_0)^{1/2}} = \det_{(ij)} \frac{\theta_e(\xi_i - \xi_j')}{\theta_e(\vec{0}) E(\xi_i, \xi_j')} \tag{7.92}
$$

is naturally interpreted as Wick's theorem:

$$
\left\langle \prod_{i=1}^{n} \tilde{\psi}(\xi_i) \prod_{j=1}^{n} \psi(\xi_j') \right\rangle_e = \frac{\theta_e(\vec{0})}{(\det \partial_0)^{1/2}} \det_{(ij)} G_e^{(1/2)}(\xi_i, \xi_j'), \tag{7.93}
$$

where

$$
G_e^{(1/2)}(\xi_i, \xi_j') = \frac{\langle \tilde{\psi}(\xi)\psi(\xi')\rangle_e}{\det_e \bar{\partial}_{1/2}} = \frac{\theta_e(\xi - \xi')}{\theta_e(\vec{0}) E(\xi, \xi')}, \tag{7.94}
$$

and

$$\det_e \bar{\partial}_{1/2} = \frac{\theta_e(\vec{0})}{(\det \bar{\partial}_0)^{1/2}} \tag{7.95}$$

are the fermionic propagator and determinant.

Proceed now to the case of arbitrary $j \in \mathbb{Z}/2$. The simplest way to work out the answer is to change the variables:

$$b(\xi) = \Omega_{j-1/2}(\xi)\tilde{\psi}(\xi), \qquad c(\xi) = \Omega_{j-1/2}^{-1/2}(\xi)\psi(\xi), \tag{7.96}$$

where $\Omega_{j-1/2}(\xi)$ is the holomorphic $(j - \frac{1}{2})$-differential with zeros, located at the points $Q_1, \ldots, Q_{n_j}$; $n_j = (2j - 1)(p - 1)$. It is obvious, that the OPE for such b and c has the proper form

$$b(\xi)c(\xi') = \frac{1}{\xi - \xi'} + \text{regular terms}. \tag{7.97}$$

The ordinary norms of the fields b and c correspond to the following norms for $\tilde{\psi}$ and ψ:

$$||\delta b||^2 = ||\Omega_{j-1/2}\delta\tilde{\psi}||^2 = \int \rho^{1-j}|\Omega_{j-1/2}\delta\tilde{\psi}|^2; \qquad ||\delta c||^2 = \int \frac{\rho^j|\delta\psi|^2}{|\Omega_{j-1/2}|^2}. \tag{7.98}$$

(The metric is supposed to be in conformal gauge, $g(\xi, \bar{\xi}) = \rho(\xi, \bar{\xi})d\xi d\bar{\xi}$.) Thus, integration over regular fields b and c is equivalent to the integration over $\tilde{\psi}$ possessing poles at the points $Q_1, \ldots, Q_{n_j}$, and ψ with zeros at the same points. Thus, we observe the following relation between integration measures:

$$\mathcal{D}b\mathcal{D}c = \mathcal{D}\tilde{\psi}\mathcal{D}\psi \prod_{i=1}^{n_j} \frac{\psi(Q_i)}{\left[\Omega'_{j-1/2}(Q_i)\right]^{1/(2j+1)}}, \tag{7.99}$$

where

$$\Omega_{j-1/2}(\xi) = \Omega'_{j-1/2}(Q_i)(\xi - Q_i) + o\left[(\xi - Q_i)^2\right]. \tag{7.100}$$

The action of b, c-fields is:

$$S = \int d^2\xi(b\bar{\partial}c) = \int d^2\xi(\tilde{\psi}\Omega_{j-1/2})\bar{\partial}(\psi\Omega_{j-1/2}^{-1}) = \int d^2\xi(\tilde{\psi}\bar{\partial}\psi), \tag{7.101}$$

and we obtain the following equality:

$$\left\langle \prod_{i=1}^{m} b(x_i) \prod_{j=1}^{n} c(y_j) \right\rangle$$

$$= \left\langle \prod_{i=1}^{n_j} \frac{\psi(Q_i)}{\left[\Omega'_{j-1/2}(Q_i)\right]^{1/(2j+1)}} \prod_{\mu=1}^{m} \Omega_{j-1/2}(x_\mu)\tilde{\psi}(x_\mu) \prod_{\nu=1}^{n} \frac{\psi(y_\nu)}{\Omega_{j-1/2}(y_\nu)} \right\rangle. \tag{7.102}$$

The charge conservation in fermionic theory leads to the following restriction: $m = n + n_j$, or

$$m - n = (2j + 1)(p - 1). \qquad (7.103)$$

The norms (7.98) are not exactly the standard norms on the bundles of j- and $(1 - j)$-differentials, which look like

$$||b||^2 = \int |\underbrace{b_{\xi \cdots \xi}}_{j\ pa3}|^2 (g^{\xi \bar{\xi}})^j \sqrt{\det g}\, d^2\xi = \int \rho^{1-j} |b|^2 d^2\xi; \qquad ||c||^2 = \int \rho^j |c|^2 d^2\xi;$$

$$||\tilde{\psi}||^2 = \int \rho^{1/2} |\tilde{\phi}|^2 d^2\xi; \qquad ||\psi||^2 = \int \rho^{1/2} |\psi|^2 d^2\xi. \qquad (7.104)$$

To make (7.98) and (7.104) the same, let us choose the metric g and the $(j - 1/2)$-differential $\Omega_{j-1/2}$ as follows:

$$g = \left| \nu_*(\xi) \right|^4, \qquad \Omega_{j-1/2} = \nu_*(\xi)^{2j-1}. \qquad (7.105)$$

Then we obtain the following answer for correlators of b, c-fields in the metric $g = |\nu_*(\xi)|^4$:

$$\left\langle \prod_{\mu=1}^{n+n_j} b(x_\mu) \prod_{\nu=1}^{n} c(y_\nu) \right\rangle_e = \frac{\prod_{\mu<\mu'}^{n+n_j} E(x_\mu, x_{\mu'}) \prod_{\nu<\nu'}^{n} E(y_\nu, y_{\nu'})}{(\det \bar{\partial}_0)^{1/2} \prod_{\mu,\nu} E(x_\mu, y_\nu)}$$

$$\times \left(\frac{\prod_\mu \sigma_*(x_\mu)}{\prod_\nu \sigma_*(y_\nu)} \right)^{2j-1} \left(\frac{\prod_{\alpha<\beta}^{p-1} E(R_\alpha^*, R_\beta^*)}{\prod_{\alpha=1}^{p-1} \nu'_*(R_\alpha^*)} \right)^{(2j-1)^2}$$

$$\times \theta_e \left(\sum_{\mu=1}^{n+n_j} \vec{x}_\mu - \sum_{\nu=1}^{n} \vec{y}_\nu - (2j-1)\vec{\Delta}_* \right). \qquad (7.106)$$

Let us now discuss how a transformation from one theta-characteristic to another may be performed (to * as a particular case). We use the same trick—a change of variables,

$$\tilde{b}(\xi) = b(\xi) f_{e,\tilde{e}}(\xi), \qquad \tilde{c}(\xi) = c(\xi) f_{e,\tilde{e}}^{-1}(\xi), \qquad (7.107)$$

where $f_{e,\tilde{e}}$ is defined by the condition that $\tilde{b}(b)$ has boundary conditions associated with the characteristic $\tilde{e}(e)$. The explicit expression is

$$f_{e,\tilde{e}}(\xi) = \prod_{i=1}^{N} \frac{E(\xi, Q_i)}{E(\xi, P_i)}; \qquad \sum_{i=1}^{N} \vec{P}_i - \sum_{i=1}^{N} \vec{Q}_i = \frac{\vec{\varepsilon} - \vec{\tilde{\varepsilon}}}{2} + \frac{\vec{\delta} - \vec{\tilde{\delta}}}{2} T. \qquad (7.108)$$

264   A.Yu. Morozov and A.M. Perelomov

By changing the variables in (7.106) according to (7.107), we obtain:

$$
\left\langle \prod_{\mu=1}^{n+n_j} b(x_\mu) \prod_{\nu=1}^{n} c(y_\nu) \prod_{\mu=1}^{n+n_j} f(x_\mu) \prod_{\nu=1}^{n} f(y_\nu)^{-1} \right.
$$

$$
\left. \times \prod_{i=1}^{N} \frac{b(P_i)\nu_*(P_i)^{1-2j} c(Q_i)\nu_*(Q_i)^{2j-1}}{\left([f^{-1}(P_i)]'\right)^{1/2} \left(f'(Q_i)\right)^{1/2}} \right\rangle_e
$$

$$
= \frac{\prod_{\mu<\mu'}^{n+n_j} E(x_\mu, x_{\mu'}) \prod_{\nu<\nu'}^{n} E(y_\nu, y_{\nu'})}{(\det \bar{\partial}_0)^{1/2} \prod_{\mu,\nu} E(x_\mu, y_\nu)} \cdot \left(\frac{\prod_\mu \sigma_*(x_\mu)}{\prod_\nu \sigma_*(y_\nu)} \right)^{2j-1}
$$

$$
\times \theta_e \left(\sum_{\mu=1}^{n+n_j} \vec{x}_\mu - \sum_{\nu=1}^{n} \vec{y}_\nu - (2j-1)\vec{\Delta}_* \right) \cdot e^{-\frac{i\pi}{4}(\delta-\tilde{\delta})T(\delta-\tilde{\delta})} F\{R_\alpha^*\}.
$$

$$(7.109)$$

The formula

$$
\theta\begin{bmatrix} \vec{\delta} \\ \vec{\varepsilon} \end{bmatrix} \left(\vec{z} + \frac{\vec{\tilde{\varepsilon}} - \vec{\varepsilon}}{2} + \frac{\vec{\tilde{\delta}} - \vec{\delta}}{2}T \right) = e^{-\frac{i\pi}{4}(\delta-\tilde{\delta})T(\delta-\tilde{\delta})} \theta\begin{bmatrix} \vec{\delta} \\ \vec{\varepsilon} \end{bmatrix}(\vec{z}) \qquad (7.110)
$$

has been used here. The last two factors in (7.109) are compensated by
Quillen's anomaly (Quillen [1985]) associated with the transformation (7.107).
Thus, we obtain

$$
\left\langle \prod_{\mu=1}^{n+n_j} b(x_\mu) \prod_{\nu=1}^{n} c(y_\nu) \right\rangle_e = \frac{\prod_{\mu<\mu'}^{n+n_j} E(x_\mu, x_{\mu'}) \prod_{\nu<\nu'}^{n} E(y_\nu, y_{\nu'})}{(\det \bar{\partial}_0)^{1/2} \prod_{\mu,\nu} E(x_\mu, y_\nu)}
$$

$$
\times \left(\frac{\prod_\mu \sigma_*(x_\mu)}{\prod_\nu \sigma_*(y_\nu)} \right)^{2j-1} \theta_e \left(\sum_{\mu=1}^{n+n_j} \vec{x}_\mu - \sum_{\nu=1}^{n} \vec{y}_\nu - (2j-1)\vec{\Delta}_* \right) \cdot F\{R_\alpha^*\}.
$$

$$(7.111)$$

Let us comment separately on the case of $j = 1$. When $n = 0$ and $j = 1$,
the theta-function in (7.111) vanishes. This only indicates, that when $j = 1$,
there are additional zero modes of the fields b and c in the functional integral.
In this case, the smallest possible value of n in (7.111) is 1, and instead of
(7.111) we have

$$
\left\langle \prod_{\mu=1}^{n+p} b(x_\mu) \prod_{\nu=1}^{n+1} c(y_\nu) \right\rangle_*
$$

$$
= \frac{\prod_{\mu<\mu'}^{n+p} E(x_\mu, x_{\mu'}) \prod_{\nu<\nu'}^{n+1} E(y_\nu, y_{\nu'})}{\prod_{\mu,\nu} E(x_\mu, y_\nu)}
$$

$$
\times \frac{\prod_\mu \sigma_*(x_\mu)}{\prod_\nu \sigma_*(y_\nu)} \cdot \frac{\theta_* \left(\sum_{\mu=1}^{n+p} \vec{x}_\mu - \sum_{\nu=1}^{n+1} \vec{y}_\nu - \vec{\Delta}_* \right)}{(\det \bar{\partial}_0)^{1/2}} \cdot F\{R_\alpha^*\}. \qquad (7.112)
$$

Consider now (7.91) and (7.106) from the point of view of bosonization (7.85):

$$b = \nu_*^{2j-1} \exp(i\phi_{(j)}),$$
$$c = \nu_*^{1-2j} \exp(-i\phi_{(j)}). \tag{7.113}$$

We may derive these formulas directly from (7.86) by the following shift of the field $\phi_{(1/2)}$:

$$\phi_{(j)} = \phi_{(1/2)} - i(2j-1) \cdot \log\left|\nu_*(\xi)\right|^2. \tag{7.114}$$

After such a shift of the integration variable in the functional integral, we have:

$$\left|\left\langle \prod_{\mu=1}^{m} b(x_\mu) \prod_{\nu=1}^{n} c(y_\nu) \right\rangle\right|^2 = \int \mathcal{D}\phi\, e^{\tilde{S}(\phi)} \prod_{\mu=1}^{m} e^{i\phi(x_\mu)} \left|\nu_*(x_\mu)\right|^{2(2j-1)}$$
$$\times \prod_{\nu=1}^{n} e^{-i\phi(y_\nu)} \left|\nu_*(y_\nu)\right|^{-2(2j-1)}, \tag{7.115}$$

where the shifted action is

$$\tilde{S}(\phi) = -\frac{1}{2\pi} \int d^2\xi \left[|\partial\phi|^2 - \frac{i(2j-1)}{2} \mathcal{R}^{|\nu_*|^4}\phi + \frac{(2j-1)^2}{4} \mathcal{R}^{|\nu_*|^4} \Delta^{-1} \mathcal{R}^{|\nu_*|^4} \right]. \tag{7.116}$$

Keeping in mind that the integration is over the fields taking values in a circle, we obtain the following answer:

$$\left|\left\langle \prod_{\mu=1}^{m} b(x_\mu) \prod_{\nu=1}^{n} c(y_\nu) \right\rangle\right|^2 = \delta\big[m - n - (2j-1)(p-1)\big] \cdot \int \mathcal{D}\phi\, e^{\tilde{S}(\phi)}$$

$$\times \prod_{\alpha=1}^{p-1} \frac{e^{-i(2j-1)\phi(R_\alpha^*)}}{\left|\nu_*'(R_\alpha^*)\right|^{2(2j-1)}} \prod_{\mu=1}^{m} e^{i\phi(x_\mu)} \left|\nu_*(x_i)\right|^{2(2j-1)} \prod_{\nu=1}^{n} e^{-i\phi(y_\nu)} \left|\nu_*(y_\nu)\right|^{-2(2j-1)}$$

$$= \sum_{\text{even } e} \left| \frac{\prod_{\mu<\mu'}^{n+n_j} E(x_\mu, x_{\mu'}) \prod_{\nu<\nu'}^{n} E(y_\nu, y_{\nu'})}{\prod_{\mu,\nu} E(x_\mu, y_\nu)} \right.$$

$$\times \left(\frac{\prod_\mu \sigma_*(x_\mu)}{\prod_\nu \sigma_*(y_\nu)} \right)^{2j-1} \left(\frac{\prod_{\alpha<\beta}^{p-1} E(R_\alpha^*, R_\beta^*)}{\prod_{\alpha=1}^{p-1} \nu_*'(R_\alpha^*)} \right)^{(2j-1)^2}$$

$$\times \left. \frac{\theta_e\left(\sum_{\mu=1}^{n+n_j} \vec{x}_\mu - \sum_{\nu=1}^{n} \vec{y}_\nu - (2j-1)\vec{\Delta}_*\right)}{(\det \bar{\partial}_0)^{1/2}} \right|^2 \exp\left(-\frac{c_j}{12\pi} S_L(|\nu_*|^4)\right). \tag{7.117}$$

Here $S_L(g)$ stands for the Liouville action, and the coefficient $2c_j = 2(6j^2 - 6j + 1)$ (the central charge of the Virasoro algebra for j-differentials) is a sum of two contributions:

$$-\frac{(2j-1)^2}{8\pi} + \frac{1}{2} \cdot \frac{2}{24\pi} = -\frac{2(6j^2 - 6j + 1)}{24\pi} = -\frac{2c_j}{12\pi}. \qquad (7.118)$$

The second item on the left-hand side of (7.118) comes from the general formula

$$\det{}' \Delta_j = \det N_j \det N_{1-j} \left|\det \bar{\partial}_j\right|^2 \exp\left(-\frac{2c_j}{24\pi} S_L(g)\right) \qquad (7.119)$$

in the case of $j = 0$.

Taking (7.119) into account, we see that (7.117) is in agreement with (7.106).

In conclusion, it is useful to stress that the bosonization prescription works well with *any* metric on a Riemann surface (not necessarily singular).

7.5. β, γ-Systems with Arbitrary (Half)-integer Spin

β, γ-systems are the analogs of b, c-systems but possess another kind of statistics: They are bosons. Originally, β, γ-systems appeared in the role of superghost fields in the theory of Neveu–Schwarz–Ramond superstrings (Friedan, Martinec and Shenker [1986]). The same systems appear in the bosonization of various conformal field theories, in particular that of the Wess–Zumino–Novikov–Witten model (Gerasimov, Marshakov, Morozov, Olshanetskij and Shatashvili [1989]). The theory of β, γ-systems with arbitrary spin j is considered in Verlinde and Verlinde [1987], Atick and Sen [1987a,b], Morozov [1988], Semikhatov [1989], and Losev [1989]; below we give a brief review of these results.

First let us discuss the generic features of β, γ-systems and their conformal blocks. Since they only differ from b, c-systems by their opposite statistics, the determinants for β, γ-systems are inverse of those for b, c-fields. The central charge has opposite value: $+2c_j$ instead of $-2c_j$. The functional integrals on the surfaces with boundaries, which had the form of $\Phi\{b, c\} \sim \exp(b\mathcal{K}c)$, do not change: $\Phi\{\beta, \gamma\} \sim \exp(\beta\mathcal{K}\gamma)$, however the integration over boundary conditions now gives $1/\det K$ instead of $\det K$. Exact technical consideration differs from these qualitative remarks by an accurate account of zero modes of fields β and γ. This is even more important than in the case of b, c-systems, since now zero modes give rise to divergences rather than to zeros of functional integrals.

At first glance, the integral $\int D\beta D\gamma e^{\int \beta\bar{\partial}\gamma}$ may seem absolutely senseless, because for $j \neq \frac{1}{2}$, there are holomorphic j and $1 - j$-differentials on any closed Riemann surface. This, however, means nothing other than that only correlators of β, γ-fields of some special form are allowed, such that zero modes do not arise in the integral. All other correlators may be considered vanishing. This reasoning suggests that for $j > \frac{1}{2}$, at least as many as $n_j = (2j-1)(p-1)$ delta-functions, $\prod_{a=1}^{n_j} \delta(\beta(Q_a))$, should be inserted into the functional integral in order to ascertain vanishing of the β-field at n_j points Q_a and exclude n_j holomorphic j-differentials from the functional integration domain over β. In order to get rid of all zero modes in this way, it is necessary that the n_j equations

$$\beta(Q_a) = 0, \quad a = 1 \ldots n_j, \tag{7.120}$$

possess no solutions of the form

$$\beta(\xi) = \sum_{a=1}^{n_j} s_a B_a^{(j)}(\xi) \tag{7.121}$$

with holomorphic j-differentials $B_a^{(j)}(\xi)$. In other words, the determinant $\det_{(ab)} B_a^{(j)}(Q_b)$ should be nonvanishing. If it vanishes, a pole appears in the functional integral, considered as a function of the Q_a's:

$$\left\langle \prod_{a=1}^{n_j} \delta(\beta(Q_a)) \right\rangle_{\beta,\gamma} \sim \frac{1}{\det_{(ab)} B_a^{(j)}(Q_b)} \sim \frac{1}{\langle \prod_{a=1}^{n_j} b(Q_a) \rangle_{b,c}}. \tag{7.122}$$

(We used the fact that the correlator of the b,c-fields is proportional to the same determinant composed of zero modes $B_a^{(j)}$, see (4.15).)

The δ-functions in (7.122) may be substituted by something more familiar if the "bosonization" formalism for β,γ-systems is used (Friedan, Martinec and Shenker [1986]), similar to that for b,c-systems considered in the previous section. In order to change fermions for bosons, it is natural to change the scalar field ϕ in the bosonization formulas (7.16) for $i\phi$. This is equivalent to the change of $\alpha_0 = \frac{2j-1}{2\sqrt{2}}$ in the action (7.39) for

$$\alpha_0 = i\frac{2j-1}{2\sqrt{2}} \quad (\alpha_0 = i\lambda/\sqrt{2}). \tag{7.123}$$

However, this time the central charge

$$1 - 24\alpha_0^2 = 1 + 12\lambda^2 = 1 + 3(2j-1)^2 = 2c_j + 2 \tag{7.124}$$

differs from the proper value of $+2c_j$ by 2. Another problem is that the fields $e^{+\phi}$ and $e^{-\phi}$, which should serve as bosonic analogs of β and γ like in (7.85), possess inadequate operator expansion. The situation may be improved if an auxiliary b,c-system with spin 1 is introduced, which is conventionally denoted by (ξ, η). ξ and η are Grassmannian fields with spins 0 and 1, respectively. The central charge of the ξ, η-system is $-2c_{-1} = -2$ and corrects (7.124) to a proper value. Bosonization formulae look like

$$\beta = \partial\xi \cdot e^{-\phi}\nu_*^{2j-1}, \qquad \gamma = \eta \cdot e^{\phi}\nu_*^{1-2j}; \tag{7.125}$$

or

$$\xi = H(\beta), \qquad \eta = \partial\gamma \cdot \delta(\gamma), \tag{7.126}$$

$$e^{\phi} = \nu_*^{2j-1}\delta(\beta), \qquad e^{-\phi} = \nu_*^{1-2j}\delta(\gamma). \tag{7.127}$$

Here $H(\ )$ stands for a Heaviside function, its derivative being a δ-function. The b, c-fields ξ and η can be bosonized themselves:

$$\xi = e^{i\sigma}, \qquad\qquad \eta = e^{-i\sigma} \tag{7.128}$$

$$\beta = i\partial\sigma \cdot e^{-(\phi - i\sigma)}\nu_*^{2j-1}, \qquad \gamma = e^{(\phi - i\sigma)}\nu_*^{1-2j}. \tag{7.129}$$

The stress tensor

$$T_{\beta,\gamma} = j\beta\partial\gamma - (1-j)\gamma\partial\beta = T_\phi + T_\sigma = (\partial\phi)^2 + i(2j-1)\partial^2\phi + (\partial\sigma)^2 + \partial^2\sigma \tag{7.130}$$

corresponds to the following Lagrangian of the fields ϕ and σ:

$$\begin{aligned}
\mathcal{L} &= \frac{1}{\pi}\left[\eta\bar\partial\xi + \frac{1}{2}|\partial\phi|^2 + \frac{i}{2}\,j(2j-1)\sqrt{\det g}\,\mathcal{R}\phi\right] \\
&= \frac{1}{2\pi}\left[|\partial\phi|^2 + |\partial\sigma|^2 + \sqrt{\det g}\,\mathcal{R}\big(\sigma + i(2j-1)\phi\big)\right]. \tag{7.131}
\end{aligned}$$

Locally on the Riemann surface, i.e., from the point of view of operator algebra, the fields ϕ and σ are completely independent. However, boundary conditions, corresponding to the turns along noncontractible contours, intermix instantonic configurations of these fields, thus making evaluation of multiloop correlators with the help of bosonization procedure a bit more subtle. This is why we shall use another way for this calculation: We shall exploit the relation between b, c- and β, γ-systems.

The following expression is, in fact, unity:

$$\int D\beta D\gamma Db Dc\, e^{\int \beta\bar\partial\gamma + b\bar\partial c}\delta\big(b(z_1)\big)\delta\big(\beta(z_1)\big)\ldots\delta\big(b(z_{n_j})\big)\delta\big(\beta(z_{n_j})\big). \tag{7.132}$$

Auxiliary δ-function insertions arise because of the zero modes. Making use of the simple property of Grassmannian fields,

$$\delta\big(b(z)\big) = \frac{1}{i}\int d\varepsilon\, e^{i\varepsilon b(z)} = b(z), \tag{7.133}$$

we obtain the following result for the chiral determinant of the β, γ-system:

$$\int D\beta D\gamma\, e^{\int \beta\bar\partial\gamma}\delta\big(\beta(z_1)\big)\ldots\delta\big(\beta(z_{n_j})\big) = \frac{1}{\big[\det_{(ab)} B_a^{(j)}(z_b)\big]} \cdot \frac{1}{\mathrm{Det}\,\bar\partial_j}. \tag{7.134}$$

(Let us recall that the $\{B_a^{(j)}(\xi)\}$ denote the basis of holomorphic j-differentials, whereas $\det\bar\partial_j$ stands for the chiral determinant of the b, c-system.) Following the same line of reasoning as in the section about b, c-systems, we shall first directly evaluate the β, γ-correlators in the simplest case of $j = \frac{1}{2}$, and then obtain expressions for an arbitrary (half-)integer j by a change of variables.

Thus, we proceed to correlators with $j = \frac{1}{2}$. In this case, we shall use the same notation as for Grassmannian spinors: $\beta^{(1/2)} = \tilde{\psi}$, $\gamma^{(1/2)} = \psi$. The main operators are

$$\psi, \tilde{\psi}, \delta(\psi) = \int \frac{dp}{2\pi} e^{ip\psi}, \quad \delta(\tilde{\psi}) = \int \frac{d\tilde{p}}{2\pi} e^{i\tilde{p}\tilde{\psi}}, \quad H(\psi) = \int \frac{dp}{2\pi(p+i0)} e^{ip\psi}.$$
(7.135)

The operator with a Heaviside function plays an important role in superstring theory in the integral over odd moduli as well as in other applications. It is straightforward to check the operator expansion rules:

$$\tilde{\psi}(\xi)\delta(\psi(\xi')) \sim (\xi - \xi')\partial H(\tilde{\psi}(\xi')) + \cdots,$$
(7.136)

$$\delta(\tilde{\psi}(\xi))\delta(\psi(\xi')) \sim (\xi - \xi') \cdot 1 + \cdots,$$
(7.137)

$$H(\tilde{\psi}(\xi)\psi(\xi')) \sim \frac{1}{\xi - \xi'}\delta(\tilde{\psi}(\xi)) + \cdots.$$
(7.138)

Let us evaluate the correlator

$$\left\langle \prod_{i=1}^{n} \psi(y_i) \prod_{j=1}^{n} \delta(\psi(w_j)) \prod_{k=1}^{n+1} H(\tilde{\psi}(x_k)) \right\rangle_e.$$
(7.139)

It is easily expressible through the Green function of the β, γ-fields $\tilde{\psi}, \psi$, which is of course just the same as the Green function of analogous b, c-fields, $G_e^{(1/2)} = \frac{\theta_e(\vec{\xi} - \vec{\xi'})}{\theta_e(\vec{0})E(\xi,\xi')}$. The only thing to be used in the calculation is the integral representation of $\delta(\psi)$ and $H(\tilde{\psi})$:

$$\left\langle \prod_{i=1}^{n}\left[\frac{\partial}{\partial r_i}e^{ir_i\psi(y_i)}\right]_{r_i=0} \prod_{j=1}^{n}\int \frac{dp_j}{2\pi}e^{ip\psi(w_j)} \cdot \prod_{k=1}^{n}\int \frac{dq_k}{2\pi(q_k+i0)}e^{i\tilde{\psi}(x_k)q_k}\right\rangle_e$$

$$= \prod_i \frac{\partial}{\partial r_i} \prod_j \int \frac{dp_j}{2\pi} \prod_k \int \frac{dq_k}{2\pi(q_k+i0)}\left\{\frac{1}{\mathrm{Det}_e\,\bar{\partial}_{1/2}}\right.$$

$$\times \exp \sum_k q_k\left(\sum_i r_i G_e^{(1/2)}(y_i,x_k) + \sum_j p_j G_e^{(1/2)}(w_j,x_k)\right)\biggr\}\biggr|_{r_i=0}$$

$$= \prod_{k=0}^{n} \int \frac{dq_k}{2\pi(q_k+i0)} \prod_{i=1}^{n}\left(\sum_{k=0}^{n} q_k G_e^{(1/2)}(y_i,x_k)\right)$$

$$\times \prod_{j=1}^{n}\left(\sum_{k=0}^{n} q_k G_e^{(1/2)}(w_j,x_k)\right)\frac{1}{\mathrm{Det}_e\,\bar{\partial}_{1/2}}.$$
(7.140)

270   A.Yu. Morozov and A.M. Perelomov

Let us integrate over all the q_k besides q_0. Then we obtain

$$\frac{\prod_{i=1}^n[G_e^{(1/2)}(y_i,x_0)-\sum_{k,j=1}^{k,j=n}G_e^{(1/2)}(y_i,x_k)[G_e^{(1/2)}(x_k,w_j)]^{-1}G_e^{(1/2)}(w_j,x_0)]}{\prod_{k=1}^n[\sum_{j=1}^n[G_e^{(1/2)}(x_k,w_j)]^{-1}G_e^{(1/2)}(w_j,x_0)]\operatorname{Det}_e\bar\partial_{1/2}\det_{(j,k)}G_e^{(1/2)}(w_j,x_k)}.$$

$$(7.141)$$

Now apply the Cramer identity

$$\sum_{j=1}^n\left[G_e^{(1/2)}(x_k,w_j)\right]^{-1}G_e^{(1/2)}(w_j,x_0)$$

$$=\det\left\|\begin{matrix}G_e^{(1/2)}(x_1,w_1)&\dots&G_e^{(1/2)}(x_n,w_1)\\ \vdots&&\vdots\\ G_e^{(1/2)}(x_1,w_n)&\dots&G_e^{(1/2)}(x_n,w_n)\end{matrix}\right\|_{x_k\to x_0}$$

$$\times\det\left[G_e^{(1/2)}(x_k,w_j)\right]^{-1}.\qquad(7.142)$$

Together with the relation (7.92),

$$\det G_e^{(1/2)}(z_i,w_j)=G(z_1\dots z_n|w_1\dots w_n)$$

$$=\frac{\prod_{i<i'}E(z_i,z_{i'})\prod_{j<j'}E(w_j,w_{j'})}{\prod_{i,j}E(z_i,w_j)}\cdot\frac{\theta_e(\sum\vec z_i-\sum\vec w_i)}{\theta_e(\vec 0)};$$

$$(7.143)$$

this gives rise to the following result:

$$\left\langle\prod_{i=1}^n\psi(y_i)\prod_{j=1}^n\delta(\psi(w_j))\prod_{k=1}^{n+1}H(\tilde\psi(x_k))\right\rangle_e$$

$$=\frac{\prod_{i=1}^nG(x_0\dots x_n|y_iw_1\dots w_n)}{\prod_{k=0}^nG(x_0\dots\check x_k\dots x_n|w_1\dots w_n)\operatorname{Det}_e\bar\partial_{1/2}}.\qquad(7.144)$$

It is convenient to transform this expression by making use of the relation between the determinants

$$(\operatorname{Det}\bar\partial_0)^{1/2}\operatorname{Det}_e\bar\partial_{1/2}=\theta_e(\vec 0).\qquad(7.145)$$

The final answer is

$$\left\langle\prod_{i=1}^n\psi(y_i)\prod_{j=1}^n\delta(\psi(w_j))\prod_{k=1}^{n+1}H(\tilde\psi(x_k))\right\rangle_e$$

$$=\frac{\prod_{i=1}^n\langle b^{(1/2)}(x_0)\dots b^{(1/2)}(x_n)c^{(1/2)}(w_1)\dots c^{(1/2)}(w_n)\rangle_e}{\prod_{k=0}^n\langle b^{(1/2)}(x_0)\dots\check b^{(1/2)}(x_k)\dots b^{(1/2)}(x_n)c^{(1/2)}(w_1)\dots c^{(1/2)}(w_n)\rangle_e}.$$

$$(7.146)$$

If expressed in terms of "bosonized" fields,

$$\xi = H(\tilde{\psi}); \qquad e^{\phi} = \delta(\tilde{\psi}); \qquad \tilde{\psi} = \partial\xi \cdot e^{-\phi};$$
$$\eta = \psi \cdot \delta(\psi); \qquad e^{-\phi} = \delta(\psi); \qquad \psi = \eta \cdot e^{\phi}, \tag{7.147}$$

with the Lagrangian

$$\frac{1}{2\pi}\left[\xi\bar{\partial}\eta + |\partial\phi|^2\right], \tag{7.148}$$

(7.146) looks like

$$\left\langle \prod_{i=0}^{n} \xi(x_i) \prod_{j=1}^{n} \eta(y_j) \prod_{k} e^{q_k\phi(z_k)} \right\rangle_e$$

$$= \frac{\prod_{j=1}^{n} \theta_e(-\vec{y}_j + \sum\vec{x} - \sum\vec{y} + \sum q\vec{z})}{\prod_{i=0}^{n} \theta_e(-\vec{x}_i + \sum\vec{x} - \sum\vec{y} + \sum q\vec{z})}$$

$$\times \frac{\prod_{i<i'} E(x_i, x_{i'}) \prod_{j<j'} E(y_j, y_{j'})}{\prod_{i,j'} E(x_i, y_j) \prod E(z_k, z_{k'})^{q_k q_{k'}} \operatorname{Det}_e \bar{\partial}_{1/2}}. \tag{7.149}$$

Now proceed to the case of arbitrary spin $j \in \mathbb{Z}/2$. As was done for the b, c-systems, perform a change of variables in the functional integral

$$\beta(z) = \Omega_{j-1/2}(z)\tilde{\psi}(z), \qquad \gamma(z) = \Omega_{j-1/2}^{-1}(z)\psi(z), \tag{7.150}$$

where the holomorphic $(j - \frac{1}{2})$-differential $\Omega_{j-1/2}(z)$ has zeros at the points $Q_1 \ldots Q_{n_j}$. The integration measure is

$$D\beta D\gamma = D\tilde{\psi}D\psi \prod_{i=1}^{n_j} \delta\big(\psi(Q_i)\big)\left[\Omega'_{j-1/2}(Q_i)\right]^{1/(2j+1)}. \tag{7.151}$$

Taking $\Omega_{j-1/2} = \nu_*^{2j-1}$, we obtain the following expression for correlators of the β, γ-fields with arbitrary j and with the metric $g = |\nu_*|^4$:

$$\left\langle \prod_{i=1}^{n} \gamma(y_i) \prod_{j=1}^{n-(2j-1)(p-1)} \delta\big(\gamma(w_j)\big) \prod_{k=0}^{n} H\big(\beta(x_k)\big) \right\rangle$$

$$= \left\langle \prod_{i=1}^{n} \eta(y_i) \frac{e^{\phi(y_i)}}{\nu_*(y_i)^{2j-1}} \prod_{j=1}^{n-n_j} e^{-\phi(w_j)}\nu_*(w_j)^{2j-1} \right.$$

$$\times \left. \prod_{k=0}^{n} \xi(x_k) \prod_{\alpha=1}^{p-1} e^{-(2j-1)\phi(R_\alpha^*)}\left[\nu'_*(R_\alpha^*)\right]^{(2j-1)^2} \right\rangle$$

$$= \frac{\prod_{j=1}^{n} \theta_e(-\vec{y}_j + \sum_k \vec{x}_k - \sum_k \vec{w}_k - (2j-1)\vec{\Delta}_*)}{\prod_{i=0}^{n} \theta_e(-\vec{x}_i + \sum_k \vec{x}_k - \sum_k \vec{w}_k - (2j-1)\vec{\Delta}_*)}$$

$$\times \frac{\prod_{i<i'} E(x_i, x_{i'}) \prod_{i,j} E(y_i, w_j)}{\prod_{i,j} E(x_i, y_j) \prod_{j<j'} E(w_j, w_{j'})}$$

$$\times \left[\frac{\prod_j \sigma_*(w_j)}{\prod_i \sigma_*(y_i)}\right]^{(2j-1)} \left[\frac{\prod_{\alpha<\beta} E(R_\alpha^*, R_\beta^*)}{\prod_\alpha \nu'_*(R_\alpha^*)}\right]^{(2j-1)^2} (\mathrm{Det}\,\bar{\partial}_0)^{1/2}. \qquad (7.152)$$

Chapter 8
Multiloop Amplitudes in the Theory
of Closed Bosonic Strings

Now we have all the necessary data for devising the simplest expressions in the closed bosonic string theory.

The P-loop contribution to the scattering amplitude of N particles is given by the following formula:

$$\mathcal{A} = \int Dg_{ab}DX^\mu \exp\left[-\frac{1}{2\pi}\int \sqrt{\det g}\, g^{ab}\partial_a X^\mu \partial_b X^\mu d^2\xi\right] V_1\{g; X\}\ldots V_N\{g; X\}$$

$$(\partial_a = \tfrac{1}{2}\partial/\partial\xi^a; \quad g^{ab}\partial_a\partial_b = -\Delta = \rho\partial\bar{\partial}). \qquad (8.1)$$

Here $X^\mu(\xi)$ is a mapping of a two-dimensional world surface of genus p into d-dimensional space–time $\mathbb{R}^d$; $g_{ab}(\xi)$ is a world sheet metric. The $V_j\{g; X\}$ stand for certain functionals of $g(\xi)$ and $X(\xi)$ associated with external particles. They are known as vertex operators. For the external state with a definite d-momentum k^μ,

$$V\{g; X\} = \int v\{g(\xi); X(\xi)\}e^{ik^\mu X^\mu(\xi)}\sqrt{\det g}\, d^2\xi, \qquad (8.2)$$

with the function $v(g; X)$ depending on quantum numbers of the external state. For the simplest string excitations—tachions—$v\{g; X\}$ is independent of X. For the sake of brevity, we consider below only tachions.

8.1. The Measure Dg_{ab}

The measure in the integral over the metrics in (8.1) is induced by the norm

$$||\delta g||^2 = \int \delta g_{ab}\delta g_{a'b'} g^{aa'} g^{bb'} \sqrt{\det g}d^2\xi. \qquad (8.3)$$

All objects in (8.1) are $2d$-covariant, and $\mathcal{A}$ does not change under the simultaneous change of coordinates and metric:

$$\xi \to \zeta = \xi + \varepsilon(\xi, \bar{\xi}), \qquad (8.4)$$

$$g(\xi) = g_{ab}(\xi)d\xi^a d\xi^b \rightarrow \tilde{g}(\zeta) = \tilde{g}_{ab}(\zeta)d\zeta^a d\zeta^b = g_{ab}(\xi)d\xi^a \xi^b. \qquad (8.5)$$

Making use of this invariance, one may choose the conformal gauge for metrics,

$$g(\xi) = \rho(\xi, \bar{\xi})d\xi d\bar{\xi}. \qquad (8.6)$$

Arbitrary variation of the metric can be expressed in the form of (3.24):

$$\delta g = \tilde{\delta}\tilde{\rho}d\xi d\bar{\xi} + \rho(\bar{\partial}\tilde{\varepsilon} + \sum_{\alpha=1}^{n_2} \delta y_\alpha \eta_\alpha)d\bar{\xi}^2 + \rho(\partial\bar{\tilde{\varepsilon}} + \sum_{\alpha=1}^{n_2} \overline{\delta y_\alpha \eta_\alpha})d\xi^2, \qquad (8.7)$$

$$\tilde{\delta}\tilde{\rho} = \delta\rho + \rho\partial\tilde{\varepsilon} + \rho\overline{\partial\tilde{\varepsilon}}. \qquad (8.8)$$

Let us recall that the Beltrami differential (3.29) looks like

$$\eta_\alpha = c_{\alpha\beta}\frac{\overline{f_\beta}}{\rho} + \bar{\partial}\varepsilon_\alpha,$$

$$c_{\alpha\beta} = \langle \eta_\alpha, f_\gamma \rangle (N_2^{-1})_{\gamma\beta}; \qquad \langle \eta_\alpha, f_\beta \rangle = \int^{d^2\xi} \eta_\alpha f_\beta. \qquad (8.9)$$

Thus, the norm of variation (8.7) is equal to

$$||\delta g||^2 = \int \rho^{-1}(\tilde{\delta}\tilde{\rho})^2 d^2\xi + \int \rho|\bar{\partial}\varepsilon|^2 d^2\xi + \delta y_\alpha (cN_2 c^\dagger)_{\alpha\beta}\overline{\delta y_\beta}. \qquad (8.10)$$

Here,

$$\varepsilon = \tilde{\varepsilon} + \sum_{\alpha=1}^{n_2} \delta y_\alpha \eta_\alpha. \qquad (8.11)$$

It follows from (8.7)–(8.11) that the norm $||\delta g||^2$ is expressed in terms of $\delta\rho, \varepsilon, \bar{\varepsilon}, \delta y, \overline{\delta y}$ by means of an upper triangular matrix. Its determinant is easy to calculate, and

$$Dg = \frac{D\rho|D'\varepsilon|^2 |\prod_{\alpha=1}^{n_2} dy_\alpha|^2}{\det N_2} \left|\det_{\alpha\beta}\langle \eta_\alpha, f_\beta \rangle\right|^2 \mathrm{Det}'(\rho^{-2}\partial\rho\bar{\partial}). \qquad (8.12)$$

Note that a measure $D'\varepsilon$ rather than $D\varepsilon$ appears in (8.12). The reason is that there are vector fields $\varepsilon \neq 0$ that do not contribute to the variation of the metric (8.7). They should satisfy the equation $\bar{\partial}\varepsilon = 0$. The number n_{-1} of such "conformal Killing vectors" or holomorphic -1-differentials on closed Riemann surfaces is different from zero only for $p = 0$ ($n_{-1} = 3$) or $p = 1$ ($n_{-1} = 1$). Killing vectors are associated with some symmetry group K (in the case of $p = 0$, this is the group of rational transformations, while in the case of $p = 1$, it is that of linear transformations). Let us express holomorphic -1-differentials in terms of some basis ε_a and orthogonal coordinates v_a on the group $\varepsilon = \sum_{a=1}^{n_{-1}} v_a \varepsilon_a$. The norm of the infinitesimal vector field is

$$||\varepsilon||^2 = \int \rho^{-2}|\varepsilon|^2 d^2\xi. \qquad (8.13)$$

Thus, the full integration measure over vector fields is

$$|D\varepsilon|^2 = \left| D'\varepsilon \sum_{a=1}^{n-1} dv_a \right|^2 \det{}_{(ab)} \int \overline{\varepsilon_a}\varepsilon_b \rho^2 d^2\xi = |D'\varepsilon|^2 d\mu_K \det N_{-1}, \quad (8.14)$$

with $d\mu_K$ being an invariant measure on the group K. If one is going to integrate only K-invariant quantities, the measure may be substituted by the volume of the group $V_K = \int_K d\mu_K$:

$$|D'\varepsilon|^2 \to \frac{|D\varepsilon|^2}{V_K \det N_{-1}}, \quad (8.15)$$

and finally:

$$Dg = |D\varepsilon|^2 D\rho \left| \prod_{\alpha=1}^{n_2} dy_\alpha \right|^2 \frac{\mathrm{Det}' \Delta_{-1}}{V_K \det N_{-1} \det N_2} \left| \det{}_{(\alpha\beta)} \langle \eta_\alpha, f_\beta \rangle \right|^2. \quad (8.16)$$

The volume V_k may depend on the moduli $y, \bar{y}$, for example, in the case of $p = 1$, $V_K = \mathrm{Im}\,\tau$. Note also, that according to (4.15),

$$\det{}_{(\alpha\beta)} \langle \eta_\alpha, f_\beta \rangle \, \mathrm{Det}\, \bar\partial_2 = \int Db^{(2)} Dc^{(-1)} e^{\int b\bar\partial c} \prod_{\alpha=1}^{n_2} \int_{d^2\xi} \eta_\alpha b = \left\langle \prod_{\alpha=1}^{n_2} \int_{d^2\xi} \eta_\alpha b \right\rangle. \quad (8.17)$$

8.2. The Integral over X-Fields

The integral over X^μ has already been calculated in Chapt. 7. Let us make use of the formula (7.37); then

$$\int DX^\mu \exp\left[-\frac{1}{2\pi} \int \sqrt{\det g}\, g^{ab} \partial_a X^\mu \partial_b X^\mu d^2\xi + \sum_{I=1}^{N} k_I^\mu X^\mu(\xi_I) \right]$$

$$= \left(\frac{\det N_0}{\mathrm{Det}' \Delta_0} \right)^{d/2} \prod_{I<J}^{N} G(\xi_I, \xi_J)^{k_I^\mu k_J^\mu} \prod_{I=1}^{N} \left(\sqrt{\det g(\xi_I)} \right)^{-k_I^2/2} \prod_{\mu=1}^{d} \delta\left(\sum_{I=1}^{N} k_I^\mu \right);$$

$$G(\xi, \xi') = |E(\xi, \xi')|^2 \exp\left\{ \pi \,\mathrm{Im}(\xi - \xi') \frac{1}{\mathrm{Im}\,T} \mathrm{Im}(\xi - \xi') \right\}. \quad (8.18)$$

As a result, for tachionic vertices, we obtain from (8.1):

$$\mathcal{A} = \int |D\varepsilon|^2 D\rho \left| \prod dy \right|^2 \frac{\det \langle \eta_\alpha, f_\beta \rangle \, \mathrm{Det}' \Delta_{-1}}{V_K \det N_{-1} \det N_2} \left(\frac{\det N_0}{\mathrm{Det}' \Delta_0} \right)^{d/2} \delta\left(\sum_{I=1}^{N} k_I^\mu \right)$$

$$\times \prod_{I=1}^{N} \int^{d^2\xi} v(g(\xi_I)) \left(\sqrt{\det g(\xi_I)} \right)^{1-k_I^2/2} \prod_{I<J} G(\xi_I, \xi_J)^{k_I^\mu k_J^\mu} \quad (8.19)$$

$$= \int |D\varepsilon|^2 D\rho \left(\exp -\frac{c_{-1} - \frac{d}{2}c_0}{12\pi} S_L\{\rho\} \right) \left| \frac{\prod dy \, \det \langle \eta_\alpha, f_\beta \rangle \, \mathrm{Det}\, \bar{\partial}_{-1}}{(\mathrm{Det}\, \bar{\partial}_0)^{d/2}} \right|^2$$

$$\times \frac{\delta(\sum_{I=1}^N k_I^\mu)}{V_K (\det N_1)^{d/2}} \prod_{I=1}^N \int^{d^2\xi} v(\rho(\xi_I)) \cdot \rho(\xi_I)^{1-k_I^2/2} \prod_{I<J}^N G(\xi_I, \xi_J)^{k_I^\mu k_J^\mu}.$$

$$(8.20)$$

It is conventional to omit the volume of the diffeomorphism group, changing the amplitude for $\tilde{\mathcal{A}} \equiv \mathcal{A}/\int |D\varepsilon|^2$. In the absence of tachionic vertices, (8.19) turns into (5.14), which has been taken earlier as a definition of Polyakov's string measure.

The determinant $\det N_1$ is independent of the metric (in variance with all other $\det N_j$ with $j \neq 1$). Therefore, (8.20) is further simplified in the case of what is called the *critical* string, when $c_{-1} - \frac{d}{2}c_0 = 0$ and $1 - \frac{1}{2}k_I^2 = 0$, $v = 1$, i.e., when there is no dependence on the conformal factor $\rho(\xi)$ in (8.20). In this case, $d = 2c_{-1}/c_0 = 26$, and the Mumford measure on moduli space arises in (8.20):

$$d\mu_M = \frac{\mathrm{Det}\, \bar{\partial}_2}{(\mathrm{Det}\, \bar{\partial}_0)^{13}} \det_{(\alpha\beta)} \langle \eta_\alpha, f_\beta \rangle \prod_{\alpha=1}^{n_2} dy_\alpha, \quad (\mathrm{Det}\, \bar{\partial}_2 = \mathrm{Det}\, \bar{\partial}_{-1}). \qquad (8.21)$$

The final expression for the amplitude at $d = 26$, $k_I^2 = 2$, $v\{p\} = 1$ is:

$$\tilde{\tilde{\mathcal{A}}} = \frac{\mathcal{A}}{\int |D\varepsilon|^2 D\rho} = \delta\left(\sum_{I=1}^N k_I^\mu \right) \int_{\mathcal{M}_p} \prod_{I=1}^N \int^{d^2\xi_I} \frac{|d\mu_M|^2 \prod_{I<J}^N G(\xi_I, \xi_J)^{k_I k_J}}{V_K (\det N_1)^{13}}$$

$$= \delta\left(\sum_{I=1}^N k_I^\mu \right) \int_{\mathcal{M}_p} \prod_{I=1}^N \int^{d^2\xi_I} \left| d\mu_M \prod_{I<J}^N E(\xi_I, \xi_J)^{k_I^\mu k_J^\mu} \right|^2$$

$$\times \frac{\prod_{I<J}^N \exp \pi k_I k_J \, \mathrm{Im}(\xi_I - \xi_J) \frac{1}{\mathrm{Im}\, T} \, \mathrm{Im}(\xi_I - \xi_J)}{V_K (\det N_1)^{13}}. \qquad (8.22)$$

The product under the module sign can also be treated as a distinguished measure on the module space of curves with punctures.

Explicit formulas for the Mumford measure (8.21) can be derived as a synthesis of expression (4.15) for the chiral determinant through the correlators of the b, c-fields and formulae (7.106) for these correlators:

$$d\mu_M^{\mathrm{Can}} = \frac{\langle \prod_{\alpha=1}^{n_2} \int \eta_\alpha b \rangle_{\mathrm{spin2}} \prod_{\alpha=1}^{n_2} dy_\alpha \sqrt{\det \bar{\partial}_0}}{\left[\langle \phi(x_0) \prod_{i=1}^{n_1} W(x_i) \rangle_{\mathrm{spin1}} \left(\det_{(ij)} \omega_i(x_j) \right)^{-1} \sqrt{\mathrm{Det}\, \bar{\partial}_0} \right]^9}$$

$$\frac{[\prod_{\alpha=1}^{n_2} dy_\alpha \int^{d^2\xi} \eta_\alpha(\xi_\alpha)] \theta_*(\sum_{\alpha=1}^{n_2} \vec{\xi}_\alpha - 3\vec{\Delta}_*) [\prod_{\alpha<\beta}^{n_2} E(\xi_\alpha, \xi_\beta) \prod_{\alpha=1}^{n_2} \sigma_*(\xi_\alpha)]}{[\theta_*(\sum_{i=1}^{n_1} (\vec{x}_i - \vec{x}_0 - \vec{\Delta}_*)) \prod_{i<j}^{n_1} E(\xi_i, \xi_j) \prod_{i=1}^{n_1} (\sigma_*(x_i)/E(xi_i, x_0))/\sigma_*(x_0) \det_{(ij)} \omega_i(x_j)]^9}.$$

$$(8.23)$$

8.3. Strings Beyond Critical Dimension

For noncritical strings ($d \neq 26$), a nontrivial functional integral over the field ρ survives in (8.20), which does not allow us to represent the amplitude as a finite-fold integral over the moduli space. Let us present the results of ρ-integration in the quasiclassical approximation (they are widely believed to be in fact exact, though the proof is still lacking; see Knizhnik, Polyakov and Zamolodchikov [1988] and Alekseev and Shatashvili [1988] for a discussion of these points).

Let us decompose the field ρ into "classical" ρ_0 and "quantum" ϕ components: $\rho = \rho_0 e^\phi$ or $\sqrt{\det g} = \sqrt{\det g_0} \cdot e^\phi$. The measure in the functional integrals over ϕ should be g- (and not g_0-) invariant, i.e., it does depend on ϕ. This is what makes the Liouville theory nonlinear and complicated. In particular, the Liouville action in $\exp(-\frac{\lambda^2}{12\pi} S_L\{g\})$ gets renormalized, the parameter λ is changing. In quasiclassical (one-loop) approximation, ϕ-dependence of the measure is unessential. However, the quasiclassical approximation is applicable only in the vicinity of some critical point $\lambda = \lambda_*$.

In quasiclassical approximation, the factor of

$$\exp\left(-\frac{\lambda_*^2}{12\pi} S_L\{g\}\right) = \exp\left(-\frac{\lambda_*^2}{12\pi} S_L\{g_0\}\right)$$
$$\times \exp\left(-\frac{\lambda_*^2}{12\pi} \int \sqrt{\det g_0}\,(g_0^{ab}\partial_a\phi\partial_b\phi + 2\mathcal{R}_0\phi)\right)$$

$$(8.24)$$

should be inserted in the original formula (8.1). Moreover, in this case in (8.2),

$$V_I\{g\} = \left(\sqrt{\det g}\right)^{\lambda_* A_I} = \left(\sqrt{\det g_0}\right)^{\lambda_* A_I} e^{\lambda_* A_I \phi}. \tag{8.25}$$

After a change of the variables $\phi \to \frac{\sqrt{3}\phi}{\lambda_*}$, the following integral over ϕ arises:

$$\int D\phi \exp\left\{-\frac{1}{4\pi}\int \sqrt{\det g_0}\,\left(g_0^{ab}\partial_a\phi\partial_b\phi + \frac{2\lambda_*}{\sqrt{3}}\mathcal{R}_0\phi\right) + \sqrt{3}\sum_{I=1}^{N} A_I\phi(\xi_I)\right\}.$$

$$(8.26)$$

As we already mentioned, the rigorous analysis of Liouville theory requires evaluation of (8.26) with a $g = g_0 e^\phi$-invariant, i.e. ϕ-dependent, measure; however, in quasiclassical approximation, we can restrict ourselves to the g_0-invariant measure. Then the theory is quadratic and calculations are trivial. Let us use the results of Chapt. 7 (7.47) and obtain instead of (8.20):

$$\tilde{A} = \int \exp\left\{-\frac{\lambda_*^2}{12\pi} S_L\{g_0\}\right\} \int^{d^2\xi} \prod_{I=1}^{N} \left(\sqrt{\det g_0(\xi_I)}\right)^{\lambda_* A_I}$$

$$\times \left[\exp\left(\frac{1}{12\pi}\left(\lambda_*^2 + \frac{d-1}{2}c_0 - c_{-1}\right)\right)S_L\{g_0\}\right] \cdot \left|\frac{\prod dy\,\det\langle\eta_\alpha, f_\beta\rangle\,\mathrm{Det}\,\bar{\partial}_{-1}}{(\mathrm{Det}\,\bar{\partial}_0)^{(d+1)/2}}\right|^2$$

$$\times \frac{\prod_{\mu=1}^{d} \delta\left(\sum_{I=1}^{N} k_I^{\mu}\right) \delta\left(\sum_{I=1}^{N} A_I + \frac{1}{3}\lambda_*(p-1)\right)}{V_K (\det N_1)^{(d+1)/2}}$$

$$\times \left[\prod_{I=1}^{N} \left(\sqrt{\det g_0(\xi_I)}\right)^{1-k_I^2/2+3A_I^2-\lambda_* A_I}\right] \prod_{I<J} G(\xi_I, \xi_J)^{k_I^{\mu} k_J^{\mu} - 6A_I A_J}$$

$$\times \prod_{I=1}^{N} \prod_{n=1}^{p-1} G(\xi_I, R_n^*)^{2\lambda_* A_I} \mathcal{F}\{R^*\}. \tag{8.27}$$

The critical point λ_* is defined from the condition of the renorm-invariance of the Liouville action, i.e., the condition that the corrections in square brackets in (8.27) are equal to unity. This restriction leads to constraints of the following form:

$$\lambda_*^2 + \frac{d-1}{2}c_0 - c_{-1}, \quad \text{i.e., } \lambda_*^2 = \frac{25-d}{2}; \tag{8.28}$$

$$1 - k_I^2/2 - \lambda_* A_I + 3A_I^2 = 0,$$

$$\text{i.e., } A_I = \frac{25-d}{2}\left[1 - \sqrt{1 - \frac{24}{25-d}\left(1 - \frac{1}{2}k_I^2\right)}\right]. \tag{8.29}$$

At $d = 25$, $\lambda_* = 0$, and we come back to the case of the critical string, with the Liouville field ϕ playing the role of the 26th coordinate.

This kind of reasoning is easy to generalize for the case of arbitrary conformal theory (see David [1988] and Distler and Kawai [1988]). An alternative derivation of (8.29) for anomalous dimensions A_I is given in Knizhnik, Polyakov and Zamolodchikov [1988].

References[1]

Alekseev, A., Shatashvili, S.
[1988]   Path integral quantization of the coadjoint orbits of the Virasoro group
         and 2-d gravity. Preprint LOMI E-16-88. Leningrad: LOMI (Russian).
Alessandrini, V., Amati, D.
[1971]   Properties of dual multiloop amplitudes. Nuovo Cimento A 4, 793–844.
Alvarez-Gaume, L., Bost, J.-B., Moore, G., Nelson, P., Vafa, C.
[1987]   Bosonization for higher genus Riemann surfaces. Commun. Math. Phys.
         112, 503–552. Zbl.647.14019
Alvarez-Gaume, L., Witten, E.
[1985]   Gravitational anomalies. Nucl. Phys. B 234, 269–305.

---

[1] For the convenience of the reader, references to reviews in Zentralblatt für Mathematik (Zbl.), compiled using the MATH database, have, as far as possible, been included in this bibliography.

Arakelov, S.Yu.
[1974]   Intersection theory for divisors on the arithmetical surface. Izv. Akad.
         Nauk SSSR. Ser. Math. *38*, 1179–1192. Engl. transl.: Math. USSR, Izv.
         8 (1974), 1164–1180 (1976). Zbl.355.14002
Arbarello, E., Cornalba, M., Griffiths, P., Harris, J.
[1986]   Geometry of Algebraic Curves. Berlin: Springer-Verlag. Zbl.559.14017
Atick, J., Sen, A.
[1987a] Covariant one-loop fermion emission amplitudes in closed string theories.
         Nucl. Phys. B *286*, 189–201.
[1987b] Spin field correlators on an arbitrary genus Riemann surface and non-
         renormalization theorems in string theories. Phys. Lett. B *186*, 339–346.
Beilinson, A.A., Manin, Yu.I.
[1986]   The Mumford form and the Polyakov measure in string theory. Commun.
         Math. Phys. *107*, 359–376. Zbl.604.14016
Belavin, A.A., Knizhnik, V.G.
[1986]   Algebraic geometry and the geometry of quantum strings. Phys. Lett.
         B *168*, 201–206. Zbl.693.58043
Belavin, A.A., Knizhnik, V.G., Morozov, A.Yu., Perelomov, A.M.
[1986]   Two- and three-loop amplitudes in the bosonic string theory. Phys. Lett.
         B *177*, 324–328.
Bers, L.
[1972]   Uniformization, moduli and Kleinian groups. Bull. Lond. Math. Soc. *4*,
         257–300. Zbl.257.32012
Bismut, J.-M., Freed, D.
[1986a] The analysis of elliptic families. 1. Commun. Math. Phys. *106*, 159–176.
         Zbl.657.58037
[1986b] The analysis of elliptic families. 2. Commun. Math. Phys. *107*, 103–163.
         Zbl.657.58038
Bost, J.B.
[1987]   Fibres déterminants, déterminants régularisés et mesures sur les espaces
         de modules des courbes complexes. Sémin. Bourbaki, 39ème année. Exp.
         No. 676, Astérisque 152/153, 113–149. Zbl.655.32021
Candelas, P., Horowitz, G., Strominger, A., Witten, E.
[1985]   Vacuum configurations for superstrings. Nucl. Phys. B *256*, 46–65.
Chern, S.S.
[1956]   Complex Manifolds. Chicago: University of Chicago. Zbl.88,378
Clemens, C.H.
[1980]   A Scrapbook of Complex Curve Theory. New York, London: Plenum
         Press. Zbl.456.14016
David, F.
[1988]   Conformal field theory, coupled to 2-d gravity in the conformal gauge.
         Mod. Phys. Lett. A *3*, 819–826.
Deligne, P.
[1987]   Le déterminant de la cohomologie. Contemp. Math. *67*, 93–177.
         Zbl.629.14008
Deligne, P., Mumford, D.
[1969]   The irreducibility of the space of curves of given genus. Inst. Hautes Etud.
         Sci., Publ. Math. *36*, 75–109. Zbl.181,488
Distler, J., Kawai, H.
[1989]   Conformal field theory and 2-D quantum gravity or who's afraid of Joseph
         Liouville? Nucl. Phys. B *321*, 509.

Faltings, G.
[1984] Calculus on arithmetic surfaces. Ann. Math., II. Ser. *119*, 387–424. Zbl.559.14005

Farkas, H., Kra, I.
[1980] Riemann Surfaces. Berlin: Springer-Verlag. Zbl.475.30001

Fay, J.D.
[1973] Theta-functions on Riemann surfaces. Lect. Notes Math. *352*. Zbl.281.30013

Friedan, D., Martinec, E., Shenker, S.
[1986] Conformal invariance, super-symmetry and string theory. Nucl. Phys. B *271*, 93–165.

Gerasimov, A., Marshakov, A., Morozov, A.Yu., Olshanetsky (= Olshanetskij, M.A.), M.A., Shatashvili, S.
[1990] Wess–Zumino–Witten model as a theory of free fields. Int. J. Mod. Phys. A *5*, 2495–2589.

Goto, T.
[1971] Relativistic quantum mechanics of one-dimensional mechanical continuum and subsidiary conditions of dual resonance model. Prog. Theor. Phys. *46*, 1560–1569.

Green, M., Schwarz, J.
[1984] Anomaly cancellations in supersymmetric D-10 gauge theory and super-string theory. Phys. Lett. B *149*, 117–122.
[1985] Infinity cancellations in SO(32) superstring theory. Phys. Lett. B *151*, 21–27.

Green, M., Schwarz, J., Witten, E.
[1987] Superstring theory, Vol. 1, Vol. 2. Cambridge: Cambridge University Press. Zbl.619.53002

Griffiths, P., Harris, J.
[1978] Principles of Algebraic Geometry. New York: Wiley-Interscience Publication. Zbl.408.14001

Gross, D., Harvey, J., Martinec, E., Rohm, R.
[1985] Heterotic string theory. The free heterotic string. Nucl. Phys. B *256*, 253–291.
[1986] Heterotic string theory. The interacting heterotic string. Nucl. Phys. B *267*, 267–305.

Igusa, J.I.
[1972] Theta Functions. Berlin: Springer-Verlag. Zbl.251.14016

Kallosh, R., Morozov, A.Yu.
[1988] The Green–Schwarz action and loop calculations in superstrings. Int. J. Mod. Phys. A *3*, 1943–1958.

Knizhnik, V.G.
[1986] Analytic fields on Riemann surfaces. Phys. Lett. B *180*, 247–254. Zbl.656.58043

Knizhnik, V.G., Polyakov, A.M., Zamolodchikov, A.
[1988] Fractional structure of 2-d gravity. Mod. Phys. Lett. A *3*, 819–826.

Losev, A.
[1989] Once more on β, γ-systems. Phys. Lett. B *226*, 62.

Manin, Yu.I.
[1986] String statistical sum can be expressed through theta-functions. ZhETP Lett. *161*, 161–163 (Russian).

Moore, G.
[1986] Modular forms and two-loop string physics. Phys. Lett. B *176*, 369–373.

Morozov, A.Yu.
[1988]   Two-loop statistical sum of superstring. Yad. Fiz. *48*, 869–885 (Russian).
Morozov, A.Yu., Olshanetskij, M.A.
[1988]   Statistical sum of the bosonic string, compactified on an orbifold. Nucl. Phys. B *299*, 389–406.
Mumford, D.
[1977]   Stability of projective varieties. Enseign. Math. I. Ser. *23*, 39–110. Zbl.363.14003
[1983,1984]   Tata Lectures on Theta. I, II. Boston: Birkhäuser. Zbl.509.14049, Zbl.549.14014
Nambu, Y.
[1970]   Lectures at Copenhagen Summer Symposium. Copenhagen: NORDITA.
Polyakov, A.M.
[1981a]   Quantum geometry of bosonic string. Phys. Lett. B *103*, 207–210.
[1981b]   Quantum geometry of fermionic string. Phys. Lett. B *103*, 211–214.
Quillen, D.
[1985]   Determinants of Cauchy–Riemann operators on a Riemann surface. Funkts. Anal. Prilozh. *19*, 37–41. Engl. transl.: Funct. Anal. Appl. *19*, 31–34 (1985). Zbl.603.32016
Schumacher, G.
[1990]   The theory of Teichmüller spaces. In: Several Complex Variables VI. Encyc. Math. Sci. 69, Berlin: Springer-Verlag.
Schwarz, J.
[1985]   Superstrings: The First Fifteen Years. Vol. 1 and 2. Singapore: World Scientific.
Semikhatov, A.M.
[1989]   ZhETP Lett. *49*, 181–185 (Russian).
Shapiro, J.
[1970]   Electroweak analogue for the Virasoro model. Phys. Lett. B *33*, 361–362.
Smit, D.-J.
[1988]   String theory and algebraic geometry of moduli spaces. Commun. Math. Phys. *114*, 645–685.
Verlinde, E., Verlinde, H.
[1987]   Multiloop calculations in covariant superstring theory. Phys. Lett. B *192*, 95–102.

Subject Index

Encyclopaedia of Mathematical Sciences
Editor-in-Chief: R.V.Gamkrelidze

Analysis

Volume 13:
R.V.Gamkrelidze (Ed.)
Analysis I
Integral Representations and Asymptotic Methods
1989. VII, 238 pp. 3 figs. ISBN 3-540-17008-1

Volume 14:
R.V.Gamkrelidze (Ed.)
Analysis II
Convex Analysis and Approximation Theory
1990. VII, 255 pp. 21 figs. ISBN 3-540-18179-2

Volume 26:
S.M.Nikol'skij (Ed.)
Analysis III
Spaces of Differentiable Functions
1991. VII, 221 pp. 22 figs. ISBN 3-540-51866-5

Volume 27:
V.G.Maz'ya, S.M.Nikol'skij (Eds.)
Analysis IV
Linear and Boundary Integral Equations
1991. VII, 233 pp. 4 figs. ISBN 3-540-51997-1

Volume 19:
N.K.Nikol'skij (Ed.)
Functional Analysis I
Linear Functional Analysis
1992. V, 283 pp. ISBN 3-540-50584-9

Volume 20:
A.L.Onishchik (Ed.)
Lie Groups and Lie Algebras I
Foundations of Lie Theory. Lie Groups of Transformations
1993. VII, 235 pp. 4 tabs. ISBN 3-540-18697-2

Several Complex Variables

Volume 7:
A.G.Vitushkin (Ed.)
Several Complex Variables I
Introduction to Complex Analysis
1990. VII, 248 pp. ISBN 3-540-17004-9

Volume 8:
A.G.Vitushkin, G.M.Khenkin (Eds.)
Several Complex Variables II
Function Theory in Classical Domains. Complex Potential Theory
1993. Approx. 260 pp. 19 figs. ISBN 3-540-18175-X

Volume 9:
G.M.Khenkin (Ed.)
Several Complex Variables III
Geometric Function Theory
1989. VII, 261 pp. ISBN 3-540-17005-7

Volume 10:
S.G.Gindikin, G.M.Khenkin (Eds.)
Several Complex Variables IV
Algebraic Aspects of Complex Analysis
1990. VII, 251 pp. ISBN 3-540-18174-1

Volume 69:
W. Barth, R. Narasimhan (Eds.)
Several Complex Variables VI
Complex Manifolds
1990. IX, 310 pp. 4 figs. ISBN 3-540-52788-5